Developments in Plant Genetics and Breeding, 7

Diversity in Barley

(Hordeum vulgare)

Developments in Plant Genetics and Breeding

1A ISOZYMES IN PLANT GENETICS AND BREEDING, PART A
edited by S.D. Tanksley and T.J. Orton
1983 x +516 pp.

1B ISOZYMES IN PLANT GENETICS AND BREEDING, PART B
edited by S.D. Tanksley and T.J. Orton
1983 viii +472 pp.

2A CHROMOSOME ENGINEERING IN PLANTS: GENETICS, BREEDING, EVOLUTION, PART A
edited by P.K. Gupta and T. Tsuchiya
1991 xv + 639 pp.

2B CHROMOSOME ENGINEERING IN PLANTS: GENETICS, BREEDING, EVOLUTION, PART B
edited by T. Tsuchiya and P.K. Gupta
1991 vi + 630 pp.

3 GENETICS IN SCOTS PINE
edited by M. Giertych and Cs. Mátyás
1991 280 pp.

4 BIOLOGY OF BRASSICA COENOSPECIES
edited by C. Gómez-Campo
1999 x + 490 pp.

5 PLANT GENETIC ENGINEERING: TOWARDS THE THIRD MILLENNIUM
edited by A.D. Arencibia
2000 x + 272 pp.

6 PHYTOSFERE '99 - Highlights in European Plant Biotechnology Research and Technology Transfer
edited by G.E. de Vries and K. Metzlaff
2000 VIII + 286 pp.

7 DIVERSITY IN BARLEY (*Hordeum vulgare*)
edited by R. von Bothmer et al
2003 xx + 280 pp.

On the Cover

The image depicts morphologically diverse spikes from the Genebank collection at IPK Gatersleben, Germany. These spikes were selected from the barley multiplication plots of the Genebank in June 1997, and scanned.

Developments in Plant Genetics and Breeding, 7

Diversity in Barley
(Hordeum vulgare)

Editors

Roland von Bothmer *Swedish University of Agricultural Sciences, Sweden*

Theo van Hintum *Wageningen University and Research Centre, The Netherlands*

Helmut Knüpffer *Institute of Plants Genetics and Crop Plant Research, Germany*

Kazuhiro Sato *Research Institute for Bioresources, Japan*

2003

ELSEVIER

Amsterdam - Boston - Heidelberg - London - New York - Oxford
Paris - San Diego - San Francisco - Singapore - Sydney - Tokyo

ELSEVIER SCIENCE B.V.
Sara Burgerhartstraat 25
P.O. Box 211, 1000 AE Amsterdam, The Netherlands

First edition 2003

Library of Congress Cataloging in Publication Data
A catalog record from the Library of Congress has been applied for.

British Library Cataloguing in Publication Data

Diversity in barley (Hordeum Vulgare)
1.Barley - Variation 2.Barley - Genetics
I.Bothmer, Roland von
633.1'67

ISBN: 0 444 505857

♾ The paper used in this publication meets the requirements of ANSI/NISO Z39.48-1992 (Permanence of Paper).
Printed in The Netherlands.

Preface

'Diversity in Barley' describes and analyses the germplasm that is the key to barley improvement so that new and diverse needs and purposes are met. It collates information on a broad spectrum of diversity created through evolution, mutation, breeding and selection, cytogenetics and biotechnology. This book serves as an extensive source of data on different aspects of diversity in barley and provides key references to further detailed information.

The genetic basis of diversity in barley was recorded before the rediscovery of Mendel's famous results. In the Museum of The Swedish Seed Association, a table was found with barley ears from an F_2 generation grown in the nursery at Svalöv in 1887. The ears presented a segregation of three gene pairs: v/V, b/B, k/K. Neergaard, who made the cross and studied the recombination of characters in F_2 was definitely interested in the diversity in barley. Pehr Bolin confirmed the crosses in his "gene pool" of barley grown from 1890-1896. He found that the F_1 generation was uniform and that some characters were "dominating" while others were "hidden". In F_2, however, possible combinations of characters from the parents were formed in numbers that could be calculated.

Very few plants have had their traits studied and subsequently genetically revealed as intensively as barley. Among crop plants barley is in the top 3 or 4 in this category. Barley is a self-fertilizing diploid, which facilitates intensive genetic analyses and breeding activities. It has only seven pairs of chromosomes, which can be visually identified and has been subject to extensive cytogenetic analyses. Barley has been the object of a very long history of mutation induction, both by chemicals and irradiation, producing one of the best collections of mutants among plants and creating much diversity for breeding and genetic studies. Moreover, the huge collections of chromosome mutants, translocations, deletions and duplications, have also created a significant degree of new diversity and have been invaluable in genetic analyses. Investigations at the biochemical and molecular levels are rapidly expanding and barley is one of the front runners in the development of relevant knowledge among plants. One outcome of these investigations is the already large collection of Quantitative Trait Loci (QTLs) and DNA markers for genetic and breeding studies. It is the world's leading crop plant for brewing and distilling.

Thus, barley can be considered to be a **model plant** among crop plants and especially the Triticeae whose analyses now and in the future can help to advance knowledge in many other plants, especially those used for food, fibre and medicine. For instance, recent findings reveal that the DNA of barley, rice, wheat and rye have major similarities. Future research on the DNA of barley can expand the knowledge of DNA and the basis for innumerable traits in these crops. The information contained in this book will be of great importance and useful for workers analysing traits and their diversity in many plant species.

This book examines many aspects of knowledge of the barley plant – its genetics and cytogenetics, breeding behaviour, ecology, physiology, morphology, taxonomy, biochemistry, molecular biology, agronomic value, seed and plant quality and environmental interactions, with emphasis on biotic and abiotic stress. It focuses on the diversity found in all the traits, their origins, their future analyses and enhancement, and prospects for the use of diversity to provide a greater understanding of the genetics. This is important for the improvement of the barley crop plant for all its current and future uses such as food for livestock, man or others, brewing and distilling, and medicine or health care products.

In addition to the extensive description of the germplasm diversity in barley, it emphasizes the efforts to preserve the germplasm and where it can be found in various genebanks and in the Core Collection.

The editors have carefully selected key scientists in the various fields to coordinate and develop the chapters. They have provided authors with a clear objective and format to follow in order to provide this key reference work on barley. "Diversity in Barley" will serve as a monumental source of information on barley for years to come.

Robert A. Nilan, Ken J. Kasha and Arne Hagberg

Genetic diversity in barley illustrated by a segregating population from the Swedish Seed Association at Svalöv in 1887.

Contributors

Bjørnstad, Åsmund (Chapter 7)
Department of Horticulture and Crop Sciences, Agricultural University, N-1432 Ås, Norway
Bothmer, Roland von (Chapters 1, 2 and 14)
Department of Crop Science, Swedish University of Agricultural Sciences, Box 44, SE-230 53 Alnarp, Sweden
Castro, Ariel (Chapter 10)
Department of Crop and Soil Science, Oregon State University, Corvallis, OR, 97331, USA
Cattivelli, Luigi (Chapter 9)
Experimental Institute for Cereal Research, Section of Fiorenzuola d'Arda, Via S. Protaso, 302, I-29017 Fiorenzuola d'Arda (PC), Italy
Corey, Ann (Chapter 10)
Department of Crop and Soil Science, Oregon State University, Corvallis, OR, 97331, USA
Fischbeck, Gerhard (Chapters 2 and 3)
Department of Agronomy and Plant Breeding, D-85350 Freising-Weihenstephan, Germany
Franckowiak, Jerome D. (Chapter 5)
Department of Plant Sciences, North Dakota State University, Fargo, ND, 58105, USA
Graner, Andreas (Chapter 7)
Institute of Plant Genetics and Crop Plant Research (IPK), Corrensstraße 3, D-06466 Gatersleben, Germany
Habekuß, Antje (Chapter 8)
Bundesanstalt für Züchtungsforschung an Kulturpflanzen, Institut für Epidemiologie und Resistenz, Theodor-Römer-Weg 1-4, D-06449 Aschersleben, Germany
Hammer, Karl (Chapter 4)
Universität Kassel, Fachbereich 11, Fachgebiet Agrarbiodiversität, Steinstraße 19, D-37213 Witzenhausen, Germany
Hayes, Patrick (Chapter 10)
Department of Crop and Soil Science, Oregon State University, Corvallis, OR, 97331, USA
Henson, Cynthia (Chapter 10)
Cereal Crops Research Unit, USDA-ARS, 501 Walnut St., Madison, WI 53705, USA and Department of Agronomy, University of Wisconsin, Madison, WI 53706, USA
Hintum, Theo van (Chapters 1, 12, 13 and 14)
Centre for Genetic Resources, The Netherlands (CGN), PO Box 16, NL-6700 AA Wageningen, The Netherlands
Jones, Berne L. (Chapter 10)
Cereal Crops Research Unit, USDA-ARS, 501 Walnut St., Madison, WI 53705, USA and Department of Agronomy, University of Wisconsin, Madison, WI 53706, USA
Kling, Jennifer (Chapter 10)
Department of Crop and Soil Science, Oregon State University, Corvallis, OR, 97331, USA
Knüpffer, Helmut (Chapters 1, 4, 13 and 14)
Institute of Plant Genetics and Crop Plant Research (IPK), Corrensstraße 3, D-06466

Gatersleben, Germany
Komatsuda, Takao (Chapter 2)
National Institute of Agrobiological Sciences, Kannondai, Tsukuba, 305-8602, Japan
Konishi, Takeo (Chapter 7)
294 Okada, Mabi-cho, Kibi-gun, Okayama 710-1311, Japan
Kopahnke, Doris (Chapter 8)
Bundesanstalt für Züchtungsforschung an Kulturpflanzen, Institut für Epidemiologie und Resistenz, Theodor-Römer-Weg 1-4, D-06449 Aschersleben, Germany
Kovaleva, Olga (Chapter 4)
N.I. Vavilov Institute of Plant Production, Bolshaya Morskaya 42-44, 190 000 St. Petersburg, Russia
Künzel, Gottfried (Chapter 6)
Institute of Plant Genetics and Crop Plant Research (IPK), Corrensstraße 3, D-06466 Gatersleben, Germany
Linde-Laursen, Ib (Chapter 6)
Botanical Section, Department of Ecology, The Royal Veterinary and Agricultural University, Rolighedsvej 21, DK-1958 Frederiksberg C (Copenhagen), Denmark
Lundborg, Tomas (Chapter 9)
Department of Crop Science, Swedish University of Agricultural Sciences, SE-230 53 Alnarp, Sweden
Lundqvist, Udda (Chapter 5)
SvalöfWeibull AB, SE-268 81 Svalöv, Sweden
Marquez-Cedillo, Luis (Chapter 10)
Department of Crop and Soil Science, Oregon State University, Corvallis, OR, 97331, USA
Mather, Diane (Chapter 10)
Department of Plant Science, McGill University, Ste Anne de Bellevue, QC H9X 3V9 Canada
Matus, Ivan (Chapter 10)
Department of Crop and Soil Science, Oregon State University, Corvallis, OR, 97331, USA
Menting, Frank (Chapter 12)
Centre for Genetic Resources The Netherlands (CGN), P.O. Box 16, NL-6700 AA, Wageningen, The Netherlands
Munck, Lars (Chapter 11)
The Royal Veterinary and Agricultural University, Rolighedsvej 30, DK-1958 Frederiksberg C, Denmark
Ordon, Frank (Chapter 7)
Institute of Crop Science and Plant Breeding I, Justus-Liebig-University, D-35390 Giessen, Germany
Proeseler, Gerhard (Chapter 8)
Bundesanstalt für Züchtungsforschung an Kulturpflanzen, Institut für Epidemiologie und Resistenz, Theodor-Römer-Weg 1-4, D-06449 Aschersleben, Germany
Romagosa, Ignacio (Chapter 9)
Centre UdL-IRTA – Universitat de Lleida, Alcade Rovira Roure 177–25198 Lleida, Spain
Rossi, Carlos (Chapter 10)
Department of Crop and Soil Science, Oregon State University, Corvallis, OR, 97331, USA
Sato, Kazuhiro (Chapters 1, 2, 4, 8, 10 and 14)
Barley Germplasm Center, Research Institute for Bioresources, Okayama University,

Kurashiki, 710-0046, Japan
Stanca, A. Michele (Chapter 9)
Experimental Institute for Cereal Research, Section of Fiorenzuola d'Arda, Via S. Protaso, 302, I-29017 Fiorenzuola d'Arda (PC), Italy
Taketa, Shin (Chapter 6)
Department of Life Sciences, Faculty of Agriculture, Kagawa University, Ikenobe, Kagawa 761-0795, Japan
Takeda, Kazuyoshi (Chapter 9)
Barley Germplasm Centre, Research Institute for Bioresources, Okayama University, Kurashiki 710-0046, Japan
Terentyeva, Irina (Chapter 4)
N.I. Vavilov Institute of Plant Production, Bolshaya Morskaya 42-44, 190 000 St. Petersburg, Russia
Terzi, Valeria (Chapter 9)
Experimental Institute for Cereal Research, Section of Fiorenzuola d'Arda, Via S. Protaso, 302 I-29017 Fiorenzuola d'Arda (PC), Italy
Walther, Ursula (Chapter 8)
Bundesanstalt für Züchtungsforschung an Kulturpflanzen, Institut für Epidemiologie und Resistenz, Theodor-Römer-Weg 1-4, D-06449 Aschersleben, Germany
Weibull, Jens (Chapter 8)
Swedish Biodiversity Centre, PO Box 54, SE-230 53 Alnarp, Sweden
Yasuda, Shozo (Chapter 2)
Research Institute for Bioresources, Okayama University, Kurashiki, 710-0046, Japan

Contents

To the memory of Professor Ryuhei Takahashi

(September 8, 1910 - May 8, 1999)

for his pioneering research on genetic diversity in barley

Section I

Introduction

Diversity in Barley *(Hordeum Vulgare)*
Roland von Bothmer et al (Editors).

Chapter 1

Barley diversity – an introduction

Roland von Bothmer[a], Kazuhiro Sato[b], Helmut Knüpffer[c] and Theo van Hintum[d]

[a]Department of Crop Science, Swedish University of Agricultural Sciences, Box 44, SE-230 53 Alnarp, Sweden
[b]Barley Germplasm Center, Research Institute for Bioresources, Okayama University, Kurashiki, 710-0046, Japan
[c]Institute of Plant Genetics and Crop Plant Research (IPK), D-06466 Gatersleben, Germany
[d]Plant Research International B.V., Centre for Genetic Resources, The Netherlands (CGN), PO Box 16, NL-6700 AA Wageningen, The Netherlands

Biodiversity – a matter of global concern

Genetic diversity is one of the main resources sustaining human life on this planet. Without it, crops would not be able to adapt to changing biotic and abiotic conditions, they would soon disappear and leave us hungry. Food security largely depends on the availability and utilisation of this diversity, and there are very large economical interests at stake both in the food and the non-food sectors. It is a combination of these elements, which makes genetic diversity a commodity of strategic importance for countries and companies. However, due to the activities of man over the last few centuries, plant and animal species are dying out at a rate which has increased 1000-fold (Soulé, 1991). Plant populations are vanishing or becoming impoverished with regard to their genetic variation, the factor which would ensure their long-term survival and a necessary prerequisite for evolutionary changes.

The importance of conservation and utilisation of biodiversity has been widely recognised in recent years, and genetic resources have become a major issue of global concern and a driving force for political decisions. The issues of extinction, genetic erosion and possibilities for long-term conservation were the topics of the world conference in Rio de Janeiro in 1992 and the Convention of Biodiversity (CBD) which, since 1993, has been ratified by 180 countries (CBD, 2001). At one time, genetic diversity was considered to be a common heritage of mankind, but that is no longer the case since the Rio convention states national ownership over the native biodiversity. All countries are, however, interdependent with regard to the access to, and utilization of plant genetic resources. The particular issues concerning access to, utilisation and benefit sharing of the exploitation of plant and animal genetic resources are still unsolved problems. Therefore, a multinational agreement is urgently needed in order to secure increased efforts for conservation as well as for utilisation of plant genetic resources to broaden the genetic

---

Bothmer, R. von, K. Sato, H. Knüpffer and Th. van Hintum, 2003. Barley diversity – an introduction. In: R. von Bothmer, Th. van Hintum, H. Knüpffer and K. Sato (eds), Diversity in Barley (*Hordeum vulgare*), pp. 3-8. Elsevier Science B.V., Amsterdam, The Netherlands.

basis for breeding material.

The Rio declaration led to intensified activities of the FAO Commission on Genetic Resources, with for example, the initiative to report on the Current State of the World (SW; FAO, 1996a) based on information from individual countries. The Global Plan of Action (GPA) and the Leipzig Declaration (1996) stressed the issues of conservation and utilisation even further (FAO, 1996b). Based on the SW report, we now have a fairly good view of the general problems and status in the world, but the detailed knowledge of variation patterns, status of conservation of individual crops are far from being satisfactorily reviewed.

Genetic diversity also plays an important role in scientific research. To understand properly how plants grow and multiply, how populations develop and how species evolve, we need to improve our knowledge of distribution and function of genetic diversity. Since barley is one of the major crops in the world, it is natural that this interesting species is the focus of a book on diversity.

Importance of barley

Barley has a long history as a domesticated crop, as one of the first to be adopted for cultivation. Migration of people together with their seed crops led to a major diversification and adaptation to new areas, and the crop is now virtually found worldwide. Conscious selection of desired genotypes by farmers at an early stage, together with natural selection, increased the diversity and created the rich genepool source of variation found today in local varieties. These landraces also formed the basic material for modern plant breeding, which started some 150 years ago. The development of new technology and methods increased the genetic diversity even further and turned barley into the universal, highly diverse crop it is today.

The importance of the crop cannot be overestimated. Today it is cultivated on a total area of 55 million hectares (USDA-FAS, 2001), and is grown and used in fertile as well as in marginal areas under extreme conditions, including at altitudes up to 4,500 m in the Himalayas, in seasonally flooded areas in SE Asia, and in arid regions of the Mediterranean. Barley thus shows a very wide spectrum of adaptation. Over the centuries, barley has been planted for many different purposes. It was initially used as source of human food and animal feed. As a human staple food it has persisted until today in large regions in the mountainous areas of Central Asia, in southwest Asia and northern Africa including Ethiopia, where barley is still used for bread, porridge and "tsamba" (flour of roasted barley with black tea and yak butter). Its importance as an animal feed has increased over the years and barley is now one of the most important feed crops in temperate areas. Early on in the history of agriculture, man invented the processes of malting and brewing. For example, barley beer was one of the most important drinks in ancient Egypt. The processes have been refined and beer produced from pure malt as well as good malting barley varieties are currently in demand all over the world. Not to mention other beverages based on distillation, such as whisky and "barley water" (Central Asia) for which the special flavour and malting characteristics of barley are utilised. Special uses are the characteristic barley tea of Japan and Korea, pearled barley for mixing with rice in SE Asia, and "barley grass" as a functional food. Moreover, due to its specific characteristics, barley has also received attention in religious ceremonies or other rituals since ancient times.

One of the aims of this book is to review the present state of knowledge on diversity in a particular crop of global importance - barley. Over the last century, barley has been the subject of a great deal of scientific work, but, so far, no collective attempt has been made to produce a general review of the diversity of this important crop. Such a survey will eventually identify gaps in our current knowledge or in existing germplasm collections. It is our intention that this book will initiate further exploration and research.

Barley has the advantage, apart from being an important agricultural crop for food and feed, that it has also been used virtually worldwide as a model species for biological research. It is a diploid species with a low chromosome number (2n=14) possessing large chromosomes, and hence, a large genome. Over many decades, starting as early as in the 1930s, it has been the target of intense research on mutagenesis, mutagens, and mutants, particularly in Japan, Sweden and the USA (cf. Lundqvist, 1992). Pioneering works in the USA, Sweden and Germany have been dedicated to cytogenetics and chromosomal rearrangements, thus creating a fundamental knowledge of chromosomal structures (Hagberg *et al.*, 1961; Hagberg, 1986). Other fields of importance have been the basic work on morphological and adaptational differentiation mainly by Russian scientists (Sinskaya, 1969; Trofimovskaya, 1972; Lukyanova *et al.*, 1990), and on principles of evolution and domestication (cf. Zohary and Hopf, 1993; Bothmer *et al.*, 1995). Barley has also been a target organism for more practical approaches such as conservation and gene-banking (cf. Hintum and Visser, 1995), creation of core collections (cf. Hintum and Haalman, 1994; Knüpffer and Hintum, 1995), and plant breeding methodology for self-pollinating crops (Sunesson, 1963). Over the last decade, a great number of studies have been directed at research on molecular markers and genetic diversity for conservation and utilisation.

This book makes one thing very clear: there is a large diversity in barley, but there is possibly an even larger diversity in barley research. These two levels of diversity provide an excellent starting point for a book.

Components of the book

This book describes various aspects of the genetic diversity found in barley. It starts with a section on the origin of the current diversity. Chapters 2 and 3 put the genus *Hordeum*, its various species and macro-geographical groups into a genetic and historical context. They also explain how the current genetic diversity was shaped in nature and by man. The evolutionary and the domestication processes are described together with the early migration of the crop, and the results of breeding efforts in ancient as well as in modern times.

The diversity can be measured or assessed by a number of different methods, each showing a specific pattern, and it may often be difficult to compare sets of data obtained by different techniques. How can we compare the importance of a molecular marker study with a study of disease resistance or yield? The section includes detailed descriptions of various aspects of barley diversity. It starts with a, nowadays, rather unusual way of looking at the morphological and adaptational diversity in barley. Using the ecogeographical differentiation of the crop as a starting point, the resulting morphological peculiarities of the locally adapted material are described in the Vavilovian tradition (Chapter 4). Chapter 5 describes natural and induced mutations, which have received particular attention by the molecular geneticists, as a number of induced mutations allow physiological processes to be unravelled. The cytogenetic diversity pattern includes the variation found in chromosomal rearrangements, such as translocations, deletions, trisomics, heterochromatic regions, all of which have been thoroughly studied in barley over many years (Chapter 6). A wealth of new and interesting techniques for surveying the diversity have been developed over the last few decades, including biochemical and molecular markers, and numerous studies have contributed to our knowledge of the genetic variation patterns (Chapter 7). The diversity in response to stress is dealt with in two chapters, one on biotic and one on abiotic stress (Chapters 8 and 9, respectively). The first is a comprehensive survey of diseases and pests, and the genetic diversity to resist them. Chapter 9 lists abiotic stresses and the mechanisms and genetic background found in barley to adapt to, or tolerate them. The section is concluded by a discussion on quality characters and particularly the QTL variation in barley, what is known and what can be expected to be found (Chapter 10).

After describing the genetic diversity, another section will look at how genetic diversity can be utilised. It starts with a visionary chapter, a source of ideas about how information on genetic diversity in barley can be efficiently gathered, and how a holistic multivariate approach could result in new information on genotypes hidden behind complex phenotypes, viz. the new area of bioinformatics (Chapter 11). Barley is one of the most frequently collected crops, and barley accessions are widely distributed in genebanks. Chapter 12 sheds some new light on the diversity conserved and how to identify gaps and redundancies in existing collections. In the last part of the section, one chapter (13) is devoted to the International Barley Core Collection (BCC), a research tool which summarises the genetic diversity in barley in an accessible, optimised collection of barley lines, landraces and wild species in a growing base of knowledge about the material.

The concluding chapter presents a view on barley diversity, discusses various issues arising from the information presented in the book, composes the various chapters and searches for consensus and discrepancies.

Conventions

A few conventions have been used throughout the book. Most of them are of a typographical and editorial nature, such as the style of references and the way names of cultivars are written. Other matters are less trivial.

For reasons of convenience an attempt was made to use a single taxonomic classification system consistently throughout the book (cf. Bothmer *et al.*, 1995). This implies that, for example, the closest wild ancestor of cultivated barley is referred to as *Hordeum vulgare* L. ssp. *spontaneum* (C. Koch) Thell., or simply ssp. *spontaneum*, and all cultivated barleys as *H. vulgare* ssp. *vulgare*.

The nomenclature for designating barley genes and mutations has been a matter of intense discussion and disagreement over the years; several suggestions for gene designations have been proposed. At the 7th International Barley Genetics Symposium (IBGS) in Saskatoon, Canada in 1996, a consensus was reached where each gene is symbolised by a three-letter code. This system was proposed by Lundqvist *et al.* (1996), and their publication includes a detailed description of all known genes in barley. This system has been adopted and followed in the book.

Just like the gene symbols, the genomic constitution and designation of barley (*Hordeum vulgare*) and its wild allies in the genus *Hordeum* has been a matter of controversy over the years. Likewise at the 7th IBGS in 1996, a decision was taken that barley (*H. vulgare* s. lat.) and its secondary genepool relative (*H. bulbosum*) contain the genome named H, and the major part of other *Hordeum* species have the genome I, or still imperfectly known genomes, named Xa and Xy, respectively (Bothmer *et al.*, 1995; Linde-Laursen *et al.*, 1997). At the same meeting, the numbering sequence of the barley chromosomes was changed (1H-7H) to be congruent with the numbering system in wheat to which the barley chromosomes are homoeologous (Linde-Laursen *et al.*, 1997). These recommendations are followed throughout this book.

Whenever possible, genebank accession numbers are given of material described. These numbers can usually be recognised via a preceding letter code, such as CI for the material in the USDA collections, HOR for the material in the collection of the genebank of IPK, Gatersleben, or K for material in the collection of the Vavilov Institute, St. Petersburg. This material is, in principle, available for interested scientists and breeders.

Closing remarks

What image do we form if we cannot see the elephant, but only touch it? Will the descriptions of one person feeling the leg conflict with those of another person feeling the ear or the tusk or the trunk? This book is written in an attempt to create as complete a picture of the elephant as possible, by bringing together the descriptions of all aspects observed.

The book is a tribute to all scientists who created the basis of current knowledge of barley, naming only the two legendary scientists N.I. Vavilov (Loskutov, 1999) of the former Soviet Union and Henry Harlan (Harlan and Martini, 1936; Harlan, 1957) of the USA. Both have made fundamental contributions to the knowledge of genetic diversity in barley through their life-long compassion for collecting barley germplasm around the world. The editors would, in particular, like to dedicate this book to the memory of the "nestor" of barley genetic research, namely the late Professor R. Takahashi of Japan.

Finally, the editors wish to stress not only the important prosaic utilisation of the crop as a food, feed or malt or the exciting scientific development but also the beauty of the barley plant. What can be more beautiful than an individual, vigorous barley plant with its long and delicate awns or a whole heading field moving in the wind.

Acknowledgements

A publication aimed to review genetic diversity comprising contributions by various authors must by necessity be as diverse in thoughts and ways of presentation as reflected by the different authors. This diversity contributes to an added value to the publication. We are most grateful to authors of the different chapters for taking their tasks seriously, which has resulted in valuable, interesting and comprehensive contributions.

A publication will not be successful without a positive, patient and promoting publisher. We acknowledge Elsevier Science for being encouraging and cooperative during the process of development of the book. Thanks are due particularly to Ms. Brenda Vollers for skilful corrections of all manuscripts.

The International Barley Genetics Symposium, being the platform and providing the network for joint barley research and international collaboration, has been an important forum of great value for the finalisation of this book.

The Nilsson-Ehle Foundation, Lund, Sweden, is gratefully acknowledged for providing grants for the publication of colour illustrations.

The process of completion of this book has been promoted by our respective organizations providing working facilities and other support for the editors, and therefore we wish to acknowledge:

Department of Crop Science, The Swedish University of Agricultural Sciences, Alnarp, Sweden

Centre for Genetic Resources The Netherlands (CGN), part of Plant Research International, Wageningen, The Netherlands

Genebank Department, Institute of Plant Genetics and Crop Plant Research (IPK), Gatersleben, Germany

Barley Germplasm Centre, Research Institute for Bioresources, Okayama University, Kurashiki, Japan

References

Bothmer, R. von, N. Jacobsen, C. Baden, R.B. Jørgensen and I. Linde-Laursen, 1995. An ecogeographical study of the genus *Hordeum*. Systematic and Ecogeographic Studies on Crop Genepools, 7. IPGRI, Rome, 2nd ed.

CBD, 2001. Convention on Biological Diversity website (http://www.biodiv.org/Index.html) info January 2001.

FAO, 1996a. FAO State of the World's Plant Genetic Resources for Food and Agriculture. Food and Agriculture Organization of the United Nations, 510 p.

FAO, 1996b. Global Plan of Action for the Conservation and Sustainable Utilization of Plant Genetic Resources for Food and Agriculture and the Leipzig Declaration. Food and Agriculture Organization of the United Nations, Rome, 63 p.

Hagberg, A., 1986. Barley as a model crop. In: Barley Genetics V. Proc. 5th Int. Barley Genet. Symp., Okayama, Japan, pp. 823-831.

Hagberg, A., T. Ramage and C. Burnhman, 1961. A summary of translocation studies in barley. Crop Sci. 1: 277-279.

Harlan, H.V., 1957. One Man's Life with Barley. Exposition Press, New York.

Harlan, H.V., and M.L. Martini, 1936. Problems and Results in Barley Breeding. U.S. Dept. Agric. Yearbook of Agriculture.

Hintum, Th.J.L. van, and D. Haalman, 1994. Pedigree analysis for composing a core collection of modern cultivars, with examples from barley (*Hordeum vulgare* s. lat.). Theor. Appl. Genet. 88: 70-74.

Hintum, Th.J.L. van, and D.L. Visser, 1995. Duplication within and between germplasm collections. II. Duplication in four European barley collections. Genet. Res. Crop Evol. 42: 135-145.

Knüpffer, H., and Th.J.L. van Hintum, 1995. The Barley Core Collection – an international effort. In: T. Hodgkin, A.H.D. Brown, Th.J.L. van Hintum and E.A.V. Morales (eds), Core Collections of Plant Genetic Resources. John Wiley and Sons, UK, pp. 171-178.

Linde-Laursen, I., J.S. Heslop Harrison, K.W. Shepherd and S. Taketa, 1997. The barley genome and its relationship with the wheat genomes. A survey with an internationally agreed recommendation for barley chromosom nomenclature. Hereditas 126: 1-16.

Lukyanova, M.V., A.Ya. Trofimovskaya, G.N. Gudkova, I.A. Terentyeva and N.P. Yarosh, 1990. Flora of cultivated plants, vol. 2, part 2, Barley. Leningrad, Agropromizdat. (In Russian).

Lundqvist, U., 1992. Mutation research in barley. PhD thesis at The Swedish University of Agricultural Sciences.

Lundqvist, U., J.D. Franckowiak and T. Konishi, 1996. New and revised descriptions of barley genes. Barley Genet. Newsl. 26: 22-516.

Loskutov, I.G., 1999. Vavilov and his Institute. A history of world collection of plant genetic resources in Russia. IPGRI, Rome, 188 p.

Sinskaya, E.N., 1969. Historical geography of the cultivated flora. Leningrad, Kolos. (In Russian).

Soulé, M.E., 1991. Conservation: tactics for a constant crisis. Science 253: 744-750.

Sunesson, C.A., 1963. Breeding techniques – composite crosses and hybrid barley. In: Barley Genetics I. Proc. 1st Int. Barley Genet. Symp., Wageningen, The Netherlands, pp. 303-309.

Trofimovskaya, A. Ya., 1972. Barley (evolution, classification, breeding). Leningrad, Kolos. (In Russian).

USDA-FAS, 2001. United States Department of Agriculture – Foreign Agricultural Service website (http://www.fas.usda.gov/) info January 2001.

Zohary, D., and M. Hopf, 1993. Domestication of Plants in the Old World. Oxford Science Publications.

Section II

Origin of Barley Diversity

Diversity in Barley *(Hordeum Vulgare)*
Roland von Bothmer et al (Editors).

Chapter 2

The domestication of cultivated barley

Roland von Bothmer[a], Kazuhiro Sato[b], Takao Komatsuda[c], Shozo Yasuda[b], Gerhard Fischbeck[d]

[a]Department of Crop Science, Swedish University of Agricultural Sciences, Box 44, SE-230 53 Alnarp, Sweden
[b]Barley Germplasm Center, Research Institute for Bioresources, Okayama University, Kurashiki, 710-0046, Japan
[c]National Institute of Agrobiological Sciences, Kannondai, Tsukuba, 305-8602, Japan
[d]Department of Agronomy and Plant Breeding, D-85350 Freising-Weihenstephan, Germany

Taxonomic position of barley

Together with wheat (*Triticum aestivum* L.), rye (*Secale cereale* L.), and several important forages, like Russian wildrye (*Psathyrostachys fragilis* (Boiss.) Nevski) and crested wheatgrass (*Agropyron cristatum* (L.) Gaertn.), barley belongs to the tribe (tribus) Triticeae. This tribe represents a highly successful evolutionary branch in the grass family (Poaceae) and comprises a vast number of species and genera. The numerous wild species are thus potential gene sources for cereal breeding. The Triticeae comprises very complex modes of speciation, including polyploidy, interspecific and intergeneric hybridizations, which have resulted in a reticulate pattern of relationships. There are still major disagreements among taxonomists especially with regard to generic delimitations. No comprehensive systematic review of Triticeae has been presented in recent years (cf. Löve, 1984).

The tribe is distributed worldwide in all major temperate areas and is even present in the subtropics. Because of the large distribution area and the fact that diversity centres are confined to remote areas, which have not yet been fully explored botanically, there is still much basic taxonomic research to be done on species in the Triticeae. There is even a considerable uncertainty as to the number of species in the tribe, ranging from ca. 325 (Dewey, 1984) to ca. 500 (Löve, 1984).

Triticeae is considered to be monophyletic and is a good example of a plant group with a high degree of biological diversity including, for example, several genomes, various degrees of polyploidy, versatility in life forms, reproductive and dispersal patterns (Dewey, 1984).

The genus *Hordeum* and the genepools of barley

Barley, *Hordeum vulgare* L., is placed in *Hordeum*, which is a moderately sized genus with ca. 32 species and altogether ca. 45 taxa (Table. 2.1; cf. Bothmer *et al.*, 1995, for review and

---

Bothmer, R. von, K. Sato, T. Komatsuda, S. Yasuda and G. Fischbeck, 2003. The domestication of cultivated barley. In: R. von Bothmer, Th. van Hintum, H. Knüpffer and K. Sato (eds), Diversity in Barley (*Hordeum vulgare*), pp. 9-27. Elsevier Science B.V., Amsterdam, The Netherlands.

references). All species in *Hordeum* have a similar set of diagnostic, morphological characters, particularly with three, one-flowered spikelets at each rachis node, called a triplet. The two lateral florets are pedunculate, or sessile and may be sterile (as in two-rowed barley) or fertile (as in six-rowed barley). The glumes are setaceous or flattened and placed on the adaxial side of (and not surrounding) the spikelet.

Despite the seemingly homogeneous structure in basic morphology and specialisation, *Hordeum* shows a high degree of biological diversity. Some species are annuals often with more or less strict inbreeding, like *H. marinum* Huds., *H. murinum* L. and *H. pusillum* Nutt. Some species are perennials with a self-incompatibility system, like *H. bulbosum* L. and *H. brevisubulatum* (Trin.) Link. The majority of species are perennials with a versatile reproductive system.

Most species, like cultivated barley, are diploids (2n=2x=14), but tetraploids (2n=4x=28) and hexaploids (2n=6x=42) are also frequent (Table 2.1). Autoploidy is found in two species, *H. bulbosum* and *H. brevisubulatum*. The majority of polyploids are segmental alloploids including different variants of the basic genome I (for genome designations, see also Chapter 6). A few

Table 2.1. Taxa in the genus *Hordeum*, their distribution, chromosome numbers and life forms.

Species	Subspecies	2n	Life form*	Distribution
H. vulgare L.	2	14	A	Cultivated and E Mediterranean
H. bulbosum L.		14, 28	P	Mediterranean
H. murinum L.	3	14, 28, 42	A	Europe, Mediterranean to Afghanistan
H. pusillum Nutt.		14	A	USA, N Mexico and S Canada
H. intercedens Nevski		14	A	SW California and N Mexico
H. euclaston Steud.		14	A	C Argentina, Uruguay and S Brazil
H. flexuosum Steud.		14	A/P	Argentina and Uruguay
H. muticum Presl		14	P	W South America
H. chilense Roem. & Schult.		14	P	C Chile and W Argentina
H. cordobense Bothm. *et al.*		14	P	Argentina
H. stenostachys Godr.		14	P	Argentina, Uruguay and S Brazil
H. pubiflorum Hook. f.	2	14	P	W Argentina, Chile, Bolivia, Peru
H. comosum Presl		14	P	Chile and W Argentina
H. jubatum L.		28	P	W North America to E Russia
H. arizonicum Covas		42	A/P	S USA and N Mexico
H. procerum Nevski		42	P	C Argentina
H. lechleri (Steud.) Schenk		42	P	Chile and Argentina
H. marinum Huds.	2	14, 28	A	Mediterranean to Afghanistan
H. secalinum Schreb.		28	P	W Europe and N Africa
H. capense Thunb.		28	P	South Africa and Lesotho
H. bogdanii Wil.		14	P	C Asia
H. roshevitzii Bowd.		14	P	S Siberia, Mongolia, N China
H. brevisubulatum (Trin.) Link	5	14, 28, 42	P	Asia
H. brachyantherum Nevski	2	14, 28, 42	P	W N America to Kamchatka
H. depressum (Scribn. & Sm.) Rydb.		28	A	W USA
H. guatemalense Bothm. *et al.*		28	P	N Guatemala
H. erectifolium Bothm. *et al.*		14	P	C Argentina
H. tetraploidum Covas		28	P	S Argentina
H. fuegianum Bothm. *et al.*		28	P	S Argentina and S Chile
H. parodii Covas		42	P	S Argentina and S Chile
H. patagonicum (Haum.) Covas	5	14	P	S Argentina and S Chile

*A=annual; P=perennial; A/P=annual or weakly perennial

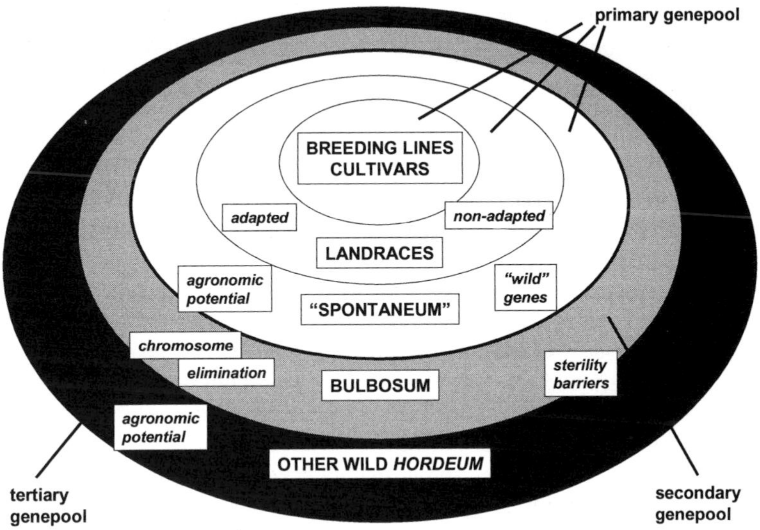

Figure 2.1. Genepools in cultivated barley (*Hordeum vulgare*).

species, like *H. secalinum* Schreb. and *H. capense* Thunb., are true alloploids including the I genome together with another genome, most probably the Xa genome from *H. marinum* (Taketa *et al.*, 1999).

Despite the close resemblance in basic morphology, the genus comprises four basic genomes as revealed by meiotic analysis of interspecific hybrids, cpDNA, karyotypes, isoenzymes, tandemly repeated DNA sequences, molecular markers and sequence analysis (Svitashev *et al.*, 1994; Bothmer *et al.*, 1995 and references therein; Marillia and Scoles, 1996; Komatsuda *et al.*, 1999). *H. vulgare* shares the basic genome H with *H. bulbosum* even though several studies point towards a clear differentiation between the two species (cf. Svitashev *et al.*, 1994). Most *Hordeum* species share variants of the basic genome I, whereas the two widespread weeds *H. marinum* and *H. murinum* possess quite distinct genomes, Xa and Ya, respectively (cf. Bothmer *et al.*, 1995, for references).

The genus is widespread and occurs in temperate areas in several types of biotopes worldwide, except in Australia. Diversity centres, here defined as areas housing the highest number of native species,are southern South America with 15 species, western North America with seven species, the Mediterranean with four species, and Central Asia with three species.

When applied to barley and its wild allies the genepool concept of Harlan and de Wet (1971) presents a very clear-cut picture (Figure 2.1). In addition to elite material, varieties, and landraces, the progenitor of domesticated barley, *H. vulgare* ssp. *spontaneum*, also belongs to the primary genepool of barley. Crossing combinations of cultivated barley with this form show no incompatibility barriers, hence there is a full capacity for gene transfer.

The secondary genepool includes only a single species, *H. bulbosum*, sharing the basic H genome with barley, but this crosses with some difficulty to the crop. However, in recent years it has been demonstrated that genes from *H. bulbosum* can be transferred to cultivated barley, thus providing a new source for breeding (Pickering, 2000, for review and references). Formerly, *H.*

bulbosum was used for the production of doubled haploids in barley breeding through chromosome elimination (Kasha and Kao, 1970; Pickering, 1984, 2000; Subrahmanyam and Bothmer, 1987).

All the remaining species of *Hordeum* are classified into the tertiary genepool. They cross with barley only with difficulty and backcrossing to the crop is even more difficult (Bothmer *et al.*, 1983; Bothmer and Linde-Laursen, 1989).

Due to the occurrence of different basic genomes, the obviously distant relationships between groups of species, and the possible polyphyletic origins of these groups, the taxonomy has been a matter of controversy for years. Primarily two genera have been considered, namely *Hordeum*, which in the narrow sense would include *H. vulgare* and sometimes also *H. bulbosum*. With this circumscription all other species are placed in *Critesion* (cf. Löve, 1984). However, the attempts, so far, to place the species into two different genera do not reflect true, consistent monophyletic phylogeny and relationships. Before any conclusive taxonomy can be achieved, reflecting a true phylogenetic pattern the treatment placing all species in a single genus, *Hordeum*, is the most practical solution and has gained general acceptance.

Studies of the wild species have, so far, mainly been directed towards taxonomy, distribution, morphological variation patterns, cytology, species relationships, and genome contents (see Bothmer *et al.*, 1995, for references). Comparatively few investigations have studied genetic diversity parameters in the wild species, apart from *H. vulgare* ssp. *spontaneum*.

The wild progenitor of barley

The immediate ancestor of cultivated barley is still abundant in nature. It was first discovered in Turkey by the German botanist Carl Koch, and described by him as a separate species, *Hordeum spontaneum*. However, based on several criteria, the progenitor form is nowadays regarded as a subspecies (ssp. *spontaneum* (C. Koch) Thell.) within the same major species, *H. vulgare* L., as cultivated barley (ssp. *vulgare*).

The centre of distribution for ssp. *spontaneum* lies in SW Asia, particularly in the Middle East. The natural distribution includes the eastern Mediterranean area with eastern Greece and Turkey, the Cyrenaica area of Libya and Egypt and the taxon extends eastwards to Afghanistan, Turkmenia and Baluchistan in West Pakistan (Giles and Bothmer, 1985; Zohary and Hopf, 1993). Brittle rachis types occurring outside this area (Morocco or Ethiopia in Africa or western China in the east) represent weedy forms and segregation products and not true wild forms (Figure 2.2).

Ssp. *spontaneum* has a large ecological amplitude. It grows in natural habitats in arid or semiarid biotopes, but also on segetal habitats on disturbed ground. It is an important annual component in open, herbaceous vegetation and it is particularly common in the summer-dry deciduous oak forests in the western part of the Middle East (Zohary and Hopf, 1993). Outside this area it occurs in drier steppes and semi-deserts representing more weedy types (Harlan, 1971). It may also be an aggressive weed in man-made habitats, in e.g. cultivated fields, edges of fields and roadsides. In the central parts of the distribution area ssp. *spontaneum* often occurs in very dense stands in large populations. In the more marginal parts of its native distribution area it is less abundant, scattered and even rare.

It may sometimes be difficult to distinguish between true, wild *spontaneum* and primitive forms and landraces of cultivated barley. The two taxa (ssp. *spontaneum* and ssp. *vulgare*) are morphologically similar but can be distinguished by a combination of a number of characters. Ssp. *spontaneum* is always two-rowed, often taller than ssp. *vulgare* of the same area, but lower types may occur in certain areas or habitats. The wild subspecies has a brittle rachis, but this character alone cannot identify the wild form since mutations and segregation products of crosses may also occur within ssp. *vulgare*. Ssp. *spontaneum* is usually more open-flowering and hence has a higher frequency of cross-pollination than the cultivated form. Outbreeding of up to 10% has been reported (Brown *et al*., 1978; Nevo, 1992). The dispersal is adapted to zoochory (seeds are attached to, for example, furs of animals) and these traits are still intact in ssp. *spontaneum* but modified under domestication in ssp. *vulgare*. The wild traits include long and tough bristles on rachis segments and on the rachilla as well as a tough (non-brittle) awn. The kernels are often shrunken, not plump, as in cultivated barley. Apart from the visible, morphological traits the wild ssp. *spontaneum* also has a number of specific adaptive traits, such as a well developed dormancy system and high drought tolerance (Nevo, 1992; Rijn *et al*., 2000).

No crossing barriers have been developed between the wild and the cultivated forms and spontaneous and artificial crosses are easily obtained (cf. Asfaw and Bothmer, 1990). However, in some populations of ssp. *spontaneum* chromosome translocations may occur, resulting in a reduced fertility in some crosses (Ahokas, 1999, see also Chapter 6). There has certainly been a high frequency of introgression in areas where the wild and the cultivated forms are in close contact. The wild form is thus an excellent source of useful alleles for barley breeding as has been demonstrated in several current national and international projects (Schönfeld *et al*., 1996; Lehmann *et al*., 1998).

Since ssp. *spontaneum* occupies a large ecological amplitude, a large diversity in several characters has been developed often related to climatic or edaphic gradients or other causes (e.g., resistance factors or biochemical compositions) across many ecological niches demonstrating a wide range of adaptability and underlying genetic diversity (cf. Nevo, 1992; Zohary and Hopf, 1993). Several studies have been performed on Middle East material in resistance traits, drought and salt tolerance as reviewed in different chapters of this book (see also Nevo, 1992, for review and references).

The domestication of barley

The identified area for the dawn of agriculture lies in the particular area of the Eastern Mediterranean called The Fertile Crescent, which comprises the arch from present day central Israel, over western Jordan, Lebanon, Syria, SE Turkey, N Iraq, and to the Zagros mountains in

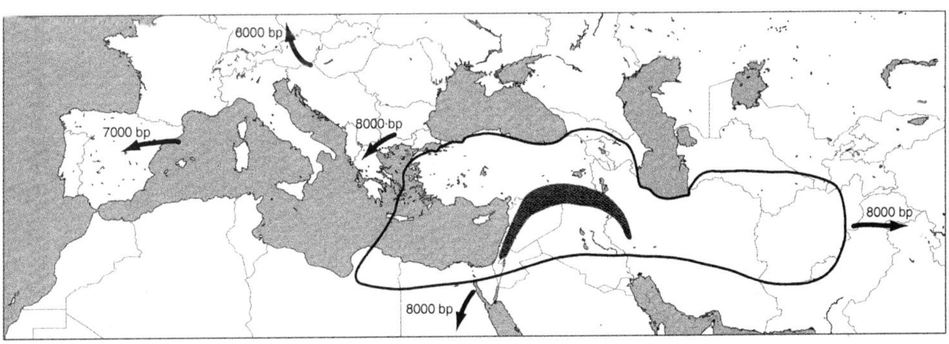

Figure 2.2. The Fertile Crescent, the area of early domestication of cultivated barley (*Hordeum vulgare* ssp. *vulgare*) in the Middle East, distribution of the wild progenitor of barley (*H. vulgare* ssp. *spontaneum*) (within solid line), and approximate time (year before present) for cultivated barley to reach different areas.

western Iran (Figure 2.2). It constitutes mainly a mountainous or hilly area with relatively dry steppes and dry woodlands (oak forests). Early settlement remains have shown that in The Fertile Crescent man went from being hunter/gatherer to become a sedentary farmer. The welfare of the early societies and the basis for our current civilisation was created by the development of agriculture and the gradual domestication of a number of plant and animal species, like einkorn and emmer wheat, barley, flax, vetches, lens, peas and goat, sheep, cattle and pigs (cf. Zohary and Hopf, 1993; Smith, 1995; Lev-Yadun *et al.*, 2000).

Barley, einkorn and emmer wheat are probably the crops presenting the best archaeological and biological evidence for the process of domestication. Remains of prehistoric cereal cultures are often composed of barley-wheat mixtures. As a result of intensive archaeological research during the second half of the 20th century, carbonised kernel imprints in pottery and mud-bricks have appeared among the remains showing the transition from the wild to a domesticated state (Zohary and Hopf, 1993; Smith, 1995; Ladizinsky, 1999).

Gathering of wild barley seeds from nature seems to have occurred in great quantities as early as 17-19 thousand years ago (Kislev *et al.*, 1992; Harlan, 1992, 1995; Zohary and Hopf, 1993; Ladizinsky, 1999). Gradually, barley was adapted to cultivation and there is clear evidence of early cultivation as well as signs of initial domestication dating from ca 10,000 years ago (Zohary and Hopf, 1993; Harlan, 1995; Smith, 1995; Ladizinsky, 1999). Evident remains of non-brittle barley in a low frequency (10-12%) are found in the 8th millennium BC, usually in mixtures with brittle ssp. *spontaneum* types. Suitable mutations were given the chance to survive in seed mixtures from non-fragile ear types soon after the beginning of barley cultivation. The earliest type to appear was two-rowed barley. Six-rowed types appeared somewhat later (ca. 9,500 years ago) and from ca. 6,000 BC naked forms occurred.

There is a strong tendency to higher percentages of cultivated wheat at sites with better soil fertility even in the early stages of agricultural activities within The Fertile Crescent. Barley regained dominance at lower soil fertility, harsher climates and shorter vegetation periods.

Theories of domestication

Over the last 150 years various ideas and theories have been presented as to the evolutionary pathways from the wild to the domesticated state in barley. The discovery of the wild ssp. *spontaneum* led to the hypothesis that cultivated, two-rowed barley was a direct derivative from this form based on their close resemblance (cf. de Candolle, 1959). However, it did not explain the occurrence of the six-rowed, cultivated barley. Due to the drastic morphological differences between two- and six-rowed barley, it was firmly believed that the two forms would represent completely separate, divergent evolutionary events. The discovery of a six-rowed form with brittle rachis in western China in the early 1930s seemed to solve this problem. The material was described by Åberg (1938) as *H. agriocrithon* and it was assumed to be the ancestor of six-rowed barley (Åberg, 1940; Bell, 1965). At that time one was not aware of the simple genetic background for row-number (see below) and considered that such a drastic morphological difference would indicate different phylogenetic pathways, resulting in separate speciation. The present opinion is that there is only one evolutionary line leading to two-rowed barley and with ssp. *spontaneum* as the only progenitor from The Fertile Crescent. Later mutations and crosses have resulted in the occurrence of six-rowed barleys, probably more than once in different areas. Similarly, brittle rachis types outside the primary distribution area of ssp. *spontaneum* are not evidence of wild forms, but they are a result of mutational events, crosses and segregation. There is obviously no selection against these types occurring in the fields of landraces since the genes for brittleness show a late expression and the crop is harvested when unripe (Bothmer *et al.*, 1990).

Nevski (1941) postulated that the common progenitor of barley had intermediately developed lateral spikelets and a fragile rachis. He considered that the lateral spikelets of *H. vulgare* ssp. *spontaneum* were too small to develop into forms with the large lateral spikelets of six-rowed barley, although the basis was not very solid because a single mutation at the *vrs1* locus causes a change from two-rowed to six-rowed (Lundqvist *et al.*, 1996). According to Helbaek (1960), *H. vulgare* ssp. *spontaneum* shows a large genetic variation, and Bakhteyev (1963) distinguished four varieties based only on the development and fertility of lateral spikelets. Takahashi and Tomihisa (1970) demonstrated that the variation of the awn and fertility of lateral spikelets were under the control of the alleles at the *vrs1* locus. They regarded that var. *proskowetzii*, showing intermediately developed lateral spikelets with elongated awns, is the most primitive form and, in line with the hypothesis by Nevski (1941) and Harlan (1968), and postulated that the form is a progenitor of cultivated barley.

It has been a matter of discussion whether barley was taken into cultivation only once in The Fertile Crescent or whether it was the subject of repeated domestication in space and time in the area (cf. Harlan, 1992). For the other original crops of The Fertile Crescent, a single domestication event is assumed. However, according to several authors, barley is the single species, which is assumed to have undergone multiple domestication events, also outside The Fertile Crescent. Especially the occurrence of brittle rachis types outside the core area of The Fertile Crescent has been accepted as evidence of more than one domestication event. More recently, some studies on markers have also been applied to the problem (Molina-Cano *et al.*, 1999). So far, no comprehensive study has unambiguously shown that more than one domestication event occurred in barley outside The Fertile Crescent (cf. Yasuda *et al.*, 1993; El Rabey and Salamini, 2000; Blattner and Badani Méndez, 2001).

Expansion of barley cultivation during prehistoric times

Within a time span of about 4,000 years the "Old World" type of agriculture, based on the original set of cereals and a few other species, expanded over most of the areas eligible for a sedentary way of life throughout Eurasia. The cultivation of wheat and barley reached Greece and Iran and moved eastwards to India very early on, around 8,000 years ago. The first barley remains found in Spain date from the 5th millennium BC and barley reached northern Germany and Southern Scandinavia ca. 6,000 years ago (Hjelmqvist, 1979; Zohary and Hopf, 1993; Ladizinsky, 1999). The expansion also included the North African coastal region of the Mediterranean basin, and moving upwards along the Nile probably reached Ethiopia ca 8,000 years ago. About 3,000 years ago cultivation of barley had reached China (Ho, 1977), possibly by seed exchange. Somewhat later, six-rowed hulled and naked barley became essential crops for feed and food supply in the ancient agriculture in Japan (Seko, 1987). The different routes of early migration are also the basis for the early differentiation into Oriental and Occidental barley (Strelchenko *et al.*, 1999).

Obviously seed mixtures were taken along the various routes. With rather slow migration rates there has been sufficient time for adaptational changes by natural selection within heterozygous populations as responses to changes in environmental conditions. Mass selection performed by the farming communities favouring preferential plant types might have played a role. Low, but consistent rates of outcrossing, such as in ssp. *spontaneum*, provided the basis for genetic recombination between different genotypes. Bottleneck effects resulting from restricted sizes of founder populations in a new environment affected the structure of the surviving population. In the end, this expansion process provided a multitude of barley landraces, each of them locally distributed and forming the basis for cultivation and large genetic diversity of this crop stretching into the 19th century.

Important traits for domestication and early migration

The domestication of barley was a gradual process with accumulation of genes particularly suited for cultivation. The selections may have been unconscious, i. e., as a result of the action of edaphic or climatic factors, or conscious, that is, as a result of deliberate choices of desired traits by man. Some traits were of particular importance for the domestication itself or connecting closely with the early cultivation systems. Studies of wild barley have shown its large genetic diversity which undoubtedly has been of importance in the domestication process (Nevo *et al.*, 1986). Also after the primary domestication advantageous traits may have accumulated such as high enzymatic activity in the cultivated forms of importance for malting. Important mutations occurred, such as adaptation to different climatic and edaphic conditions, during an early stage of cultivation which promoted the migration process of barley into other parts of the world.

Brittleness of rachis

Shattering is a character of natural adaptation in wild plants. The brittleness of rachis in barley promotes the spreading of seeds together with the rough awn, which may easily attach to animals for effective dispersal. The detached rachis segment with one spikelet in ssp. *spontaneum* (one dispersal unit) is shaped like an arrowhead which is why it is difficult to pull the seed out after it attaches somewhere. One of the most important traits for the domestication of barley is probably non-brittleness of rachis which is of benefit for an efficient harvest without loss of grains. In archaeological remains, wild forms of barley with fragile spikes were found in some quantity (Harlan, 1995). The elimination of such a deleterious character was probably the first requirement for a cultivated form.

Takahashi and Yamamoto (1949) clarified the genetic system of non-brittleness of barley, occurring in ssp. *spontaneum*. When they made crosses among cultivated barleys, brittleness of rachis occurred only in crosses between East Asian and European cultivars. Two recessive genes, *btr1* and *btr2*, each responsible for non-brittle rachis had been independently established by natural mutations in the wild progenitor which had a brittle rachis due to two dominant, complementary genes, *Btr1* and *Btr2* (Takahashi, 1987). Since the F_1 progeny of the cross between a genotype *Btr1Btr1 btr2btr2* (type E=eastern) and a *btr1btr1 Btr2Btr2* (type W=western) is brittle, the brittleness of rachis might be a key character to differentiate the germplasm groups. The double recessive genotype, *btr1btr1 btr2btr2*, is not found in any landrace material. According to Takahashi and Hayashi (1964), the two genes, *Btr1btr1* and *Btr2btr2*, are pseudo-allelic. The complementary action of *Btr1* and *Btr2* in the heterozygote suggested that these two were situated on different loci, but no recombinants were recovered in F_2 of a cross between the two different genotypes, *Btr1Btr1 btr2btr2* and *btr1btr1 Btr2Btr2*.

Takahashi and his co-workers made two-way test crosses of a world collection using two kinds of pollen donors, typical East Asian and European cultivars. The relative frequencies of type E cultivars and type W cultivars in various geographical regions showed that varieties of Oriental origin (Nepal, China, Korea and Japan) are mostly type E, while type W is very frequent in Occidental material, especially from Turkey, Europe and Ethiopia (Figure 2.3). Two-rowed barleys are mostly type W, with some exceptions. Type W is dominant in Occidental, six-rowed

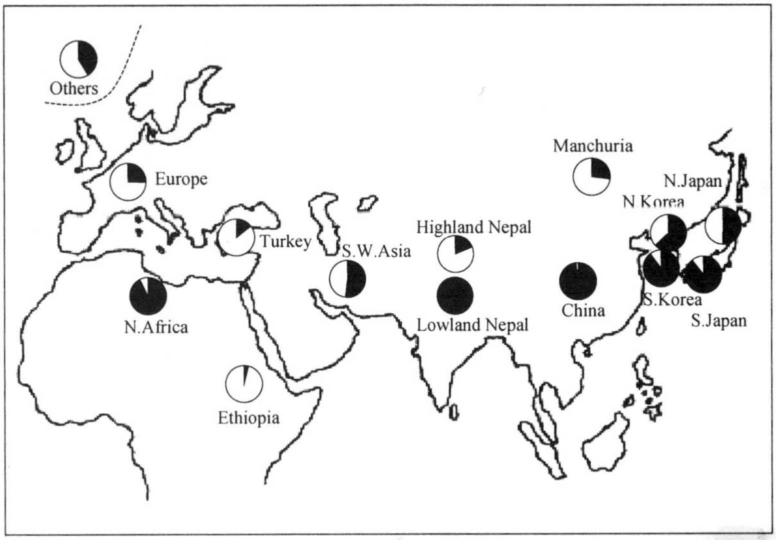

Figure 2.3. Geographical difference in the ratio of two non-brittle barley genotypes *Btr1Btr1 btr2btr2* (type E: black) and *btr1btr1 Btr2Btr2* (type W: white).

barleys but not as frequent as in the two-rowed varieties. In East Asia, varieties with type W are rather frequent in the northern parts of Japan, North Korea and Manchuria (Takahashi, 1955; 1963; Takahashi *et al.*, 1983).

Kernel row type

There is a wide variation in kernel row type with a detailed classification system (Lundqvist *et al.*, 1996). The row type is basically controlled by the gene *vrs1* and six-rowed is recessive to two-rowed (Figure 2.4.A-D). Several genes showing imperfect six-rowed, such as *vrs2* or *vrs3*, have been found mainly in artificially induced mutants. The development of lateral kernels is also controlled by the gene *int-c* (*intermedium* spike-c), which regulates the size of lateral spikelets (Figure 2.4.E). The double recessive of *vrs1* and *int-c* sometimes produces poor development of lateral spikelets in six-rowed spikes. On the other hand, *Vrs1.t* is the most dominant allele in the multiple allelic series of *vrs1* and produces the *deficiens* genotype missing lateral spikelets. *Irregulare* is randomly missing fertile lateral spikelets and it is mainly distributed in Ethiopia, Northern India and Pakistan (Takeda and Saito, 1987). The development of lateral spikelets is also influenced by growing conditions. A combination of spikelet fertility genes and an awnlessness gene *Lks1* tightly linked with *vrs1*, results in a diverse morphology of barley spikes especially in East Asian material (Takahashi, 1987).

The *vrs1* is probably the most important gene for regulation of the row number. Tanno *et al.* (1999, 2000) studied DNA sequences of some 900 bp closely linked to the *vrs1* locus, which showed that ssp. *spontaneum* has a larger variation than ssp. *vulgare*. Within ssp. *vulgare* the two-rowed form had a larger variation than the six-rowed form and the study showed that there are two distinct lineages of six-rowed barley (type I and II). Type I, including the majority of six-rowed barley in the world, had an identical DNA sequence with that of a strain of var. *proskowetzii* from Turkmenistan. Type II, being less frequent and distributed in the

Figure 2.4. Triplets in *H. vulgare* ssp. *spontaneum* (A), var. *proskowetzii* (B) and six-row cultivated (C). D shows the dominance of two-row in the F_1 of six-row (*vrs1vrs1*) and two-row (*Vrs1Vrs1*). E shows the partial fertility of the lateral spikelets of heterozygous *Vrs1vrs1* interacting with the *Int-c.h* gene.

Mediterranean region, had an identical DNA sequence as that of brittle rachis types from Morocco. The molecular study thereby showed a diphyletic origin of six-rowed barley.

Two-rowed forms analysed included four types of DNA sequences. The major type showed only one nucleotide difference with that of type I of six-rowed barley. The DNA sequence of the major type was identical to that of some materials of Iranian var. *spontaneum* (Sayed *et al.*, 2000; Tanno *et al.*, 2002), whereas the three remaining types were distributed in the Mediterranean region and Ethiopia. As the two-rowed spike is very similar in cultivated and wild forms, any introgression between the two forms can not be excluded.

A recent molecular phylogenetic study concerning the *vrs1* region showed that ssp. *spontaneum* and ssp. *vulgare* form a sister group to *H. bulbosum*. These taxa formed a sister group to *H. murinum* (Komatsuda *et al.*, 1999). Since *H. bulbosum*, *H. murinum* and many other species of the genus *Hordeum* have somewhat developed lateral spikelets, the most primitive form of ssp. *spontaneum* may be similar to the var. *proskowetzii* (Takahashi and Tomihisa, 1970).

Covered and naked kernels

Naked kernel is a single recessive character from the covered wild type. Naked barley is distributed widely in the world, but there is a higher preference for naked barleys in East Asian countries such as China, Korea and Japan and it is especially high in Tibet and the northern parts of Nepal, India and Pakistan. Since the frequency is low in the west, Vavilov (1926) considered southeastern Asia to be a centre of origin for naked barley. It has, however, become clear that naked barley was grown in Anatolia (Turkey) and in northern Europe already in ancient times (Hunter, 1952; Helbaek, 1969).

Some of the naked barleys of Oriental origin have a glutinous endosperm which segregates as a simple recessive gene to starchy endosperm (Kashiwada, 1930). In these areas, barley is used as a major human diet and naked barley is preferred. In the southwestern part of Japan, a special semi-dwarf type called "uzu", earlier covering about 80% of the whole barley acreage in Japan, is now used mainly in the naked form (Takahashi, 1951). The smaller kernel of the uzu type is preferred to cook together with rice in Japan due to the similar sizes of the two cereals.

Dormancy

Dormancy is a natural adaptational system controlling the seed germination in semi-arid areas where barley was domesticated. Takeda (1995) evaluated the dormancy of more than 4,000 cultivars and 177 wild (ssp. *spontaneum*) accessions. All wild material tested was highly dormant (see also Chapter 9). Compared to other characters which might have been important for the domestication, the genetic system of dormancy seems more complicated. Takeda (1995) reported quantitative inheritance of dormancy as a result of a diallele analysis and several QTLs from the cross of 'Harrington' × TR306. A certain level of dormancy is also useful in cultivated material to prevent pre-harvest sprouting or unnecessary starch degradation during the storage period.

Growth habit

One of the prerequisites for expansion of the cultivation area for barley must have been differentiation of spring habit. In high latitudes and in mountainous regions, barley is sown in spring to avoid damage by a severely cold winter. Accordingly, in these regions spring type cultivars are needed in order to grow and head normally. At low latitudes, on the other hand, air temperature is too high to induce vernalisation in a winter type. Spring type cultivars prevail in these regions. In mid-latitudinal regions including North Africa, southern Europe, Nepal, China

Table 2.2. Number of barley cultivars with five spring genotypes* in major barley growing areas of the world (Yasuda, 1992; Yasuda *et al.*, 1993).

Geographic regions	I *+Sgh2+*	II *+Sgh2Sgh3*	III *sgh1++*	IV *sgh1Sgh2+*	V *sgh1Sgh2Sgh3*	Total
Japan, Korea and C and S China	17		12**	22**		51
Tibet and Nepal (naked)	18	7		2		27
India	7	14		1		22
Pakistan and Afghanistan	17	54		1		72
Iran and Iraq	7	4	3			14
Turkey	6	3	10	3		22
Ethiopia	27	20		2		49
North Africa	38			6		44
Northern Japan and Korean Peninsula	4		5	15		24
North China (Manchuria)	1			13	2	16
North Europe	1			9	3	13
Central and South Europe	18	2	9	62	1	92
Russia	2	6		10		18
USA and Others	5	2		4	3	14
Total	168	112	39	150	9	478

* Genotype of winter barley: *Sgh1sgh2sgh3*. Each winter gene is shown by +
** Mostly two-rowed, exotic cultivars and some hybrid products with exotic forms

and Japan, both spring and winter barley cultivars are generally sown in autumn.

The genes *sgh1*, *Sgh2* and *Sgh3* are all regulating spring habit, and their allelic genes winter habit. Because of the epistatic effect among these genes only a single genotype, *Sgh1sgh2sgh3*, exhibits winter habit. Linkage studies of the three spring type genes, *sgh1, Sgh2* and *Sgh3*, have shown that they are located on chromosome 4H, 5H and 1H, respectively (Takahashi and Yasuda, 1956, 1958; Yasuda, 1969; Laurie *et al.*, 1995). It has also been determined that different degrees of vernalisation requirement are controlled by the multiple allelic genes *Sgh2I* and *Sgh2II* denoting the spring genes with high and moderate degree of spring habit, respectively. The *Sgh2* locus has been hypothesised to be homoeologous with the *Vrn1* locus of wheat (Karsai *et al.*, 1997).

A total of 478 spring cultivars were classified into five genotypes from seven possible spring genotypes (Table 2.2). The gene *Sgh2* is the most frequent one and is distributed worldwide, while *sgh1* is confined to the occidental region, with rare exceptions. The third gene, *Sgh3,* is less frequent than thc other two, and encountered mainly in the mountainous regions of Pakistan, Afghanistan, India and Ethiopia. There are also some geographical regularities in distribution of the different spring genotypes (Table 2.2).

Almost all strains of *Hordeum vulgare* ssp. *spontaneum* are of winter habit with the exception of a few strains which are regarded as cross products with spring cultivars (Takahashi *et al.*, 1963, 1968). Consequently, the first barley types to be domesticated might have been of winter habit type, but a dominant mutation occurred first in the *sgh2* locus, resulting in a spring barley of type I. This hypothesis is supported by the fact that the I type of spring barley is adapted to lower latitudinal regions and co-exists with the winter type. Subsequently, mutations from *Sgh1* and *sgh3* might have occurred independently in the I type, because both spring genes are mostly accompanied by *Sgh2*. Although a mutation to *sgh1* might have occurred from a winter type, it seems more probable that in Japan the genotype III might have been established after hybridisation between a winter type and type IV, since all Japanese two-rowed barleys of this genotype are known to have originated from such crosses. Type V is supposed to be the hybrid product between II and IV. The gene *Sgh3* appeared later than the gene *sgh1* and it is always accompanied by *Sgh2*. The geographical distribution of type II is restricted to mountainous regions of the Indian Subcontinent and Ethiopia.

Productivity and quality traits

Wide cultivation of barley results in increased probabilities for natural mutational events. Direct or indirect selections in cultivation have increased the ratio of genotypes preferred by humans. Productivity is apparently the largest concern for human consumption, but it is a complex character consisting of physiological changes, adaptations of the genotype and reductions of yield losses, etc. (Evans, 1993). Human needs for quality characters such as for malting purposes or human diets resulted in strong selections on the barley populations. Once categories such as malting varieties or naked semi-dwarf food barley were established, introductions of other sources of variation were restricted. In this way, some of the promising sources of variation from the wild or primitive germplasm during the process of domestication and migration might have been lost. For example, some genes with high level of enzymatic activities, of potential importance for malting, are present in ssp. *spontaneum* collections as reported by Ahokas and Naskali (1990), but not in cultivated forms.

Disease resistance

A clear host-pathogen coevolution system has been suggested in barley fungal diseases, especially in mildew resistance (Jørgensen, 1992). Even in wild *Hordeum* species, mildew resistance is a key to estimate genetic distance among species (Gustafsson and Claesson, 1988). A classical study of mildew resistance to the Japanese race13 in a world barley collection showed that *H. vulgare* ssp. *spontaneum* had a wide range of reactions and most varieties from East Asia are highly susceptible to this isolate, whereas Occidental genotypes contain higher proportions of resistant or medium-resistant forms (Nishikado *et al.*, 1951; Hiura and Heta, 1952, 1954). It is generally recognized in a gene-for-gene system that the virulence of the pathogen population might change the constitution of the resistance genes in the corresponding germplasm.

The diversity of disease resistance in ssp. *spontaneum* is wide compared to cultivated barley (Nevo, 1992). Primitive barleys or landraces have basically a fairly high level of resistance to most diseases and bottlenecks might differentiate the resistance of germplasm. For example, Barley Yellow Mosaic Virus (BaYMV) was a local barley disease in Japan until recently and has spread into other countries such as China and Germany. Yasuda and Rikiishi (1997) found a high level of resistance to BaYMV in Ethiopian germplasm. Differences between germplasms can also be found in the reaction to *Pyrenophora teres* f. *teres* in Nepalese barleys. Naked food barley, distributed in the Himalayan region, has a high level of resistance to the Canadian isolate WRS102, but hulled barley, distributed mainly in the Indian sub-continent, is quite susceptible to the isolate (Sato and Takeda, 1994; Takeda *et al.*, 2000).

Abiotic stress tolerance

A major reason for the increase of barley cultivation during the Mesopotamian age might have been higher tolerance to salt stress in barley than in wheat (Nair and Khulbe, 1990). When salinity began to increase in the irrigated land of southern Mesopotamia, wheat production declined, and a near monoculture of barley was established (Harlan, 1995). According to Mano and Takeda (1995), there is no evident geographical differentiation for salt tolerance at the seedling stage (see also Chapter 9) and most wild accessions showed better tolerance to salinity than cultivated material (Mano and Takeda, 1998). Some genotypes of ssp. *spontaneum* ripened even in 60% sea water (Nevo *et al.*, 1993), suggesting a very high salt tolerance in *Hordeum* in comparison to other cereals. Tolerance to excessive moisture is one of the key adaptive traits for cultivation in an Asian monsoon climate, especially when grown as a winter crop in paddy fields (see also Chapter 9). The tolerance should be expressed during germination and early developmental stages. Some of the most tolerant genotypes were found in East Asian accessions (Takeda and Fukuyama, 1987; Takeda, 1989).

Pigmentation

Kernel pigmentation itself is not negative for production. For example, Tibetans do not care about grain colour of hull-less barley since they mill roasted grain to make a flour dough with butter tea (tsamba). In Japan purple grain was selected for some waxy barleys in order to distinguish it from regular, non-waxy ones. However, it is not used for malting or even feeding since the pigmentation gives an unpleasant appearance. Black colouring of kernels is caused by a melanin-like substance and controlled by the gene *Blp* and purple kernels are due to the production of anthocyanin regulated by the gene *Pvc* (Lundqvist *et al.*, 1996). Black-headed, primitive barleys were found in southwestern Asia and also sparsely in Tibet and China (Takahashi, 1987). Blue aleurone layer is due to anthocyanin which is of little importance in covered barley used as feed, but apparently paid attention to when used for human consumption. This character is controlled by the two complementary genes, *Blx1* and *Blx2* (Mayler and Stanford, 1942). Most wild and primitive barleys have blue kernels, the naked varieties have predominantly white kernels, whereas more than half of the covered types of the world have blue kernels (Takahashi and Yamamoto, 1950).

Hybrid chlorosis

Inferior growth in an F_1 progeny may be due to the fact that parental genotypes are genetically distantly related. Takahashi and Hayashi (1987) found hybrid chlorosis governed by two complementary dominant genes, *Ch-a* and *Ch-e*, the latter of which was found only in an Ethiopian strain (AA12). More than a thousand accessions including ssp. *spontaneum* were crossed with AA12 to establish the chlorotic symptom of F_1 plants with the gene *Ch-a*. Varieties with the *Ch-a* gene are frequent in East and SW Asia, while the frequency of *Ch-a* is low in Occidental regions, such as Turkey, Europe and North Africa. This pattern probably reflects the early geographical differentiation in barley.

Traits neutral to migration

Some traits seem to be neutral and without having been subject of natural selection during early migration. However, these traits show an obvious large diversity in the gerplasm. Changes in DNA sequences, such as molecular or isozyme markers, are good examples of neutral traits and can be used for the estimation of relationships among accessions (see Chapter 7). Some of the kernel characters, such as rachilla and lodicule hair lengths, are used as visible markers to identify malting barley varieties, indicating their neutral performances in practical use. As an exception, Jensen (1989) reported the linkage between a short rachilla hair (*srh*) and a quantitative trait locus (QTL) controlling kernel weight. In this case, the allelic ratio of *srh*/*Srh* in the barley population might be affected by the selection of kernel weight.

Reaction to some chemicals might be neutral to selection since the crop had never been in contact with these chemicals during its early migration. Takeda (1996) found a reaction sensitivity to the insecticide diazinon and that sensitive genotypes were widely distributed in the Middle East but rare in East Asia. Takeda and Chang (1996) screened the phenol staining reaction on awns and husks and found that only 51 accessions in a world collection out of 9,000 were reaction-less. These neutral traits might be mutated, included in the barley population and kept at a certain ratio.

Conclusions and outlook

Through evolutionary processes, domestication, migration correlated with adaptation to new environmental conditions, and conscious selection by early farmers a wealth of genetic diversity was created in barley. These dramatic changes developed as a result of human activity over a period of 10,000 years. A large number of essential characters were irreversibly changed when barley became a cultivated crop, due to intense selection in the early phases of domestication. The last step for the development of the present day state of genetic diversity took place during the last century due to intensive breeding based on a wide array of new methods (see also Chapter 3).

During the time span from the early domestication up to now, particular distribution patterns of genetic diversity have been substantiated, such as the differentiation of oriental and occidental types based on the brittle rachis genes, differentiation of two- and six-rowed types, development of secondary diversity centres (such as Ethiopia, and western China) outside the centre of origin in the Fertile Crescent. Later breeding, exchange of material, and the use of exotic gerplasm have made the genetic diversity pattern even more complicated.

The diversity developed over millennia resulted in a great number of locally adapted landraces in practically all temperate parts of the world. This rich genetic variation was the basis for modern plant breeding and a great deal of the genetic variation in landraces is certainly present in our modern varieties. However, the industrialized society in particular has faced a severe genetic erosion of its older, well adapted germplasm. How much of this diversity has vanished or how large the depauperization was, is not known, and should be the target for further studies. In certain areas, such as Europe, almost all older material has been lost and is not represented in genebanks. In other areas, such as in Ethiopia, southwestern or Central Asia, the cultivation of landraces is still practiced and genebank collections have been created in recent years (see Chapter 12). Due to the high numbers of genebank accessions, gaps in the current collections are presently difficult to identify.

Still not all details of the domestication process and early migration of cultivated barley have been clarified. Further studies of the genetics behind individual traits are necessary as well as investigations of evolutionary patterns based on biosystematics and molecular markers. Molecular techniques based on ancient, archaeological material and future studies of modern material from different areas will reveal more details elucidating several, as yet unsolved, questions about domestication and evolution of barley, such as the old question of mono- or polyphyletic origin.

References

Åberg, E., 1938. *Hordeum agriocrithon* nova sp., a wild six-rowed barley. Ann. Agric. Coll. Sweden 6: 159-216.

Åberg, E., 1940. The taxonomy and phylogeny of *Hordeum* L. sect. Cerealia Ands. With special reference to Thibetan barleys. Symbolae Bot. Upsaliensis 4: 1-156.

Ahokas, H., 1999. On the territorial distribution of chromosomal interchanges in wild barley, *Hordeum vulgare* ssp. *spontaneum*. Barley Genet. Newsl. 29: 40-41.

Ahokas, H., and L. Naskali, 1990. Geographic variation of α-amylase, β-amylase, β-glucanase, pullulanase and chitinase activity in germinating *Hordeum spontaneum* barley from Israel and Jordan. Genetica 82: 73-78.

Asfaw, Z., and R. von Bothmer, 1990. Hybridization between landrace varieties of Ethiopian barley (*Hordeum vulgare* ssp. *vulgare*) and the progenitor of barley (*H. vulgare* ssp. *spontaneum*). Hereditas 112: 57-64.

Baktheyev, F., 1963. Origin and phylogeny of barley. In: Barley Genetics I. Proc. 1st Int. Barley Genet. Symp., Wageningen, The Netherlands, pp. 1-18.

Bell, G.D.H., 1965. The comparative phylogeny of the temperate cereals. In: J.B. Hutchinson (ed), Essays on Crop Plant Evolution. Cambridge University Press, pp. 70-101.

Blattner, F.R., and A.G. Badani Méndez, 2001. RAPD data do not support a second centre of barley domestication in Morocco. Genet. Res. Crop Evol. 48: 13-19.

Bothmer, R. von, J. Flink, N. Jacobsen, M. Kotimäki and T. Landström, 1983. Interspecific hybridization with cultivated barley. Hereditas 99: 219-244.

Bothmer, R. von, N. Jacobsen, C. Baden, R.B. Jørgensen and I. Linde-Laursen, 1995. An ecogeographical study of the genus *Hordeum*. Systematic and Ecogeographic Studies on Crop Genepools, 7. IPGRI, Rome, 2nd ed. 129 p.

Bothmer, R. von, and I. Linde-Laursen, 1989. Backcrosses to cultivated barley (*Hordeum vulgare* L.) and partial elimination of alien chromosomes. Hereditas 111: 145-147.

Bothmer, R. von, C. Yen and J.L. Yang, 1990. Does wild, six-rowed barley, *Hordeum agriocrithon* really exist? Plant Genet. Res. Newsl. 77: 17-19.

Brown, A.H.D., D. Zohary and E. Nevo, 1978. Outcrossing rates and heterozygosity in natural populations of *Hordeum spontaneum* Koch in Israel. Heredity 41: 49-62.

De Candolle, A., 1959. Origin of Cultivated Plants. 2nd ed. Hafner, New York. (Translated from the 1896 edition).

Dewey, D.R., 1984. The genome system of classification as a guide to hybridization with perennial Triticeae. In: J.P. Gustafson (ed), Gene Manipulation in Plant Taxonomy. Plenum Publ. Corp., New York, pp. 209-279.

El Rabey, H., and F. Salamini, 2000. Domestication history of barley and phylogenetic relationships in the genus *Hordeum*. In: Barley Genetics VIII. Proc. 8th Int. Barley Genet. Symp., Adelaide, Australia, pp. 32-36.

Evans, L.T., 1993. Crop Evolution, Adaptation, and Yield. Cambridge Univ. Press, Cambridge.

Giles, B., and R. von Bothmer, 1985. The progenitor of barley (*Hordeum vulgare* ssp. *spontaneum*) – its importance as a gene source. J. Swedish Seed Assoc. 95: 53-61.

Gustafsson, M., and L. Claesson, 1988. Resistance to powdery mildew in wild species of barley. Hereditas 108: 231-237.

Harlan, J.R., 1968. On the origins of barley. In: Barley: Origin, Botany, Culture, Winterhardiness, Genetics, Utilization. USDA Handb. 338. U.S. Gov. Print Office, Washington, DC.

Harlan, J.R., 1971. Agricultural origins: centers and noncenters. Science 174: 468-474.

Harlan, J.R., 1992. Crops and Man. American Society of Agronomy, Madison, Wisconsin, 2nd ed.

Harlan, J. R., 1995. Barley. In: J. Smartt and N.W. Simmonds (eds), Evolution of Crop Plants, 2nd ed. Longman Scientific and Technical, pp. 140-147.

Harlan, J.R., and J.M.J. de Wet, 1971. Toward a rational classification of cultivated plants. Taxon 20: 509-517.

Helbaek, H., 1960. Ecological effects of irrigation in ancient Mesopotamia. Iraq 22: 186-196.

Helbaek, H., 1969. Plant collecting, dry-farming, and irrigation agriculture in prehistoric Deh, Luran. Mem. Mus. Anthrop. Univ. Michigan 1: 383-426.

Hiura, U., and H. Heta, 1952. Studies on the disease resistance of barley varieties II. Nogaku Kenkyu 40: 89-95. (In Japanese).

Hiura, U., and H. Heta, 1954. Studies on the disease resistance of barley varieties V. Nogaku Kenkyu 41: 145-156. (In Japanese).

Hjelmqvist, H., 1979. Beiträge zur Kenntnis der prähistorischen Nutzpflanzen in Schweden. Opera Bot. 47: 1-58.

Ho, P.T., 1977. The indigenous origin of Chinese agriculture. In: C.A. Reed (ed), Origin of Agriculture, Mouton, The Hague, pp. 418-423.

Hunter, H., 1952. The Barley Crop. London, Crosby Lockwood and Son.

Jensen, J., 1989. Estimation of recombination parameters between a quantitative trait locus (QTL) and two marker gene loci. Theor. Appl. Genet. 78: 613-618.

Jørgensen, J.H., 1992. Sources and genetics of resistance to fungal pathogens. In: P.R. Shewry (ed), Barley: Genetics, Biochemistry, Molecular Biology and Biotechnology. CAB International, Wallingford, UK, pp. 441-457.

Karsai, I., K. Meszaros, P.M. Hayes and Z. Bedo, 1997. Effects of loci on chromosomes 2(2H) and 7(5H) on developmental patterns in barley (*Hordeum vulgare* L.) under different photoperiod regimes. Theor. Appl. Genet. 94: 612-618.

Kasha, K.J., and K.N. Kao, 1970. High frequency of haploid production in barley (*Hordeum vulgare* L.). Nature 225: 874-876.

Kashiwada, S., 1930. On the glutinous vs. non-glutinous characters and their inheritance in barley. Proc. Crop Sci. Soc. Japan 2: 193-194. (In Japanese).

Kislev, M.E., D. Nadel and I. Carmi, 1992. Grain and fruit diet 19,000 years old at Ohalo II, Sea of Galilee, Israel. Rev. Paleobot. Palynol. 73: 161-166.

Komatsuda, T., K.I. Tanno, B. Salomon, T. Bryngelsson and R. von Bothmer, 1999. Phylogeny in the genus *Hordeum* based on nucleotide sequences closely linked to the *vrs1* locus (row number of spikelets). Genome 42: 973-981.

Ladizinsky, G., 1999. Plant Evolution under Domestication. Kluwer Acad. Publishers. Dordrecht.

Laurie, D.A., N. Pratchett, J.H. Bezant and J.W. Snape, 1995. RFLP mapping of five major genes and eight quantitative trait loci controlling flowering time in a winter × spring barley (*Hordeum vulgare* L.) cross. Genome 38: 575-585.

Lehmann, L.C., R. Jönsson and M. Gustafsson, 1998. Identification of resistance genes to powdery mildew isolated from *Hordeum vulgare* ssp. *spontaneum* and land races of barley. J. Swedish Seed Assoc. 108: 94-101.

Lev-Yadun, S., A. Gopher and S. Abbo, 2000. The cradle of agriculture. Science 288: 1602-1603.

Löve, A., 1984. Conspectus of the Triticeae. Feddes Repert. 95: 425-521.

Lundqvist, U., J.D. Franckowiak and T. Konishi, 1996. New and revised descriptions of barley genes. Barley Genet. Newsl. 26: 22-516.

Mano, Y., and K. Takeda, 1995. Varietal variation and effects of some major genes on salt tolerance in barley seedlings. Bull. Res. Inst. Bioresources, Okayama Univ. 3: 71-81.

Mano, Y., and K. Takeda, 1998. Genetic resources of salt tolerance in wild *Hordeum* species. Euphytica 103: 137-141.

Marillia, E.F., and G.J. Scoles, 1996. The use of RAPD markers in *Hordeum* phylogeny. Genome 39: 646-654.

Mayler, J.I., and E.H. Stanford, 1942. Color inheritance in barley. J. Am. Soc. Agr. 34: 427-436.

Molina-Cano, J.L., E. Moralejo, E. Igartua and I. Romagosa, 1999. Further evidence supporting Morocco as a centre of origin of barley. Theor. Appl. Genet. 98: 913-918.

Nair, K.P.P., and N.C. Khulbe, 1990. Differential response of wheat and barley genotypes to substrate-induced salinity under North Indian conditions. Experimental Agric. 26: 221-225.

Nevo, E., 1992. Origin, evolution, population genetics and resources for breeding of wild barley *Hordeum spontaneum* in the Fertile Crescent. In: P.R. Shewry (ed), Barley: Genetics, Biochemistry, Molecular Biology and Biotechnology. CAB International, Wallingford, UK, pp. 19-43.

Nevo, E., A. Beiles and D. Zohary, 1986. Genetic resources of wild barley in the Near East: structure, evolution and application in breeding. Biol. J. Linn. Soc. 27: 355-380.

Nevo, E., T. Krugman and A. Beiles, 1993. Genetic resources for salt tolerance in the wild progenitors of wheat (*Triticum dicoccoides*) and barley (*Hordeum spontaneum*) in Israel. Plant Breed. 110: 338-341.

Nevski, S.A., 1941. Beiträge zur Kenntnis der wildwachsenden Gersten im Zusammenhang mit der Frage über den Ursprung von *Hordeum vulgare* L. und *Hordeum distichon* L. (Versuch einer Monographie der Gattung *Hordeum*). Trudy Bot. Inst. Akad. Nauk SSSR, ser. 1, 5: 64-255.

Nishikado, Y., R. Takahashi and U. Hiura, 1951. Studies on the disease resistance in barley I. Ber. Ohara Inst. landw. Forsch. 9: 411-423.

Pickering, R.A., 1984. The influence of genotype and environment on chromosome elimination in crosses between *Hordeum vulgare* × *H. bulbosum*. Plant Sci. 34: 153-164.

Pickering, R., 2000. Do the wild relatives of cultivated barley have a place in barley improvement? In: Barley Genetics VIII. Proc. 8th Int. Barley Genet. Symp., Adelaide, Australia, pp. 223-230.

van Rijn, C.P.E., I. Heersche, Y.E.M. van Berkel, E. Nevo, H. Lambers and H. Poorter, 2000. Growth characteristics in *Hordeum spontaneum* populations from different habitats. New Phytol. 146: 471-481.

Sato, K., and K. Takeda, 1994. Sources of resistance to net blotch in barley germplasm. Bull. Res. Inst. Bioresources, Okayama Univ. 2: 91-102.

Sayed-Tabatabaei, B.E., K. Tanno, F. Javadi and T. Komatsuda, 2000. Genetic variation of DNA markers linked to the *vrs1* locus in two-rowed barley from Iran. In: Barley Genetics VIII. Proc. 8th Int. Barley Genet. Symp., Adelaide, Australia, pp. 93-94.

Schönfeld, R.M., A. Ragni and G. Fischbeck, 1996. RFLP-mapping of three new loci for resistance genes to powdery mildew (*Erysiphe graminis* f. sp. *hordei*) in barley. Theor. Appl. Genet. 93: 48-56.

Seko, H., 1987. History of barley breeding in Japan. In: Barley Genetics V. Proc. 5th Int. Barley Genet. Symp., Okayama, Japan, pp. 915-922.

Smith, B.D., 1995. The Emergence of Agriculture. Scientific American Library, New York.

Strelchenko, P., O. Kovalyova and K. Okuno, 1999. Genetic differentiation and geographical distribution of barley germplasm based on RAPD markers. Genet. Res. Crop Evol. 46: 193-205.

Subrahmanyam, N.C., and R. von Bothmer, 1987. Interspecific hybridization with *Hordeum bulbosum* and development of hybrids and haploids. Hereditas 106: 119-127.

Svitashev, S., T. Bryngelsson, A. Vershinin, C. Pedersen, T. Säll and R. von Bothmer, 1994. Phylogenetic analysis of the genus *Hordeum* using repetitive DNA sequences. Theor. Appl. Genet. 89: 801-809.

Takahashi, R., 1951. Studies on the classification and the geographical distribution of the Japanese barley varieties II. Correlative inheritance of some quantitative characters with the ear type. Ber. Ohara Inst. landw. Forsch. 9: 383-398.

Takahashi, R., 1955. The origin and evolution of cultivated barley. Advances in Genetics 7, Academic Press, New York, pp. 227-266.

Takahashi, R., 1963. Further studies on the phylogenetic differentiation of cultivated barley. In: Barley Genetics I. Proc. 1st Int. Barley Genet. Symp., Wageningen, The Netherlands, pp. 19-26.

Takahashi, R., 1987. Genetic features of east Asian barleys. In: Barley Genetics V. Proc. 5th Int. Barley Genet. Symp., Okayama, Japan, pp. 7-20.

Takahashi, R., and J. Hayashi, 1964. Linkage study of two complementary genes for brittle rachis in barley. Ber. Ohara Inst. landw. Biol., Okayama Univ. 12: 99-105.

Takahashi, R., and J. Hayashi, 1987. Studies on chlorotic plants of barley by dominant complementary genes and geographical distribution of the genes concerned. In: Barley Genetics V. Proc. 5th Int. Barley Genet. Symp., Okayama, Japan, pp. 139-144.

Takahashi, R., J. Hayashi, U. Hiura and S. Yasuda, 1968. A study of cultivated barleys from Nepal, Himalaya and North India with special reference to their phylogenetic differentiation. Ber. Ohara Inst. landw. Biol., Okayama Univ. 14: 85-122.

Takahashi, R., J. Hayashi, S. Yasuda and U. Hiura, 1963. Characteristics of the wild and cultivated barleys from Afghanistan and its neighboring regions. Ber. Ohara Inst. landw. Biol., Okayama Univ. 12: 1-23.

Takahashi, R., and Y. Tomihisa, 1970. Genetic approach to the origin of two wild forms of barley, *lagunculiforme* Bacht. and *proskowetzii* Nabelek (*Hordeum spontaneum* C. Koch emend. Bacht.). In: Barley Genetics II. Proc. 2nd Int. Barley Genet. Symp., Pullman, WA, USA, pp. 51-62.

Takahashi, R., and J. Yamamoto, 1949. Studies on the classification and the geographic distribution of barley varieties, VIII. Nogaku Kenkyu 38: 81-90. (In Japanese).

Takahashi, R., and J. Yamamoto, 1950. Studies on the classification and the geographic distribution of barley varieties, XII. Nogaku Kenkyu 39: 25-32. (In Japanese).

Takahashi, R., and S. Yasuda, 1956. Genetic studies of spring and winter habit of growth in barley. Ber. Ohara Inst. landw. Biol., Okayama Univ. 10: 245-308.

Takahashi, R., and S. Yasuda, 1958. Genetic studies on heading date in barley. In: K. Sakai, R. Takahashi and H. Akemine (eds), Studies on the Bulk Method of Plant Breeding. Yokendo, Tokyo, pp. 44-64. (In Japanese).

Takahashi, R., S. Yasuda, J. Hayashi, T. Fukuyama, I. Moriya and T. Konishi, 1983. Catalogue of Barley Germplasm Preserved in Okayama University, 217 p.

Takeda, K., 1989. Varietal variation of flooding tolerance in barley seedlings, and its diallel analysis. Jap. J. Breeding 39 (Suppl. 1): 174-175.

Takeda, K., 1995. Varietal variation and inheritance of seed dormancy in barley. In: K. Noda and D.J. Mares (eds), Pre-harvest Sprouting in Cereals 1995. Center for Academic Societies, Osaka, Japan, pp. 205-212.

Takeda, K., 1996. Inheritance of sensitivity to the insecticide diazinon in barley and the geographical distribution of sensitive varieties. Euphytica 89: 297-304.

Takeda, K., and C.L. Chang, 1996. Inheritance and geographical distribution of phenol reaction-less varieties of barley. Euphytica 90: 217-221.

Takeda, K., and T. Fukuyama, 1987. Tolerance to pre-germination flooding in the world collection of barley varieties. In: Barley Genetics V. Proc. 5th Int. Barley Genet. Symp., Okayama, Japan, pp. 735-740.

Takeda, K., and W. Saito, 1987. Character expression and inheritance of 'Irregulare' spike in barley. Nogaku Kenkyu 61: 195-220. (In Japanese with English summary).

Takeda, K., H. Yoshino, K. Kato, K. Sato, H. Tsujimoto and H. Tsuyuzaki, 2000. Genetic assay and study of crop germplasm in and around China (II). A report of Grant-in-Aid for Scientific Research, Monbusho, Japan, pp. 37-71.

Taketa, S., H. Ando, K. Takeda and R. von Bothmer, 1999. Detection of *Hordeum marinum* genome in three polyploid *Hordeum* species and cytotypes by genomic *in situ* hybridization. Hereditas 130: 185-188.

Tanno, K., F. Takaiwa, S. Ota and T. Komatsuda, 1999. A nucleotide sequence linked to the *vrs1* locus for studies of differentiation in cultivated barley (*Hordeum vulgare* L.). Hereditas 130: 77-82.

Tanno, K., S. Taketa, K. Takeda and T. Komatsuda, 2002. A DNA marker closely linked to the *vrs1* locus (row type gene) indicates multiple origins of six-rowed cultivated barley (*Hordeum vulgare* L.). Theor. Appl. Genet. 104: 54-60.

Vavilov, N.I., 1926. Studies on the origin of cultivated plants. Bull. Appl. Bot., Genet. Plant Breeding USSR 16: 1-248. (In Russian with English summary).

Woodward, R.W., 1941. Inheritance of a melanin-like pigmentation in the glumes and caryopses in barley. J. Agr. Research 63: 21-28.

Yasuda, S., 1969. Linkage and pleiotropic effects on agronomic characters of the genes for spring growth habit. Barley Newsl. 12: 57-58.

Yasuda, S., 1992. Differentiation and geographical distribution of spring genes in barley. Iden 46: 26-30. (In Japanese).

Yasuda, S., J. Hayashi and I. Moriya, 1993. Genetic constitution for spring growth habit and some other characters in barley cultivars in the Mediterranean coastal regions. Euphytica 70: 77-83.

Yasuda, S., and K. Rikiishi, 1997. Screening of the world barley collection for resistance to barley yellow mosaic virus. Barley Genet. Newsl. 28: 64-66.

Zohary, D., and M. Hopf, 1993. Domestication of Plants in the Old World. Oxford Science Publications.

Diversity in Barley *(Hordeum Vulgare)*
Roland von Bothmer et al (Editors).

Chapter 3

Diversification through breeding

Gerhard Fischbeck

Department of Agronomy and Plant Breeding, D-85350 Freising-Weihenstephan, Germany

Introduction

The large diversity present in the wild progenitor of barley (*H. vulgare* ssp. *spontaneum*) as well as the early domestication process created the prerequisites for a wider distribution of the crop. Early selection, both conscious and unconscious, mutation events, recombination, segregation and random factors in prehistoric times led to adaptation of the material and its distribution to many new areas (see also Chapter 2). The process of diversification has continued throughout history, and modern plant breeding has introduced a new dimension to it.

Expansion of barley cultivation through history

The discovery of the New World by Columbus and subsequent explorers opened the door to a new phase of reciprocal exchange of cultivated plant species between the Old and the New World. Although wheat eventually became the most important plant to be introduced from the Old into the New World, barley was among the first crop species that settlers took with them following the Spanish conquest. It found its way into local farming systems practised in the mountainous highlands of Meso and South America. Again, it was Spanish settlers who introduced barley of North African origin into California (Poehlman, 1959), where it formed the basis for the 'Coast' type of six-rowed, large-seeded spring barley cultivars characteristic of this area.

Mass movements to North America during the 18th century introduced seeds from many countries, especially Central, Northern and Eastern Europe. Barley was not sufficiently competitive to become a major crop in North American farming systems, but in some more restricted areas it became an important cash crop, mainly for malting. It was developed from better-adapted seed sources such as 'Oderbrucker' and 'Manchuria', introduced by immigrants from Germany and Russia (Poehlman, 1959). They formed the basis for six-rowed spring barley cultivars with excellent malting quality grown in the Northern spring barley cultivation region in the US and extended into the Prairie provinces of Canada. In more restricted ways, winter barley became established in Ontario, Canada, and in the south-eastern part of the US; it was known as 'Tennessee Winter' and could be traced back to European sources. It was a reintroduction from Canada that later played an important role in the early phases of winter barley breeding in

---

Fischbeck, G., 2003. Diversification through breeding. In: R. von Bothmer, Th. van Hintum, H. Knüpffer and K. Sato (eds), Diversity in Barley (*Hordeum vulgare*), pp. 29-52. Elsevier Science B.V., Amsterdam, The Netherlands.

Europe.

As early as 1819, the US government instructed its foreign office employees to collect seeds and plants potentially useful for cultivation in the US. In 1898, the 'Office of Foreign Plant Introduction' was established at the US Department of Agriculture (Poehlman, 1959), and a systematic approach was launched to collect, introduce and evaluate plant materials from relevant crop species from all parts of the world. At the same time, the 'Bureau of Applied Botany', which later became the N.I. Vavilov Institute of Plant Production, was founded in St. Petersburg, Russia, with similar aims.

At the turn of the 19th to the 20th century, barley cultivation was extended to Australia based on two-rowed spring landrace selections from the UK, and it was not until this period that two-rowed spring barley also became established in Japan (Seko, 1987). Another wave of introductions of two-rowed spring barley initiated malting barley production in the Northwest of the USA, based on 'Bethge XIII', a German selection from the 'Hanna' landrace. Furthermore, malting barley production was introduced into the Pampa regions in South America, often based on well-known selections from Australia and Europe.

Present stage of barley cultivation

Cultivation of barley occupied a world total of about 39 million ha between 1934 and 1938, compared to 145 million ha for wheat, 82 million ha for maize, 56 million ha for oats, 49 million ha for rice and 42 million ha for rye (FAO, 1947). In accordance with the general increase in cereal acreage after World War II, and being more favoured than some other cereal species, barley reached its top world acreage of about 81.2 million ha during 1979/81 (Table 3.1), occupying the fourth position after wheat, rice and maize. During the last two decades, the world acreage decreased by 20 million ha, a substantial increase only being noted for countries in West Asia and North Africa (WANA, ICARDA's mandate region), mainly Turkey. Thus, barley reached the fifth position following wheat, rice, maize and soybeans. The reductions in barley acreage are mainly explained by a general decrease in agricultural activities in the USSR follower states, and policy decisions to reduce cereal surplus production in the European Union and the US. It is also based on competition with other cereals such as wheat, maize and triticale, for production of feed grain in more favourable growing conditions. On the other hand, barley cultivation remained untouched or even increased in rain-fed cereal areas under semiarid conditions as well as in regions or under conditions that require very short reproduction cycles, and/or tolerance to drought and salinity.

Besides adaptation to soil, climate and husbandry conditions, cultivation of barley depends

Table 3.1. World acreage of barley cultivation and trends of change in important barley growing areas (from FAO, Production Yearbooks, 1985 and 1998), million ha.

Region	1979 / 1981	1998	Difference
Western Europe	15.90	11.41	-4.49
Eastern Europe	4.17	3.26	-0.91
Former USSR	33.46	19.59	-13.87
WANA region	11.56	13.24	+1.68
Far East	3.61	2.61	-0.99
North America	7.84	6.64	-1.20
M & S America	0.72	0.85	+0.12
Australia	2.54	2.96	+0.42
World total	81.17	61.70	-19.47

Table 3.2. Estimates of utilisation of the barley crop (information from national reports).

Region	No. of national reports	Percentage of harvested grains used for			
		Feed	Malt	Food	Export
West Europe	8	40-90	5-40	0-2	10-40
East Europe	6	70-85	6-20	0-5	
Russia, Ukraine, Belarus	3	68-89	6-12	2-6	
Baltic States	3	80-96	2-10	1-10	
Kazakhstan, Uzbekistan, Kyrgystan	3	90-94	3-6	2	
Turkey	1	78	18	2	
North America	2	50-60	10-40	25*	
M & S America	6	10-52	10-98	0-60	
Australia	1	30	3		65

*Canada

on methods and preferences in crop utilisation. Since a comparable body of official statistics is not available, a survey has been carried out to provide estimates from barley scientists and breeders working in the major barley growing countries (Table 3.2). Animal feed grain occupies the dominant position between 60 and 95% of the harvested grains in almost all countries.

Production of malting barley ranks second. It requires high quality grains that are not easy to produce everywhere, but maintains or even increases its central position to supply the malting industry's requirements. For such reasons, malting barley production shows distinct differences even between countries within the same region. In most barley growing countries in Meso and South America, production of malting barley even occupies the prime position. Relatively high percentages for malting barley production are reported (in decreasing order) for the USA, Western, and Eastern European countries. Barley production in some countries (Australia, France and Canada) depends on very high export percentages, while Saudi Arabia, Japan, China and the USA are listed for high barley imports (Gauger, 1997). In addition, export of barley malt from Europe and Canada, and import to South America (Brazil) and East Asia (Japan) is also increasing.

With the exception of Ethiopia, Peru and probably Tibet, where barley production is still mainly used for food, this ancient purpose of barley cultivation has decreased to less than 10%, often even less than 1% of the barley crop. Adherence to traditional dishes plays an important role, while modern trends to include barley in breakfast cereals and other food substances have apparently not yet made a noticeable impact.

In some areas, vegetative parts of the barley plant are commonly used for grazing, practised during the tillering stage in the case of shortage of forage grasses in semiarid regions (Near East and Australia), while in many countries in Eastern Europe and most of the USSR follower states as well as in the WANA region, straw from the harvested barley crop is often used for animal feed.

Landraces

A landrace represents the equilibrium between heterogeneous and heterozygous genotypes within a population of a crop that is maintained by continuous multiplication under a given set of climatic, soil and husbandry conditions.

With self-pollinating crops such as barley, discrimination between homozygous genotypes of a donor population may take place within only a few generations if subjected to a new set of growing conditions. Even low levels of outcrossing and occasional admixtures of genotypes

Table 3.3. Estimates of cultivation of landraces (information from national reports).

Country	Percentage	Acreage (1,000 ha)*
Spain	very low	
France	very low	
Italy	up to 25 (r)	20
Yugoslavia	10-30 (r)	4
Russia	2 (r)	20
Ukraine	1 (r)	3
Kazakhstan	1-2	20
Uzbekistan	0-70 (rf)	10
Peru	10	15
Bolivia	10	10
Subtotal		102
Turkey	10-15	452
WANA region**		
Near East	85	3,970
North Africa	85	3,320
Ethiopia	85	760
Subtotal		8,500

* calculated from the 1998 barley acreage
** general estimate for the WANA region (C. Einfeldt, pers. comm.)
(r) – in remote areas; (rf) – under rain-fed conditions

from non donor populations well adapted to the prevailing growing conditions will contribute to slowly proceeding changes in the existing equilibrium. In this way, dynamic elements are introduced into the genetic constitution of a landrace.

Landraces became established in all regions where barley was part of the farming systems practised up to the 19th century. The cultivation of barley landraces still continues on a considerable proportion of the barley acreage (Table 3.3), since they have maintained the dominant position until recently close to the central region of its origin but also in other countries where similar harsh growing conditions and relevant farming systems prevail. This caused the Consultative Group on International Agricultural Research (CGIAR) to establish an International Centre of Agricultural Research for the Dry Areas (ICARDA) in Aleppo (Syria) with barley as a major mandate crop. The WANA region includes Ethiopia, where the wide range of growing conditions for barley – fields between 1,500 m and 4,000 m above sea level, different moisture regimes and soil conditions – have caused a wider range of variation within and between landraces than elsewhere (Berhane *et al.*, 1997). This so impressed and puzzled the famous N.I. Vavilov that he came to consider Ethiopia as the birthplace of cultivated barley.

Landraces that are still commonly grown in Syria seem to be rather uniformly restricted to either white or black kerneled two-rowed types. Nevertheless, as revealed by more detailed studies, they contain a broad range of genetic diversity within regional landrace populations as well as specific adaptation to local diversification in the overall set of environmental stress that dominates barley cultivation everywhere in the region (Weltzien and Fischbeck, 1990). In other countries of the WANA region such as Northern Africa and Iraq, landraces of the six-rowed ear type prevail and still represent the original landrace types inherited from the traditional farming systems.

In addition, beyond the WANA region, barley landraces are still being grown in a considerable number of countries (Table 3.3). Generally, they are restricted to more remote, often mountainous regions that are not affected or far less affected by the general trend to more intensified cereal culture. This applies to the European countries where sometimes only traces of landraces are still being cultivated, as well as to regions where landraces occupy at least 10% of the barley acreage, e.g., Turkey and some of the USSR follower states, and to the relics of barley landraces developed from early introductions into South America.

The remaining landrace populations will probably no longer contain all the genetic diversity of other populations already replaced by barley cultivars. However, they certainly do represent cases of *in situ* conservation of genetic diversity as long as the traditional farming systems survive and are not subjected to dramatic changes excluding the heritage of landraces from further multiplication.

The replacement of the original landraces began gradually in the United Kingdom during the first half of the 19th century, when 'improved' seed stocks were selected by farmers with specific interests from various plant types: 'Archer', 'Spratt' and 'Chevallier'. If the yield was higher or of better quality, they attracted interest and became more widely distributed. During the second half of the 19th century, such seed stocks also reached the continent, and the story behind them encouraged an increasing number of farmers, mainly owners of larger farms, to join the seed business by exerting further selections within the stands of imported seeds as well as from local landraces. This empirical basis gave rise to most of the private and cooperative breeding stations, which continue to play a major role in agricultural plant breeding activities in European countries.

Landraces were replaced within a shorter time and to a much greater extent in the centres of agricultural activity in central and north-western Europe, where they virtually disappeared during the early decades of the 20th century. They have largely been lost, since no broad-based efforts were undertaken at that time to collect and store representative seed samples. In more remote, often marginal and mountainous regions, landraces of barley have survived much longer, although uncontrolled mixtures with cultivars from neighbouring regions may have occurred (Schachl, 1976).

N.I. Vavilov has set the first example for systematic collection of landraces including barley from the huge territories of the former USSR and from other countries in his search for 'centres of diversity'. Other European countries, such as Spain and Italy, where replacement of barley landraces was delayed into the second half of the 20th century, have taken advantage of the situation and have prevented extinction and loss by systematically collecting seed samples and storing them in national genebanks.

Barley breeding

The purely empirical base on which breeding of agricultural crops developed is characterised by the specific attention paid to conspicuously different plant types occurring within local landrace populations together with separate harvest and seed multiplication.

Seed multiplication from selected plants

Along with other efforts to improve agricultural production in order to provide food for a more rapidly growing population, and also in response to side-effects from industrialisation and railroad transportation, it was in the United Kingdom that already during the first half of the 19th century, farm owners in Scotland and in England observed their cereal fields for off-types, and multiplied the separately harvested seed. They did this mainly for wheat, but also for barley and oats (Reitemeier, 1905). Similar efforts to provide better seed stocks for the improvement of barley production laid the basis for barley breeding activities in many European countries (Table

Table 3.4. Survey of barley breeding activities (information from national reports).

Country	Year of begin	Active breeding stations: public sector	Active breeding stations: private sector	Total no. of cultivars*	No. of dominating cultivars**
Spain	1940	3	1	176	
Germany	1880	2	18	110	8
France	1900	2	12	252	9
U. K.	1824	1	5	109	6
Denmark		-	3	68	
Finland	1906	-	2	23	
Sweden	1900	-	1	53	7
Italy	1917	3	3	92	9
Austria	1880	-	4	59	
Norway				13	
Greece				30	
Poland	1880	7	-	35	4
Czech Republic	1872	-	5	35	4
Slovakia	1895	-	3	32	
Romania	1958	2	-	12	3
Hungary	1940	2	?	47	10
Bulgaria	1925	3	-	10	3
Yugoslavia	1957	3	-	48	14
Russia	1903	23	?	125	11
Ukraine	1910	9	-	37	8
Belarus	1927	6	-	25	5
Lithuania	1924	1	-	14	1
Latvia	1950	3	-	11	
Estonia	1909	1	-	10	5
Kazakhstan	1924	9	1	29	7
Uzbekistan	1924	3		7	3
Kyrgystan	1928	1		8	3
Turkey	(1930)	6	?	8	1
Iran				6	
Syria				3	
Iraq				4	
Pakistan				7	
Morocco				10	
Algeria				3	
Tunisia				5	
Libya				6	
Ethiopia				8	
Japan	1891	7	3	46	
Argentina	1917	2	5	10	2
Brazil	1950	1	2	10	1
Peru	1968	1	1	8	
Uruguay	1968	2	3	10	3
Mexico	1960	3	-	5	
USA	1915	8	1	>34	7
Canada				(25)	4
Australia	1950	2	-	17	1

* official list; ** occupying more than 50% of barley acreage

3.4).

The interest in obtaining cereal seeds from different regions within Europe and in comparing their performance increased considerably during the 19th century. During the second half of the 19th century, the success in 'seed improvement' that had been obtained in the UK encouraged farmers in France, Germany, the Danube monarchy and some other neighbouring countries of continental Europe to initiate similar activities. In many cases seed samples from well-known UK sources were introduced for comparison with seed samples from domestic sources, but higher yield potential and later specific quality characteristics were often thought to be correlated with certain ear types or kernel characteristics represented by the imported sources. Mass selection for such types was practised continuously based on the assumption of continuous effects on further improvement, and multiplied seed was sold to other farmers. It is on this highly empirical basis that cereal breeding developed and gradually began to replace the former habit of self-multiplication of cereal landraces on farms or within the region.

Two types of barley landraces, i.e., 'Archer' and 'Spratt', dominated barley cultivation in the UK; the late maturing 'Archer' and the stiff-strawed 'Spratt', both yielding rather coarse grains of low malting quality (Briggs, 1978). As early as 1824, an off-type plant was selected from 'Archer', and gave rise to the 'Chevallier' type of spring barley that yielded well-shaped grains of superior malting quality; it was, however, highly susceptible to lodging because of its tender straw. Later on, original 'Archer' type landrace populations gave rise to improved seed materials distributed in Denmark ('Tystofte Prentice') and also in Ireland ('Irish Archer'), while selections from 'Spratt' became known as dense-eared and stiff-strawed 'Imperial' or 'Goldthorpe' barley on the continent.

During the last decades of the 19th century, barley exports from Moravia, called 'Hanna barley' according to the region where it was cultivated, became famous because of high malting quality. 'Hanna barley' has not only been bought by brewing companies in neighbouring European countries, but has also been resown by several of the already established farmer-breeders, subjected to their individual methods of selection and multiplication and later on sold as Hanna-type improved seed (Hillmann, 1910).

With increased experience from comparative field trials, executed by broad-based agricultural organisations such as 'Deutsche Landwirtschafts-Gesellschaft' (DLG), it became clear that improved seed of specific ear or kernel type did not guarantee better yields in all cases, and limitations in adaptation either due to different soil and climatic conditions or to differences in crop husbandry became apparent. This raised more interest in adaptational differences and possible yield potentials stored within local landrace materials. It did not take very long before 'improved landraces' appeared on the barley seed market and were able to compete successfully with the selections made from seed stocks of introductions such as 'Chevallier', 'Archer', 'Hanna Goldthorpe' and 'Imperial' types of spring barley. For example, in a summary of the status of plant breeding using agricultural crops in Germany in 1910 (Hillmann, 1910), a total of 38 breeding stations are listed that were actively engaged mainly in breeding spring barley (34). The majority of them (25) were working to improve regional landraces at least by mass selection, but most commonly by single-plant selection for plant type, albeit without large-scale progeny tests.

In this way, the early phase of barley seed exchange, within and between different European countries, and the efforts to provide improved seed stocks from continuous selection within available populations that characterised the empirical development of cereal breeding efforts in Europe throughout the 19th century, does not only mark the onset of decreasing diversity due to the progressive replacement of landraces, but also includes an element of increased diversity from outside sources.

Cross-breeding and genetic recombination

The early farmer-breeders who tried to produce improved cereal seeds have not only done so by practising different methods of selection. Even P. Shireff, the "nestor" of wheat breeding in Europe, is reported to have been working with artificial crosses, as have his contemporary colleagues in the UK (Reitemeier, 1905). The same story is reported for early cereal breeders in France (Vilmorin) and Germany (Rimpau). However, because attention was focussed on uniformity for specific plant types multiplied and sold as improved seed stocks, the time-consuming increase in variation during segregating generations greatly hampered the adoption of cross-breeding for cereal improvement. The potential for genetic recombination was not understood, largely because of the lack of a reliable theoretical base for heritability. In fact, even after the rediscovery of the Mendelian laws of inheritance and the broad-based attention they received in science, many of the cereal breeders who had meanwhile become established in Europe retained their empirical methods of seed improvement, and many of them eventually gave up their business as cross-breeding began to make its impact.

Crop superiority obtained by cross-breeding for the cultivation of two-rowed spring barley in Europe became apparent with a number of cultivars named 'Isaria', 'Kenia', 'Herta', 'Rika' and 'Proctor' (Figure 3.1). Year of release was as early as 1924 and as late as 1952. They all originated from crosses between cultivars that had been selected from landraces during the preceding phase of seed improvement. Parent origin could be traced back to one single landrace population, as in the case of 'Isaria', but also to geographically very different populations as in the case of 'Kenia' (Figure 3.1). However, the majority of later releases were regularly descended from genetic recombination between parents of different landrace origin.

Some of the early European landrace selections have also been used in early phases of non-European barley breeding programmes, viz., 'Chevallier' in Japan (Sato, pers. comm.), 'Archer' or 'Chevallier' in Australia (Baumer and Cais, 2000) and 'Bethge XIII' in the USA (Baumer and Cais, 2000).

If pedigrees of the releases from cross-breeding are studied (Fischbeck, 1992), a general

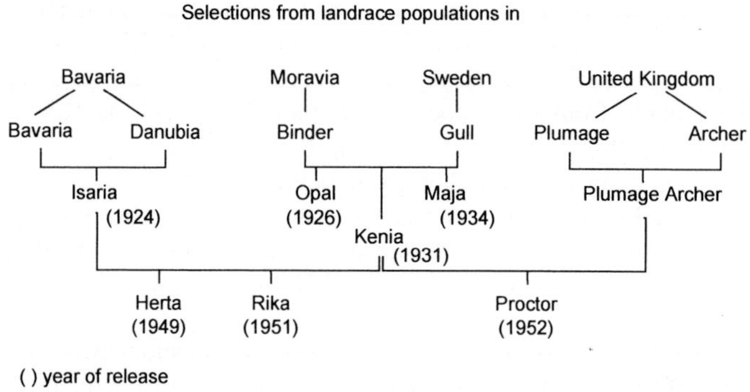

Figure 3.1. Pedigree of spring barley cultivars derived from early cycles of cross breeding (Aufhammer *et al.*, 1958).

tendency is noted that a continuously broader genetic base has been used in successive crosses, giving rise to highly successful cultivars as can be judged from the use of parents from either different geographic origin, or other breeding programmes. This is exemplified in Figure 3.2 by the pedigree of 'Alexis' and 'Trumpf', both selected from crosses between an unnamed breeder's strain and well-known earlier cultivars. 'Alexis' has been of major importance for the production of high quality malting barley in several European countries for more than a decade, while 'Trumpf' served successfully as a donor for high malting quality and favourable agronomic traits in many European barley breeding programmes. Both cultivars include exotic sources of germplasm, mainly for the improvement of disease resistance, as well as induced mutations from well-adapted European cultivars. However, apart from the more general trend towards broadening the genetic base with successive breeding cycles there always have been and still are cases of success with cultivars of spring barley that have originated from crosses between (apparently) more closely related parents. This already applied to 'Isaria' (Figure 3.1), and it is observed in the origin of 'Ingrid', a Swedish cultivar released in 1959 from a cross between closely related parents (Figure 3.3), that has dominated cultivation of spring barley in

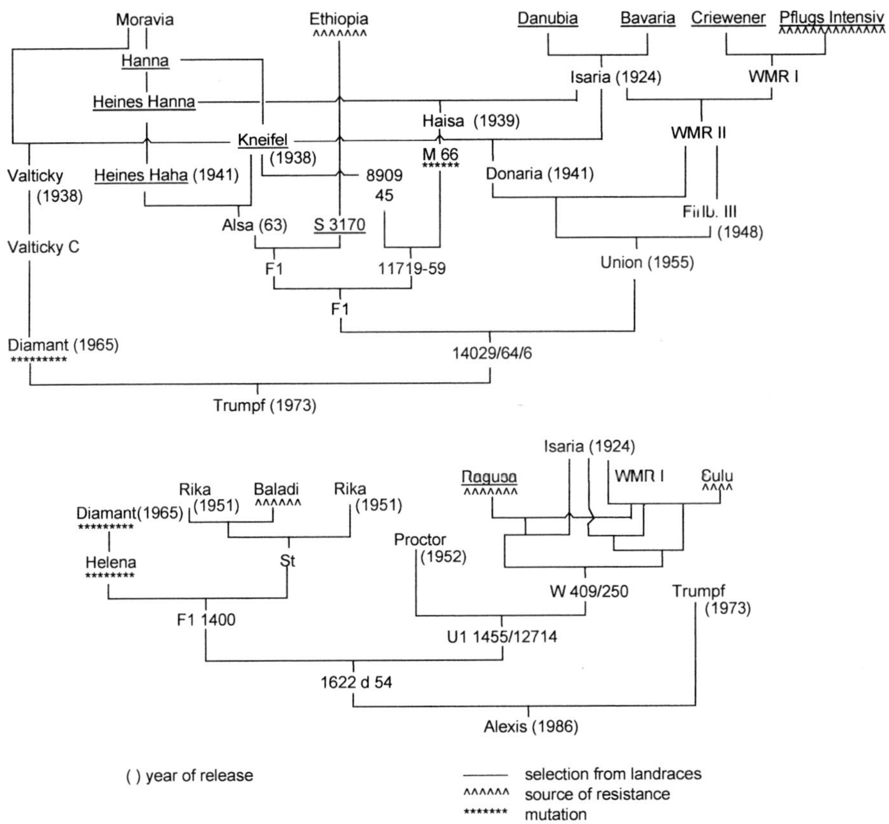

Figure 3.2. Pedigrees of the spring barley cultivars 'Trumpf' and 'Alexis' (D. Lau and J. Breun, pers. comm.).

Scandinavia in subsequent years. More recently, Wych and Rasmusson (1983) and Horsley *et al.* (1995) have documented continuous progress in a narrow based barley breeding programme (Figure 3.3), while Rasmusson and Phillips (1997) have discussed possible sources of '*de novo*' variation as a source of genetic diversity in the Minnesota barley breeding programme.

As exemplified by the pedigrees in Figures 3.1 and 3.2, landrace sources from Moravia ('Hanna') have been crucial in cross-breeding with two-rowed barley in Europe, since the aim was largely the improvement in malting quality, and Hanna-types of barley were used as reliable donors of high malting quality. Only recently, the influence of the Hanna-Moravian malting

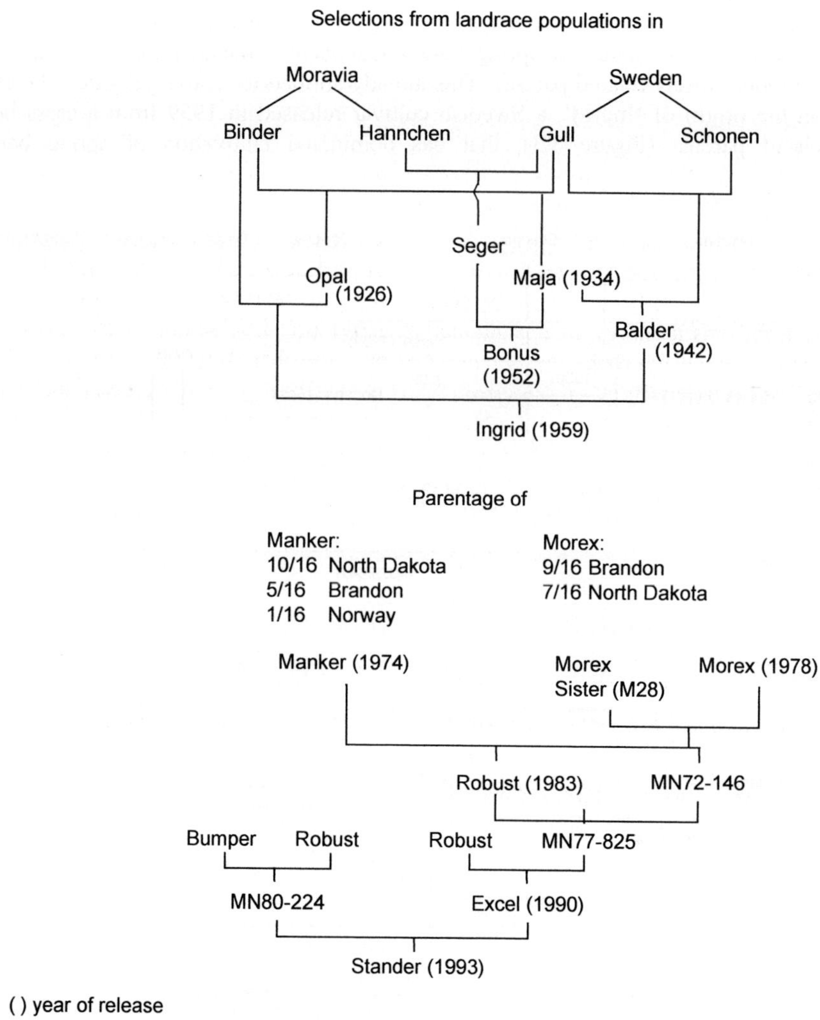

Figure 3.3. Pedigree examples for narrow-based barley breeding programmes ('Ingrid' from Arias *et al.*, 1983; 'Robust' and 'Stander' from Rasmusson and Philipps, 1997).

quality complex has been reinforced by quality characteristics conveyed into 'Trumpf' from a mutant line ('Diamant') that again traces back to the Moravian landrace.

Compared to spring barley, until recently the acreage sown with winter barley played only a minor role in Europe, not only because of lower levels of winter hardiness, but also due to the requirement for early sowing dates which could not be met on a larger scale before full mechanisation had been achieved during the second half of the 20th century. For such reasons, in winter barley cross-breeding has not been practised at the same level of intensity as with spring barley, and selections from landraces have maintained larger shares in winter barley cultivation. Winter barley cross-breeding in Europe emerged from a narrower base of endemic landraces, and in part was substituted by reintroductions from Canada (Figure 3.4). Furthermore, crosses with spring barley played an important role (Figure 3.5) and significant progress was reached with progenies from crosses, which used partners from other countries, such as 'Vogelsanger Gold' in Eastern Europe and 'Ager' and 'Astrix' in South-Eastern Europe.About 86% of the contemporary barley acreage is occupied by cultivars (Table 3.3). Most frequently they originated from several cycles of cross-breeding, and it appears that the process of barley breeding itself has become the prime source for maintenance and development of an immediately available range of diversity, which has fostered a continuous flow of genetic recombinations sufficiently well endowed to serve the needs of continuous crop improvement. Introgressions of exotic germplasm is noted with increasing frequency, but until recently has played a secondary role, mainly in improvement of crop protection against pests and diseases.

For such reasons a survey of the number of active breeding stations was carried out for countries or regions in which barley is cultivated on more than 100,000 ha of arable land. In addition, an attempt has been made to obtain the total number of barley cultivars currently offered to farmers, according to official lists.

A total of 189 breeding stations are actively engaged in barley breeding. The majority is publicly funded forming part of a national programme for agricultural research; also, agricultural research institutions often participate in barley breeding (Table 3.4). More than 70 private plant breeding enterprises sustain barley breeding programmes. These are mainly located in West

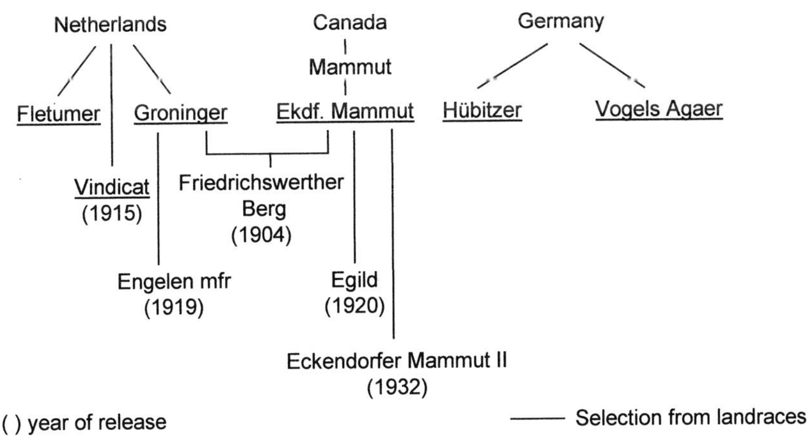

Figure 3.4. Landrace origin and pedigree of early cultivars of six-rowed winter barley (Aufhammer, 1982).

European countries, but also in South America and Japan, where large-sized breweries often maintain breeding stations for improvement of locally adapted malting barley cultivars. In addition, a considerable number of seed firms and collaborative farmers' organisations operate in probably all West European countries, for seed multiplication, having contractual arrangements with (often foreign) breeders, and sometimes maintain seed multiplication for old cultivars (for which plant breeders' rights have been terminated, and also for landrace selections) if there is local interest. One frequently comes across such situations in Italy and in Spain, as well as in other countries without active breeding stations.

About 1,700 registered cultivars serve an acreage of about 49 million ha of barley cultivation (Tables 3.4 and 3.5). The overall ratio of yearly acreage per barley cultivar amounts to about 30,000 ha, which is not a very high figure and leaves much room for diversification in

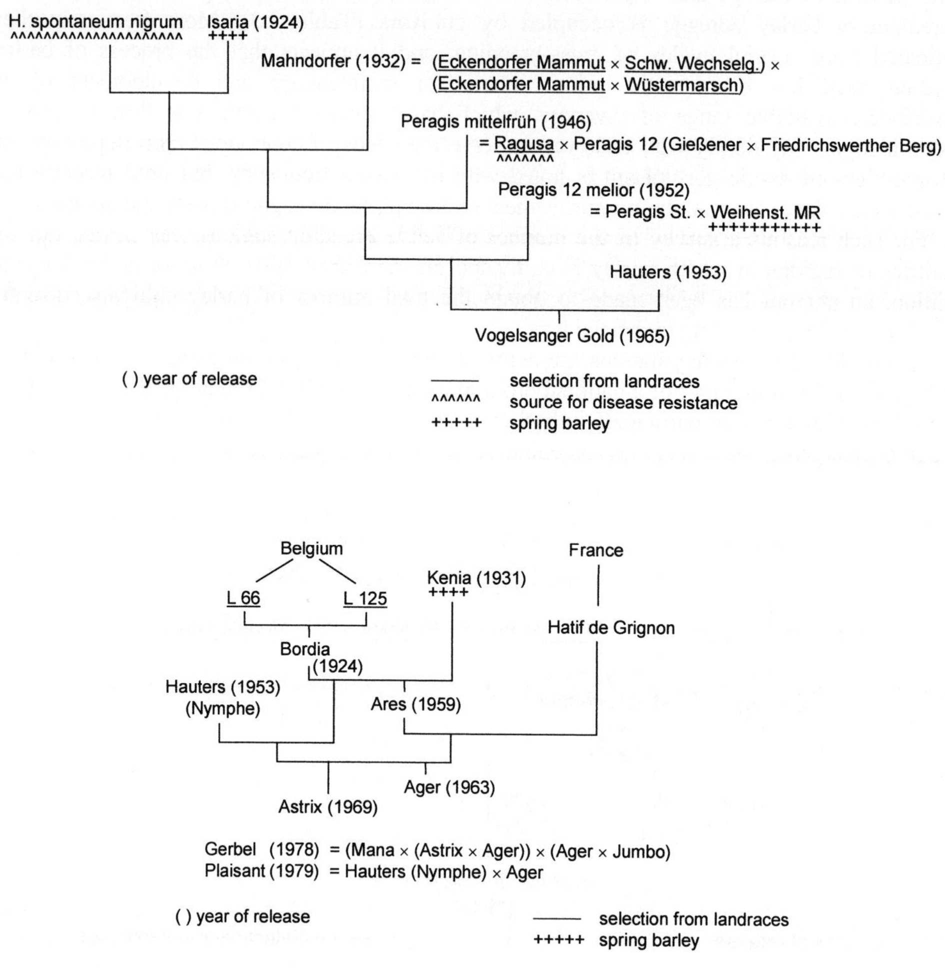

Figure 3.5. Pedigrees of the winter barley cultivars 'Vogelsanger Gold' and 'Astrix' ('V. Gold', Wienhues, pers. comm.; 'Astrix' from Aufhammer, 1982).

Table 3.5. Barley acreage per cultivar (calculated from data in Tables 3.1 and 3.4, in 1,000 ha). Total barley acreage for countries included: 49.3 million ha, overall average 29,800 ha per cultivar. Dominant cultivars (assuming 70% coverage of available acreage): available acreage for countries included: 28.9 million ha, overall average 199,000 ha per cultivar.

Region	for total number of cultivars (official list)	for dominating cultivars
West Europe	11.3	105.0
East Europe	13.9	49.0
Baltic States	24.8	78.0
Russia, Ukraine, Belarus	87.3	476.0
Kazakhstan, Uzbekistan, Kyrgystan	45.8	103.0
Japan	1.2	
Turkey	415.0	2,324.0
Near East*	23.9	
North Africa*	24.9	
Ethiopia*	16.7	
M & S America	19.7	46.0
North America	112.6	423.0
Australia	118.2	2,076.0

*Landrace acreages of Turkey and countries of the WANA region not included

Table 3.6. Barley cultivars occupying very large acreages (information from national reports).

Cultivar	Country	Year of release	Barley type*	Acreage/year (million ha)	Pedigree	See Fig.
'Tokak'	Turkey	1937	2rf	> 3	Landrace selection	
'Clipper'	Australia	1968	2rs	> 1	'Proctor' × 'Prior A'	3.1
'Aramir'	Netherlands	1972	2rs	≈ 0.8	'Volla' × 'Emir'	3.6
'Moskovsky 121'	Russia	1977	2rs	> 2	'Maja' × 'Viner' (Viner: Sel. from Kirow landrace)	3.1
'Harrington'	Canada	1981	2rs	> 1	('Klages' × 'Gazelle' × 'Betzes') × 'Centennial'	3.7
'Schooner'	Australia	1982	2rs	> 1	'Clipper' × ('Proctor' × CI 3567) (CI 3567=landrace from Egypt)	3.1
'Odessky 100'	Ukraine	1984	2rs	> 1	(Hiproly × nutans 244) × (medicum × 'Slavatich')	
'Alexis'	Germany	1986	2rs	≈ 0.8	Br. 1622 × 'Trumpf'	3.2
'Donetzky 8'	Ukraine	1988	2rs	≈ 1	(T552 × T541) × Local Turkey	
'Robust'/ 'Stander'	USA	1983/93	6rs	≈ 1	'Manker' × 'Morex' resp. MN80-224 × 'Excel'	3.3
'Igri'	Germany	1976	2rw	≈ 0.8	'Malta' × 'Ingrid'	3.8
'Plaisant'	France	1979	6rw	≈ 0.5	'Ager' × 'Nymphe'	3.5
'Cyclon'	Russia	1983	6rw	≈ 0.5	'Vogelsanger Gold' × 'Poisk'	3.5

*: 2r – two-rowed; 6r – six-rowed; f – facultative; s – spring; w – winter

the cultivation of barley if, for example, greater use was made of the advantages from growing cultivar mixtures (Wolfe, 1992). The majority of countries surveyed provided the numbers of barley cultivars that occupy a dominant position. From about 41 million ha represented (with a total of about 1,200 cultivars on the official list), about 145 cultivars (12%) belong to the dominant group (occupying at least 50% of the barley acreage). If it is assumed that this group actually represents about 70% of the acreage, an average of close to 200,000 ha/year will be sown with each cultivar, which still leaves room for diversification. However, there is a large degree of regional differentiation. A general decrease in the acreage sown with dominating cultivars is noted in countries where different types of barley (e.g., spring/winter barley) are cultivated. Much higher acreages sown with dominating cultivars are reported for countries with very high barley acreages sown with the same type of barley cultivars to produce cash crops of malting barley. Another factor that affects genotypic differentiation in barley cultivation relates to the time span that a given cultivar will retain its dominant position. With a high degree of competition between mainly private plant breeding enterprises in Western Europe, this period is presently confined to about five to ten years with a tendency to further reduction, while in Russia as well as in Australia, dominant cultivars have maintained their position sometimes for two decades or even longer.

Some barley cultivars have exceeded acreages of more than 1 million ha/year (Table 3.6). The Turkish cultivar 'Tokak' certainly represents the most exceptional case. Tracing back to a landrace selection released in 1937, it still occupies more than 3 million ha in Turkey, and only recently (1997) it has been released as an improved cultivar in Iraq (ICARDA, 1997). In addition, it is presumably involved in the pedigree of 'Donetzky 8', one of the dominant cultivars in the Ukraine. The very popular Russian cultivar 'Moskovsky 121' was derived from a rather simple cross involving one type of offspring from the important cross between Moravian and Scandinavian landrace sources (Figure 3.1) and a Russian landrace selection. Other landraces from Russia dominate the pedigree of 'Odessky 100', which, in addition, is one of the rare cases for a largely grown cultivar that includes 'Hiproly' in its parentage. 'Clipper' as well as 'Schooner' together with 'Stander' and 'Robust' have been derived from rather narrow-based breeding programmes, although 'Schooner' includes an Ethiopian-based landrace selection, probably to integrate additional genes for disease resistance. In Western Europe acreages of close to 1 million ha have only been obtained with rather broad-based cultivars of spring barley that have been successful in at least three countries with larger barley acreages, favoured either by broad adaptation (Jestin, 1985) and improved disease resistance ('Aramir', Figure 3.6) or by very

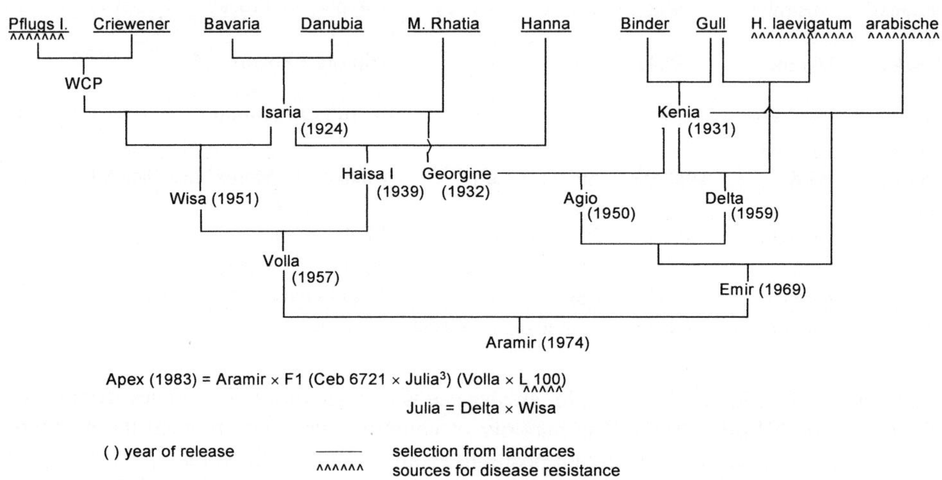

Figure 3.6. Pedigree of the spring barley cultivar 'Aramir' (from Baumer and Göppel, 1988).

favourable quality characteristics ('Alexis', Figure 3.2). 'Harrington' (Figure 3.7), a two-rowed spring cultivar that plays a major role in malting barley production in the western provinces of Canada and in some North-western states of the US, combines a broad base of well-known sources from two-rowed spring landraces from different European countries, including the heritage of the Moravian-Hanna complex combined with breeding materials from Canadian sources.

In general, winter barley cultivars have not been able to reach the very high acreages reported above for spring barley cultivars, since winter barley for a long time played only a minor role in barley production. This has changed only during the last three decades in Central and Western Europe (Jestin, 1985). Along with this increase the cultivar 'Igri', a two-rowed winter barley (Figure 3.8) has reached very high acreages in West Europe. Its pedigree combines winter and spring barley parents, while 'Plaisant' (viz. Figure 3.5), a six-rowed winter barley, grown and accepted for production of malting barley in France, once more originates from a rather narrow genetic base. 'Cyclon', a six-rowed winter barley, which is grown in the southern parts of Russia (Kuban), originates from recombination between German (Figure 3.5) and Russian winter barley cultivars.

Information about the ongoing breeding process with cultivated barley has been obtained from 32 countries. A summary referring to the breeding methods applied and source materials used is given in Table 3.7. As expected, cross-breeding together with backcrossing occupies the prime position, and in most countries, double haploids are used to shorten the segregation period, sometimes together with, or replaced by, SSD (single seed descent) procedures. Marker-assisted selection (MAS) is still in an early phase of integration into breeding programmes, and is applied in most cases to special characteristics that are qualitatively inherited but difficult to determine visually, such as resistance to Barley Yellow Mosaic Virus. Genetic transformation as a method

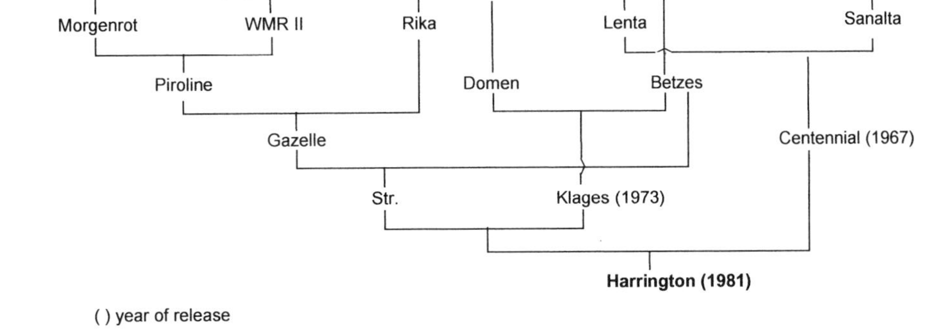

Figure 3.7. Pedigree of the spring barley cultivar 'Harrington' (from Baumer and Cais, 2000, and GRIN).

of barley breeding has been mentioned in only a few cases, still restricted to more scientific projects, whereas attempts to provide new source materials by mutagenesis are still proceeding in a considerable number of barley breeding programmes.

The information provided on the type and rank of source materials used indicates that improved breeder strains (from a given breeding programme) continue to represent the major source, closely followed by cultivars (from other breeding programmes) and introductions of cultivars (from other countries). It seems justified to interpret this information in such a way that the ongoing breeding process itself contributes the major sources for its continuation, mainly by exploitation of favourable recombinations from crosses between advanced materials. Prediction of general or specific combining ability between a given set of candidates for a crossing programme is still not very safe, and measures of genetic distances calculated from frequencies of common ancestors in the pedigree may be as indicative as the use of marker technologies (Graner *et al.*, 1994). Experience has shown that some cultivars appear rather frequently in the parentage of successful crosses without having played a prominent role in barley production (Fischbeck, 1992), but widely grown barley cultivars may only rarely show up in the pedigree of successful cultivars in subsequent generations. Actual progress has also been obtained from successful crosses between closely related parents, and the hypothesis of '*de novo*' diversity has been raised for explanation (Rasmusson and Phillips, 1997). It is for such reasons that the large number of cultivars that never join the relatively small group of dominating cultivars still provides a highly important source of successful genitors for the ongoing breeding process.

As indicated in Table 3.7, efforts to broaden the genetic base beyond the available level of improved strains and cultivars are frequently reported, but clearly rank below the exploitation of the above-mentioned sources. Mutant lines together with landrace-derived materials (most frequently for improvement of disease resistance) are employed, and ssp. *spontaneum* accessions receive growing interest as donors of more (and possibly new) genetic diversity, not only of disease resistance (Brown, 1992). A considerable number of scientific institutions, even though not actively engaged in practical breeding programmes, carry out evaluation work on landrace

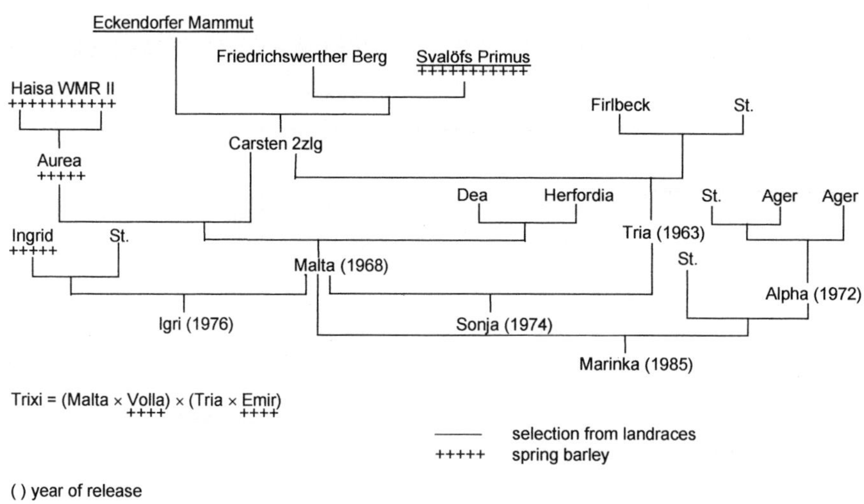

Figure 3.8. Pedigree of the two-rowed winter barley cultivar 'Igri' (from Baumer and Göppel, 1988).

materials, ssp. *spontaneum* accessions and other *Hordeum* species of the secondary genepool, initiate pre-breeding programmes with promising sources, or include new source materials in molecular research related to the barley genome.

For a number of reasons, the landraces grown in the countries of the WANA region have resisted introduction and replacement by conventionally 'improved' cultivars for a long time. In spite of the success of the recent release of 'Rihane-3', a more conventional cultivar in several countries (ICARDA, 1997), research at ICARDA has demonstrated very intricate ways of specific adaptation to the harsh growing conditions in the region (Weltzien and Fischbeck, 1990). The question has been raised whether new strategies for barley breeding in this area can be developed, which try to exploit specific rather than broad adaptation (Ceccarelli, 1994). Such strategies are in fact now employed in a 'decentralised' network of barley breeding (ICARDA, 1995). It involves an intensive flow of breeding materials between ICARDA headquarters and national programmes, which also try to involve practical farmers in evaluation work. The breeding scheme starts with landrace selections evaluated locally, proceeds to crossing work at the ICARDA centre based on promising lines from local evaluation that also may include ssp. *spontaneum* accessions, sends out segregating generations for further evaluation and selection to national programmes, and may end up with releases of mixed populations made up from specifically adapted genotypes. On a very similar basis, crop improvement activities are underway in Ethiopia (ICARDA, 1997), aiming not only to increase the yield potential of the barley crop but also to maintain an essential part of diversity stored within the Ethiopian landraces.

Measuring genetic diversity

The most conspicuous consequences of genetic diversity within a crop species are visible differences in phenotype. A large spectrum of morphological differences, such as number of kernel rows, ear density, awn formation, together with other kernel characteristics such as hulled or naked and differences in colour, are still the markers for determining botanical varieties within the cultivated *H. vulgare* ssp. *vulgare*, and it is very interesting to note that most of them are only found in cultivated landraces of barley.

Considering the range of morphological differentiation between today's cultivars, it is

Table 3.7. Survey on methods applied and source materials used from active breeding stations (information from national reports).

Methods (frequencies of mentioning)

Cross-breeding	32	Marker-assisted selection	9
Doubled haploids	21 + 4*	Mutagenesis	8
Back-crossing	20	Transformation	4

*Single seed descent

Materials (1 – 5 ranking in decreasing order of importance)

	1	2	3	4	5
Improved strains	20	9			
Cultivars	16	7			
Introductions	10	14	2		
Landraces	2	3	7	3	
Mutant lines		3	10	3	1
ssp. *spontaneum*	2	5	3		

apparent that almost only two types (*H. vulgare* convar. *distichon* and convar. *vulgare*) occupy an overall dominant position. Compared to the morphological differentiations that characterised landraces in general (Briggs, 1978) and especially those still cultivated in Ethiopia, it is apparent that genetic diversity in morphological traits has decreased considerably as a consequence of barley breeding. Even if one assumes that genetic diversity for physiological characters is not automatically correlated with morphological traits, the question remains whether barley breeding, as practised during the 20th century, has been and may be still is confounded with an inherent trend to lose genetic diversity.

Progress in analytical methodology based on protein electrophoresis has provided a new dimension to uncover genetic diversity. Electrophoretic differentiation between cereal storage proteins serves as a powerful means to recognise cultivar identity in wheat and barley on a single kernel basis, in spite of the fact that only few, albeit highly complex, loci in their genomes are involved in the expression of the electrophoretic banding patterns.

With the detection of polymorphism in electrophoretic banding patterns for large numbers of enzyme proteins (allozymes), an even more powerful methodology to detect invisible genetic variation became available, since genotypic differentiations are seized that originate from loci on many chromosomes. Finally, the development and the mapping of DNA probes from coding regions of a given genome provided practically unlimited possibilities to detect genetic diversity within a crop species, since all coding regions can be included that are marked by the available set of localised probes (Melchinger *et al.*, 1994). In practice, labour and cost considerations will

Table 3.8. Genetic diversity in landraces and different groups of European barley cultivars, determined by RFLP polymorphism at 32 loci. Data from Hatz (1997) and Backes (unpublished).

	Total	Landraces	Cultivars
N	226	50	17
H	0.381	0.363	0.379
Loci*		5 (landraces vs. cultivars)	

	Spring	Winter	two-rowed	six-rowed
N	135	89	155	69
H	0.260	0.415	0.293	0.381
Loci*	21 (spring vs. winter)		23 (two-rowed vs. six-rowed)	

	two-rowed spring	two-rowed winter	six-rowed winter
N	53	34	20
H	0.231	0.373	0.355
Loci*	18 (two-rowed spring vs. two-rowed winter)	8 (two-rowed winter vs. six-rowed winter)	
	6 (between all 3 groups)		

	two-rowed spring**		six-rowed winter**	
	before 1975	after 1975	before 1975	after 1975
N	76	53	37	20
H	0.235	0.231	0.363	0.355
Loci*	2		0	

N – number of accessions tested
H – gene diversity index (Nei, 1973)
*number of loci with significantly different allele frequencies
**year of release

set limitations, but the selection of probes for loci that are evenly distributed over the genome provides a reliable basis to obtain representative results of genetic diversity within the analysed material. Not many studies have yet been carried out with the intention of comparing changes in genetic diversity confounded with the on-going breeding process.

A study by Hatz (1997) included 226 pure line accessions of either landraces or cultivars from Central, Western and Northern Europe (Table 3.8). They were selected to represent the landrace basis of barley breeding in this region, as well as to mirror the breeding process from the early beginnings of cross-breeding to its recent stage in such a way that successful cultivars reflecting the breeding work done in different countries and for different types of barley cultivation have been chosen.

Measurement of genetic diversity was based on determination of polymorphism detected by single or low copy RFLP probes from 32 loci (2-6 loci/chromosome) and expressed by a total of 91 banding patterns obtained from treatment with different restriction enzymes. Genetic diversity, H, was calculated from allele frequencies as proposed by Nei (1973).

The overall level of genetic diversity for cultivars is very similar to that of the landrace sample included in this study (Table 3.8). Significant differences have been detected at five loci on four chromosomes only; in three cases, the cultivar sample showed a higher level of genetic diversity than the landrace sample. At the allele level (total 91 alleles), four landrace alleles have not been detected in the (larger) cultivar sample, but seven cultivar alleles did not show up in the landrace group.

Differences in genetic diversity increase substantially with the H-values calculated for major barley types. In spite of the narrower landrace base for winter barley breeding (Figure 3.4), a much lower level of genetic diversity characterises the group of two-rowed spring barley types. Highly significant differences have been detected at 21/23 RFLP loci distributed over all seven chromosomes. It is only in very few cases that a higher level of genetic diversity is attributed to the spring/two-rowed group, for which only one allele reaches frequencies >80% at 15 out of the 32 loci involved.

The recent trend to decrease spring and to increase winter barley acreages in Europe has stimulated relevant breeding activities from which two-rowed winter barley cultivars were derived as a new branch in barley cultivation. Surprisingly, genetic diversity within this latest group of barley cultivars for which no landrace basis exists in Europe reached the same level as found in six-rowed winter barley. Frequent crosses between two-rowed spring and six-rowed winter barley cultivars appear in the pedigrees of two-rowed winter barley cultivars (Figure 3.8). Apparently, this high level of genetic diversity was created within a relatively short period, and a rather large number of recombinations between the two older genepools allowed the minimum level of favourable agronomic characters required to be accepted on the official list of barley cultivars in European countries to be exceeded.

This is confirmed by a comparison of genetic diversity within the groups of two-rowed spring and six-rowed winter barley cultivars if they are subdivided in releases that occurred before 1975 and thereafter, together with two-rowed winter cultivars. Again, there are no remarkable differences in genetic diversity if earlier or later releases are compared.

It appears from this study that genetic diversity inherited from the restricted landrace base of European barley breeding has at least been maintained over a period of about 80 years of cross-breeding. However, the general picture is modified by distinct differences in genetic diversity between three major groups of barley cultivars presently cultivated in Europe. Apparently the broader landrace base of two-rowed spring barley has been counteracted by high malting quality requirements superimposed on breeding activities from the very beginning and may well be confounded with the Moravian-Hanna complex that frequently appears in the parentage of two-rowed spring barley cultivars. The multi-locus base of the large number of heritable characters that contribute to malting quality (Ullrich *et al.*, 1997) may, therefore, explain the low level of genetic diversity within this large group of European barley cultivars. On the contrary, for six-rowed winter barley breeding, the landrace base was scarce and completely absent for the most recent efforts in breeding two-rowed winter barley. In both cases, the two-rowed spring barley genepool was frequently utilised, without the restrictions imposed by malting quality traits, at least in six-rowed winter barley. As a consequence, it has been possible to establish and to maintain a definitely higher level of genetic diversity in six-rowed winter barley cultivars, and to reach the same level of diversity with the group of two-rowed winter barley cultivars within a fairly short time span. The data obtained seem to indicate, therefore, that the breeding process itself still allows many new recombinations to be selected within the existing level of genetic diversity; to a large extent these have provided a sufficient base for the progress obtained even in narrow-based crossing programmes (Rasmusson and Phillips, 1997).

Referring to Table 3.7, crosses between superior breeders strains with promising cultivars from other breeding programmes and other countries (introductions) were reported to be of major importance, and they are accompanied by a slow but continuous influx of evaluated landrace materials carrying useful genes, mainly for biotic or abiotic stress. However, when breeding two-rowed spring barley, such means have been employed for quite some time already, but apparently they have not affected the rather low level of genetic diversity in this group to a measurable extent with RFLP polymorphism at the loci involved in this study, possibly as a consequence of rather strict back-crossing.

Another aspect related to the base of genetic diversity is confounded with developments in the WANA region where cultivation of barley is just about to change from its unimproved landrace basis. The 'decentralised' system of barley breeding as developed and practised here (ICARDA, 1995) not only aims to improve the crop under the prevailing very harsh growing conditions, but also provides good chances to preserve high levels of genetic diversity between and within the seed materials which eventually may replace the original landrace populations in this region. The 'decentralised' breeding system may even be considered a promising side branch of dynamic *in situ* conservation.

Growing interest has developed in a more intensive exploitation of ssp. *spontaneum* (Fischbeck *et al.*, 1976; Ceccarelli, 1984; Brown, 1992), originally based on observations of exceptionally high diversity in genes for disease resistance. Using allozyme polymorphism, Nevo *et al.* (1986) performed comparative studies to quantify differences in diversity that are present in populations of landraces and ssp. *spontaneum* from several Near East countries (Table 3.9). No major differences are indicated for genetic diversity within and between landraces from

Table 3.9. Polymorphism at 19 allozyme loci in barley landraces and ssp. *spontaneum*. (Data from Nevo *et al.*, 1986).

	number of plants tested	number of alleles detected	loci without polymorphism
Landraces			
Europe*	178	44	10
Iran	290	40	5
H. vulgare ssp. *spontaneum* populations			
Iran	509	49	6
Israel	1,179	79	0

*Among 29 accessions 1 from India and 1 from Egypt

Europe and from Iran. With the ssp. *spontaneum* accessions from Iran, a higher level of diversity became apparent, but remained much lower than that of ssp. *spontaneum* populations collected in Israel. This indication may point to completely untapped treasuries of genetic diversity that may be stored in ssp. *spontaneum* populations from the Fertile Crescent.

Again, it is too early to draw general conclusions. The major task is to uncover and mark DNA elements that provide 'useful' diversity in recombination with the genepools available in contemporary barley cultivars. The 'advanced back-cross' method as proposed by Tanksley *et al.* (1996) has been successfully applied for efficient introgressions of useful diversity complexes from wild relatives into cultivated tomato. If this method allowed corresponding results to be obtained with barley, a new phase in barley breeding might emerge characterised by efficient gains in genetic diversity even beyond the level inherited and retained from barley landraces.

Conclusions and outlook

Within a time span of about 10,000 years, cultivation of barley has spread over all continents, and an unmeasurable number of genotypes have developed forming a wide range of landraces adapted to many different growing conditions and ways of utilisation. It is only during the last two centuries that landraces have been replaced by cultivars in a process that began gradually but gained continuously in intensity.

Twenty years ago, cultivation of barley exceeded a peak area of 81 million ha world-wide; this has now decreased to 61 million ha. The harvest is largely used for animal feed in competition with other coarse grains; malting barley ranks second with a slowly increasing market, while utilisation for food, with some exceptions, plays a minor role.

At present landraces are still being grown on an estimated area of 8.5 million ha, mainly restricted to countries in the WANA region. Information from other important barley growing countries indicates close to 1,700 registered cultivars to serve an acreage of 49 million ha. In most of these countries only a small group of cultivars (average 12%) are used to sow more than 50% of the available barley acreage, and cases are reported for cultivars occupying more than 1-3 million ha occasionally over longer periods of time.

Information from 47 countries mentions 189 active barley breeding programmes. With the exception of West European countries, barley breeding is currently mainly depending on public funding, while breeding methods largely agree on a dominant role of cross-breeding and application of doubled haploid techniques. Utilisation of marker-assisted selection techniques is at present still restricted to special programmes and genetic transformation in a more experimental phase.

The ongoing breeding process itself appears to provide the most important sources for programme continuation obtained from recombination and selection within the available genepool, and a considerable number of very successful cultivars originated from a very narrow genetic base. In addition, a slow but steady influx of additional materials from mutagenesis, landraces and ssp. *spontaneum* is noted. From the large base of landrace diversity in the WANA region a 'decentralised' system of breeding is being developed taking advantage of specific adaptation to stress conditions prevailing in different countries that at the same time should be able to conserve a high level of the existing diversity.

If an outlook on future developments can be based on provisional results obtained from attempts to determine quantitative differences in genetic diversity, it appears that the conventional breeding process can be continued without an urgent need for a general increase in the level of genetic diversity, since a review of 80 years of cross-breeding barley in Europe shows no indication of a reduction in RFLP diversity. Another picture arises if QTL methods would allow the determination as well as the efficient selection of useful introgressions of new DNA complexes, not only from landraces but even more so from untapped levels of genetic diversity that apparently exist in ssp. *spontaneum* and other wild relatives.

Acknowledgements

The diversity estimates presented in this article are largely based on information from national reports provided by cooperating scientists (see below) to whom the author would like to express his gratitude.

Contributors of national reports

Dr. T. Andras, Cereal Res. Station, Taplanszentkerezt, Hungary.
Dr. G.Arias, Embrapa, Passo Fundo, Brazil.
Dr. P. A. Atanassow, Res. Inst. Barley, Karnobat, Bulgaria.
Dr. A. Bude, RICIC, Fundulea, Romania.
Prof. Dr. O. Chloupek, Dept. Crop Sci. and Plant Breeding, Brno, Czech Republic.
Prof. Dr. H.J. Czembor, IHAR, Radzików, Poland.
Dr. M.Z. Díaz, INIFAP, Chapingo, Mexico.
Prof. Dr. St. Grib, Belarus Research Inst., Zhodino, Belarus.
Dr. A. Jahoor, Risø Nat. Laboratory, Roskilde, Denmark.
Dr. L. Jestin, INRA, Clermont-Ferrand, France.
Michael Mackay, Austr. Winter Cer. Collection, Tamworth, Australia.
Dr. J.L. Molina-Cano, UdL-IRTA, Lleida, Spain.
Dr. V.D. Navolotzky, Breeding and Genetics Inst., Odessa, Ukraine.
Prof. Dr. G.F. Nikitenko, Agr. Res. Inst. Non-Chernozem Zone, Nemchinovka, Russia.
Dr. N. Przulj, Inst. Field and Vegetable Crops, Novi Sad, Yugoslavia.
Prof. Dr. I. Rashal, Inst. Biology, Univ. Latvia, Salaspils, Latvia.
Prof. Dr. D. Rasmusson, Dept. Agron. and Plant Genetics, St. Paul, Min., USA.
M. Romero-Loli, Univ. Agraria Nacional, La Molina, Peru.
Prof. Dr. K. Sato, Barley Germplasm Centre, Kurashiki, Japan.
Dr. V. Shevtsov, ICARDA, Aleppo, Syria.
Dir. Dr. A.M. Stanca, Ist. Sperimentale Cerealicoltura, Fiorenzuola d'Arda, Italy.
Dr. M. Užík, Res. Inst. Plant Production, Piešťany, Slovakia.
Dr. H.E. Vivar, CIMMYT-ICARDA, Texcoco, Mexico.

References

Aufhammer, G., 1982. Abstammung der Wintergersten-Sorten – Ahnentafel. Bayer. Landw. Jb. 59: 536-544.

Aufhammer, G., P. Bergal and F.H. Horne, 1958. Barley varieties-EBC. Elsevier Publ. Amsterdam, New York, London.

Arias, G., L. Reiner, A. Penger and A. Mangstl, 1983. Directory of barley cultivars and lines. Ulmer, Stuttgart.

Baumer, M., and R. Cais, 2000. Abstammung der Gerstensorten. Manuscript (unpubl.). [also downloadable as PDF under: http://www.stmelf.bayern.de/lbp/info/gerste.html]

Baumer, M., and W. Göppel, 1988 Gerste, Sorten, Züchter, Ursprungsland, Zulassungsjahr, Abstammung. Manuscript (unpubl.).

Berhane, L., Y. Semeane, F. Alemayehu, H. Gebre, S. Grando, J.A.G. van Leur and S. Ceccarelli, 1997. Exploiting the diversity of barley landraces in Ethiopia. Genet. Res. Crop Evol. 44: 109-116.
Briggs, D.E., 1978. Barley. Chapman and Hall, London, pp. 39-75 and 449.
Brown, A.H.D., 1992. Genetic variation and resources in cultivated barley and wild *Hordeum*. In: Barley Genetics VI. Proc. 6th Int. Barley Genet. Symp., Helsinborg, Sweden, pp. 669-682.
Ceccarelli, S., 1984. Utilisation of landraces and *H. spontaneum* in barley breeding for dry areas. Rachis 3: 8-11.
Ceccarelli, S., 1994. Specific adaptation and breeding for marginal conditions. Euphytica 77: 205-219.
EU Kommission, 1999. Gemeinsamer Sortenkatalog für landw. Pflanzenarten, 21. Gesamtausgabe. Amtsblatt Europ. Gemeinschaften. 09. 11. 1999.
FAO, 1947, 1985 and 1998. Production Yearbooks.
Faccioli, P., M. Grossi, N. Pecchioni, G. Vale and A.M. Stanca, 1998. Barley. In: G.T. Scarascia Mugnozza and M.A. Pagnotta (eds), Italian contribution to plant genetics and breeding, pp. 247-254.
Fischbeck, G., 1992. Barley cultivar development in Europe – success in the past and possible changes in the future. In: Barley Genetics VI. Proc. 6th Int. Barley Genet. Symp., Helsinborg, Sweden, pp. 885-908.
Fischbeck, G., E. Schwarzbach, S. Sobel and I. Wahl, 1976. Types of protection against powdery mildew in Germany and Israel selected from *H. spontaneum*. In: Barley Genetics III. Proc. 3rd Int. Barley Genet. Symp., Thiemig, München, pp. 412-417.
Gauger, H.M., 1997. Statistical Digest. BVBA/SPRL, Zaventem, Belgium.
Graner, A., W.F. Ludwig and A.E. Melchinger, 1994. Relationships among European barley germplasm. II. Comparison of RFLP and pedigree data. Crop Sci. 34: 1199-1205.
Hatz, B., 1997. Untersuchungen der genetischen Diversität innerhalb der Gattung *Hordeum* mit molekularen Markertechniken. Dissetation, München Tech. Univ.
Hillmann, P., 1910. Die deutsche Pflanzenzucht. DLG, Berlin, pp. 182-190.
Horsley, R.D., P.B. Schwarz and J.J. Hammond, 1995. Genetic diversity in malt quality of North American six-row barley. Crop Sci. 35: 113-118.
ICARDA, 1995 and 1997. Germplasm program cereals. Ann. Report.
Jestin, L., 1985. Some aspects of adaptation and adaptability of barley in European conditions. Neth. J. agric. Sci. 33: 195-213.
Melchinger, A.E., A. Graner, M. Singh and M.M. Messmer, 1994. Relationship among European barley germplasm. I. Genetic diversity among winter and spring cultivars revealed by RFLPs. Crop Sci. 34: 1191-1199.
Nei, M., 1973. Analysis of gene diversity in subdivided populations. Proc. Natl. Acad. Sci. 70: 3321-3323.
Nevo, E., A. Beiles and D. Zohary, 1986. Genetic resources of wild barley in the Near East: Structure, evolution and application in breeding. Biol. J. Linn. Soc. 27: 355 380.
Poehlman, J.M., 1959. Breeding Field Crops. Harry Holt, New York.
Rasmusson, D.C., and R.L. Phillips, 1997. Plant breeding progress and gene diversity from *de novo* variation and elevated epistasis. Crop Sci. 37: 303-310.
Reitemeier, A., 1905. Geschichte der Pflanzenzüchtung. Dissertation, Univ. Breslau.
Schachl, R., 1976. Refugia of Austrian land varieties in the subalpinal region. In: Barley Genetics III. Proc. 3rd Int. Barley Genet. Symp., Thiemig, München, pp. 182-190.
Seko, H., 1987. History of barley breeding in Japan. In: Barley Genetics V. Proc. 5th Int. Barley Genet. Symp., Okayama, Japan, pp. 915-922.
Tanksley, S., S. Grandillo, T.M. Fulton, D. Zamir, Y. Eshed, V. Petiard, J. Lopez and T. Beck-Bum, 1996. Advanced backcross QTL analysis in a cross between an elite processing line of tomato and its wild relative *L. pimpinellifolium*. Theor. Appl. Genet. 92: 213-214.
Ullrich, S.E., F. Han and B.L. Jones, 1997. Genetic complexity of the malt extract trait in barley suggested by QTL analysis. J. Amer. Soc. Brew. Chem. 55: 1-4.
Weltzien, E., and G. Fischbeck, 1990. Performance and variability of local barley landraces in Near Eastern environments. Plant Breed. 104: 58-67.

Wolfe, M.S., 1992. Maintaining the value of our varieties. In: Barley Genetics VI. Proc. 6th Int. Barley Genet. Symp., Helsinborg, Sweden, pp. 1055-1067.
Wych, R.D., and D.C. Rasmusson, 1983. Genetic improvement in malting barley cultivars since 1920. Crop Sci. 23: 1037-1040.

Section III

Current Barley Diversity

Diversity in Barley *(Hordeum Vulgare)*
Roland von Bothmer et al (Editors).

Chapter 4

Ecogeographical diversity – a Vavilovian approach

Helmut Knüpffer[a], Irina Terentyeva[b], Karl Hammer[c], Olga Kovaleva[b] and Kazuhiro Sato[d]

[a]Institute of Plant Genetics and Crop Plant Research (IPK), Corrensstr. 3, D-06466 Gatersleben, Germany
[b]N.I. Vavilov Institute of Plant Production, Bolshaya Morskaya 42-44, 190 000 St. Petersburg, Russia
[c]Universität Kassel, Fachbereich 11, Fachgebiet Agrarbiodiversität, Steinstraße 19, D-37213 Witzenhausen, Germany
[d]Barley Germplasm Center, Research Institute for Bioresources, Okayama University, Kurashiki, 710-0046, Japan

Introduction

Since the initial domestication, the diversity in barley has increased during later cultivation (cf. Chapters 2 and 3). Systematic collecting of the diversity on a worldwide scale started in the early 20th century. Ecogeographical studies of cultivated plant species were initiated by Vavilov and have a long tradition in Russia (e.g., Flaksberger, 1935 on wheat; Orlov, 1936 on barley; Vavilov, 1940a on various crops). In western countries, similar studies were carried out more recently, often using marker techniques (e.g., Maxted *et al.,* 1995, 2000).

Vavilov (1926) developed his theory of centres of origin based on detailed studies of morphological variation. He used the term "main areas of origin of cultivated plants", and discussed the problems connected with Darwin's term "centre of origin" (Vavilov, 1940b). According to Vavilov, dominant alleles would predominate in the centre of origin of a species, whereas recessive alleles would prevail in the periphery.

Among the seven major centres of origin of cultivated plants, Vavilov (1926) initially postulated two centres of primary diversification for cultivated barley: Northeast Africa (mountainous areas of Ethiopia) with its wealth of two-rowed barleys, and East Asia (China, Japan, the Tibetan plateau and some adjacent areas), where naked, hooded, awnless and short-awned forms predominate. Later, Vavilov (1940b, 1957) revised his view and concluded that the Near East, with its extraordinary diversity of wild and cultivated species of wheat, rye, barley and other crops, was the main centre of origin, the above-mentioned centres being secondary (see also Chapter 2).

Vavilov's (1957) botanical-geographical approach to the study of the world's diversity in agricultural crops forms the basis for the agro-ecological classification of cultivated barley. It takes into account the evolution of ecotypes and the differentiation of morphological, physiological, resistance and other characteristics.

Knüpffer, H., I. Terentyeva, K. Hammer, O. Kovaleva and K. Sato, 2003. Ecogeographical diversity – a Vavilovian approach. In: R. von Bothmer, Th. van Hintum, H. Knüpffer and K. Sato (eds), Diversity in Barley (*Hordeum vulgare*), pp. 53-76. Elsevier Science B.V., Amsterdam, The Netherlands.

The barley world collection of the N.I. Vavilov Institute of Plant Production (VIR) in St. Petersburg comprises approximately 20,000 accessions of *Hordeum vulgare* ssp. *vulgare* from all over the world. For many years, multi-site characterisation and evaluation studies have been carried out in a network of experimental stations. The characteristics reported here for the various agro-ecological groups in barley are based on studies of germplasm collected mainly in the first half of the 20th century. In areas where the traditionally grown landraces have been replaced by new varieties, the descriptions may not always apply to present-day conditions.

While Orlov (1936) distinguished three main types of cultivated barley with twelve geographical sub-types, based on morphological and physiological characters, Bakhteev (1953) recognised 31 agro-ecological groups from the old world, by also taking agronomic traits into account. Bakhteev's classification was based on multi-site evaluations of almost 17,000 accessions during the period 1933 to 1940.

The most recent infraspecific morphological classification system for cultivated barley (Lukyanova *et al.,* 1990) includes 218 botanical varieties, grouped under two subspecies and four convarieties. This treatment also includes the 192 varieties listed by Mansfeld (1950) under five convarieties. The classification is widely used in germplasm collections and provides a key to a number of morphological characters of the spike and the caryopsis.

In the United States and Canada, work similar to that in Russia was carried out (e.g., Harlan, 1914, 1918; Cowan, 1936; Åberg, 1944; Wiebe and Reid, 1961). Åberg and Wiebe (1946) quote Harlan who noted that any evaluation of morphological and physiological characters, whether for taxonomic or for breeding purposes, must be based on field observations at several experimental sites and in different years. Harlan had the rare opportunity of developing and confirming his ideas by growing varieties for 35 years under a large variety of climatic conditions in the United States, and by studying their native habitats in numerous foreign countries. A large number of diversity studies in barley related to agro-ecological conditions have also been carried out in Japan by Takahashi and his co-workers and successors (e.g., Takahashi, 1955, 1987; Takahashi *et al.,* 1983).

This chapter intends to make studies of Russian and Soviet researchers on the agro-ecological classification of cultivated barley, which have been published mainly In Russian, accessible to the Western researcher. The present description of centres of diversity and their subdivision into agro-ecological groups follows the Russian "Flora of Cultivated Plants – Barley" (Lukyanova *et al.,* 1990), which reflects the views of Vavilov (1926, 1957, 1967) and his scholars (Orlov, 1936; Bakhteev, 1953, 1960; Sinskaya, 1969; Trofimovskaya, 1972).

Characters used for the delimitation and description of agro-ecological groups

A number of morphological, adaptive and agronomic characters are used in the classification into agro-ecological groups (Bakhteev, 1953). All traits are evaluated using standards and measured under various conditions at trial sites throughout the former USSR.

Earliness is the duration of the period from seedling emergence to ripening stage in days. This character includes five classes and ranges from very early-ripening (55-62 days) to very late-ripening (more than 92 days). The most early-ripening barleys occur in the former USSR, North Africa, India, Pakistan, Syria, Israel, and Northern Scandinavia. Very late forms are characteristic for the Pyrenean peninsula, Central Europe, Western Caucasus and Japan. Early barleys from Southern countries retain this character even under deviating environmental conditions, while early northern forms, as a rule, lose their earliness when grown in southern conditions (Bakhteev, 1953).

Vernalisation requirement is determined as the reduction in number of days to maturity in material vernalised before sowing, as compared to untreated controls. The scale ranges from low (reduction by 0-2 days) to very high vernalisation requirement (up to 15-17 days and more), divided into four classes. A very high vernalisation requirement occurs in so-called semi-winter and winter forms, i.e., untreated plants usually do not form spikes at all, whereas vernalised plants ripen normally. Semi-winter forms can be sown both in spring and autumn; they are similar to spring forms: when sown in spring, they head and ripen normally but need slightly more time to mature than spring barleys, whereas real winter barleys do not head and ripen at all when sown too late in spring.

Spring and winter habit: Spring forms predominate throughout the area of barley cultivation. However, in the Caucasus region, lowland Middle Asia, Central Europe, Balkan, the Mediterranean, India, Pakistan, China, Japan, winter and semi-winter forms also exist. Winter barleys are cultivated in regions with mild winters and abundant autumn precipitation. They are not usually grown under severe winters and in highlands.

Reaction to daylength is defined as the difference in number of days to heading between southern (short-day) and northern (long-day) growing conditions. In Northern Caucasus, the daylength in spring and summer ranges from ca. 13 to 15.5 hours, while on the Kola peninsula it reaches 24 hours daylight, resulting in a daylength difference of up to 9-11 hours. Daylength response is measured in seven classes, ranging from daylength-indifferent (no reaction) to very strong reaction (>25 days of heading delay) in Southern conditions.

Plant height ranges from 30-35 cm (dwarfs) up to above 120 cm ("giants"). The character contains seven classes. Generally, barley plants are 80-100 cm tall. Extremely tall forms (>100 cm) are grown under abundant moisture and warmth in Transcaucasia, the Far East, West, South and South-western Europe. Endemic forms with short plants are found in hot and dry regions (Middle and Southwest Asia, North Africa), or under warm and moist climate (China and Japan).

Density of foliage is a compound trait observed during heading, including the number of leaves per culm (low: 3-4, high: 5-6), leaf length (short: <15 cm, long: >30 cm), leaf width (narrow: <9 mm, broad: >20 mm), and number of culms per plant (low: 1-2, high: 6-7). The density of foliage is low, medium or high, and is a measure for the total leaf area of a plant.

Straw thickness is measured in the centre of the second internode. It ranges from thin (diameter <2.5 mm) to thick (up to 4 mm), with 3 classes.

Growth shape varies from completely erect to prostrate (culms more or less adpressed to the ground).

Spike length: short (<5 cm) to very long (>11 cm), divided into four classes.

Spike density: Very dense spikes are characteristic for endemic barleys of East Asia and the Pyrenean peninsula (abundant precipitation combined with high temperatures), Afghanistan, Syria and Israel (moisture deficiency combined with heat). In most other areas, lax spikes predominate.

Row type: Six-rowed barleys are predominant in East Asia, North Africa, and on Mediterranean islands, as well as in highlands and at the northern edges of barley cultivation. Two-rowed barleys are less cold-tolerant and less early, and generally not grown above 2,000-2,500 m a.s.l.

Kernel cover: Hulled barleys are commonly distributed, whereas naked barleys, due to their greater cold-hardiness, are predominantly found in mountainous regions where barley often is the only cereal. Naked barleys from lowland regions are more susceptible to diseases and give lower yields than hulled forms, and cannot compete with other bread cereals such as wheat and rye.

Awn length and type: Long-awned forms are common, except in East Asia (China and Japan), where short-awned and awnless forms adapted to monsoon climates predominate. Hooded barleys are concentrated in the Tibetan plateau, Mongolia, Nepal and adjacent regions, where they are adapted to highlands with a strong continental climate. This trait is often correlated with naked grains.

Thousand-grain weight (TGW): small (<30 g) to very large (>45 g). Grain size is linked with spike density. Barley forms with the largest grains (Transcaucasia, Arabia, Mediterranean countries and Western Europe) usually have lax spikes, whereas the smallest grains (Eastern Siberia, Japan and China) are found in dense barleys.

Rigidity of spike: The concept of rigidity, which is strongly correlated with geography and ecology, was first introduced by Orlov (1936). This trait has three grades: rigid (lignified), intermediate, soft/tender (herbaceous). It is a complex trait consisting of the following interrelated characters: spike (rough *vs.* soft), awns (rough, thick, not breaking off at ripening *vs.* tender, soft, breaking off before ripening), caryopses (elliptic, large *vs.* oblong, small), threshability (hard, difficult to remove awns and glumes from grains *vs.* grain easily delimited from awns and glumes), ligule (rough, thick-hulled *vs.* thin-hulled), rachis (tough, erect *vs.* soft, nodding), reaction to moisture (drought-resistant, xerophytic *vs.* moisture-dependent, mesophytic), reaction to insolation and light (high sunlight requirements, strong insolation *vs.* medium sunlight requirements, long-day plants preferring diffuse light). The rigid type occurs in the Mediterranean, Central, South and East Asia, the intermediate forms grow in Europe, Southern and Central regions of the former Soviet Union (except Central Asia), while the soft type is adapted to the Northern circumpolar regions.

Spike colour: Dark-coloured spikes are concentrated in Eastern Africa.

Glume width: Broad glumes (>1 mm) only occur in Ethiopia, Syria, Spain, Armenia and Afghanistan, while narrow glumes (<1 mm) are common in most regions.

Drought resistance is determined at different experimental stations under conditions of soil and air drought. If not specified, "drought resistant" means combined resistance to soil and air drought; it is measured in five classes, from very susceptible (all plants damaged) to highly resistant (plants remain green). Barleys from Eastern Siberia and adjacent regions of Mongolia and North-East China are most resistant to spring drought, due to their ability to slow down their development during early stages, thus surviving spring drought periods characteristic for Eastern Siberia. Most resistant to summer drought (both air and soil) are the non-irrigated lowland barleys of Southwest to Middle Asia. A remarkable resistance to air drought is found in barleys of Armenia, the Apsheron peninsula (Azerbaijan), Crimea, Northern Caucasus and steppe regions, Egypt, Tunisia, Algeria, Cyprus and Asia Minor. Barleys from mountainous and northern areas are susceptible to drought.

Winter hardiness is only determined for semi-winter and winter forms in Northern Caucasus region (Kuban station) at the northern border of commercial winter barley cultivation, both under natural conditions and in growth chambers. The trait varies from low (0-30% surviving) to high (85-90% surviving) divided into in three classes.

Lodging forms predominate in the former USSR, the Tibetan plateau and Mongolia. Barleys of Japan, India, Pakistan, Arabia, Ethiopia, the Mediterranean and Western Europe, as well as the Steppe agro-ecological group are highly lodging-resistant. Non-lodging barleys occur both in dry and moist climatic conditions.

Reaction to low and high temperatures is evaluated by multi-site trials under various climatic conditions. The plants have their highest temperature requirements between the heading and the ripening stage. Therefore, forms accomplishing this development under extreme

Northern conditions (where the mean air temperature is below 17 °C) in 25-30 days are classified as having low temperature requirements. Plants which need more than 40 days to achieve this development, or which do not ripen at all under Northern conditions, have very high temperature requirements. The character has four classes.

Reaction to diseases and pests is evaluated under field and/or laboratory conditions. The scale ranges from 0 (free of symptoms or no infection) to 4 (strong infection). Barleys most resistant to rusts, loose smut, net blotch, powdery mildew and other fungal diseases are found mainly in the Mediterranean, Ethiopia, Arabia, Transcaucasia, India and Pakistan, while the most susceptible forms occur in Northern Russia, Middle and East Asia.

Centres of diversity and agro-ecological groups

The descriptions of agro-ecological groups are arranged according to the centres of diversity. The distribution of the Old World groups is illustrated in Figure 4.1.

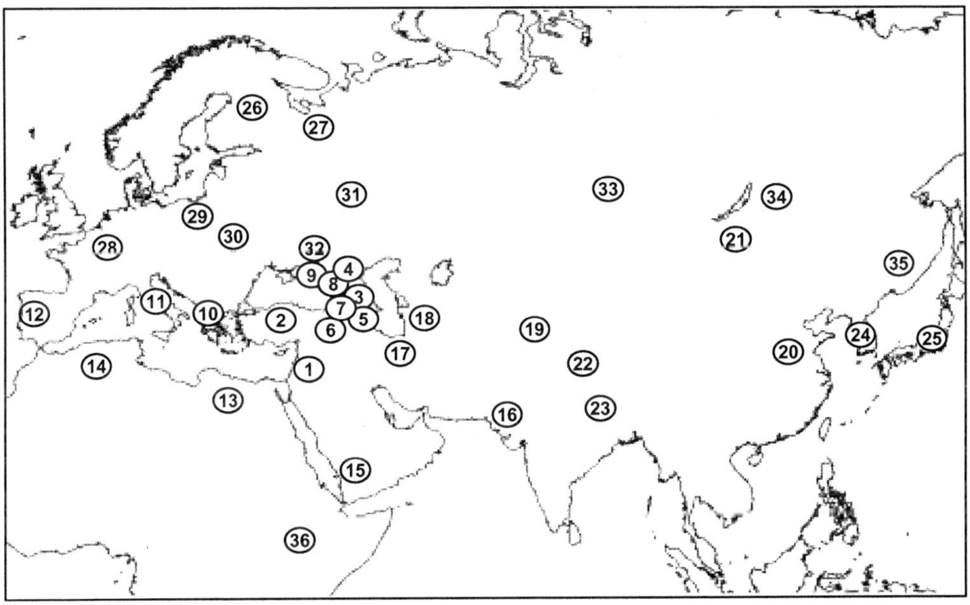

Figure 4.1. Agroecological groups in the Old World (after Bakhteev, 1953): 1 – Syrian-Palestinian Group; 2 – Anatolian Group; 3 – Azerbaijanian-Daghestanian Group; 4 – Daghestanian Group; 5 – Apsheron Group; 6 – Armenian Group; 7 – Armenian-Georgian Group; 8 – Mountainous-Caucasian Group; 9 – West-Caucasian Group; 10 – Greek-Macedonian Group; 11 – Apennine Group; 12 – Pyrenean Group; 13 – Egyptian Group; 14 – North African Group; 15 – Arabian Group; 16 – Indian Group; 17 – Iranian Group; 18 – Iranian-Turkestanian Group; 19 – Pamir-Badakhshan Group; 20 – Chinese Group; 21 – Mongolian-Tibetan Group; 22 – Tibetan Winter Group; 23 – Nepalese Group; 24 – Korean Group; 25 – Japanese Group; 26 – Northern Group; 27 – North European Naked Group; 28 – West European Group; 29 – Central European Group; 30 – West-Ukrainian Group; 31 – Steppe Group; 32 – North Caucasian Group; 33 – West Siberian Group; 34 – East Siberian Group; 35 – Far-Eastern Group; 36 – Ethiopian Group.

Near Eastern Centre

It is now well established that the Fertile Crescent was the primary region where domestication of barley took place (see also Chapter 2). Vertical zonality and the diversity of environmental and climatic factors led to an extraordinary ecological and morphological differentiation of barley in the Near East. Two-rowed hulled and naked forms with different grain colours, but also six-rowed barleys with lax or dense spikes developed here. Landraces cultivated for centuries in Asia Minor developed drought- and salt-tolerance and resistance to various diseases (Zhukovskiy, 1933). A six-rowed awnless variety with dense spikes, pyramidal in shape (var. *dundar*), reported from Asia Minor, has been considered to be a form endemic to Japan that was transferred accidentally to Asia Minor (Orlov, 1936), but it has also been regarded as one of the most ancient forms and as a link between Occidental and Oriental barleys (Trofimovskaya, 1972; see also Chapter 2). The relative isolation and the mosaic of diverse climatic and soil conditions in the Caucasus region led to the formation of several distinct agro-ecological groups adapted to specific environmental conditions. Drought- and salt-tolerant forms were found as were types with resistance to rusts (*Puccinia* spp.).

Syrian-Palestinian Group

This includes hulled two- and six-rowed spring and semi-winter barleys from the mountainous, pre-mountainous and pre-desert regions of present-day Syria, Israel and Jordan. The area is characterised by mild winters and a concentration of precipitation in late autumn to spring. The summers are dry and hot. Generally, barley is sown in winter, but spring sowing is also possible. Barleys from this region are short, and characterised by compact, very rigid spikes and a strong waxy cover on culm and leaves. They are early-ripening and show tolerance to both air and soil drought. They have a low vernalisation requirement and are daylength-indifferent.

Anatolian Group

A large diversity in two-rowed barleys is found in Turkey (Orlov, 1932; Zhukovskiy, 1933). Hulled two- and six-rowed spring barleys are cultivated in the mountainous and pre-mountainous regions of inner Anatolia, an area characterised by a dry climate and insufficient precipitation; barley is usually sown in spring. Hulled six-rowed winter barleys are grown in the coastal regions of Western Anatolia with a more humid climate, mild winters and hot summers. The spikes are mostly soft, with large grains. The hulled spring barley forms have a very strong waxy cover on culm, spike and leaves, whereas the waxy cover is expressed to a lesser extent in the winter forms and absent in the naked ones. A high drought resistance was found (Bakhteev, 1953; Trofimovskaya, 1969). The hulled two-rowed spring barleys are early-ripening, short and not lodging. The six-rowed spring and winter barleys are medium-ripening, the winter barleys showing higher resistance to fungal diseases than the spring ones. The naked forms are short, with erect broad and rigid spikes; they are early-ripening. The vernalisation requirements vary between forms, showing medium reaction to daylength. Severe attacks by stem-concealed pests are characteristic for this group. Old local forms from Turkey with a high grain protein content have been reported (Zhukovskiy, 1933).

Azerbaijanian-Daghestanian Group

This ecotype of hulled six-rowed winter and semi-winter forms is distributed in pre-mountainous and lowland regions of Azerbaijan and Daghestan, and in pre-mountainous areas of Karabakh. The winters are relatively mild, except for the northern regions of Daghestan, where they are stronger than in Azerbaijan. Winter sowings (up to 1000 m a.s.l.) predominate, but spring forms are also grown. Barleys from this group resemble both Iranian and Turkish types. The spike is strongly nodding during ripening, and the awns have large sharp bristles, similar to the mountainous-Caucasian type. The caryopsis is very large. The vernalisation requirement is above medium, the daylength response medium. These barleys are tall, late-ripening, susceptible to lodging, and have a high density of foliage. They are susceptible to powdery mildew, smut and net blotch.

Daghestanian Group

The region includes the highlands of Daghestan and Ossetia, and partly of Georgia. These areas receive sufficient moisture during the relatively short vegetation period. The complex climatic and soil conditions have led to a diversity of naked two-rowed, mostly spring forms with large caryopses and a high density of foliage. They have a low vernalisation requirement, but a high sensitivity to daylength. They are tolerant to short-term spring frosts and do not require high temperatures during ripening. The barleys are similar to those from Azerbaijan. Especially the winter forms show a high susceptibility to lodging, and barleys from this region are particularly susceptible to fungal diseases. Naked forms have an especially high protein content (Bogdanova *et al.,* 2001).

Apsheron Group

Mainly six-rowed hulled local winter barleys were grown in the driest lowland zones of Azerbaijan, particularly on the Apsheron peninsula and in the Mugani region. This group is characterised by large caryopses, early ripening and high resistance to lodging, drought and fungal diseases. The plants are short, with low winter hardiness, needing high temperatures during grain filling and ripening. They do not require vernalisation; the reaction to daylength is medium.

Armenian Group

This region covers dry areas of Armenia around Erevan, and the territory of Turkish Armenia. The precipitation during summer is insufficient; therefore, barley is usually sown in autumn (Bakhteev, 1953). The morphological diversity of the dominating two-rowed hulled forms is low (Agadzhanyan, 1928; Stoletova, 1929-1930; Gandilyan, 1980). The plants are short, early ripening with erect spikes with a low tendency to lodging. These barleys are resistant to both air and soil drought, and tolerant to short-term spring frosts, but require high temperatures for ripening. The vernalisation requirement is very low, and they are neutral to daylength. They show high resistance to smuts and leaf stripe. A two-rowed variety with wide outer glumes has been found, a characteristic common to Ethiopian barleys. Some local spring barleys have good malting properties (Minasyan, 1961).

Armenian-Georgian Group

This group is confined to the mountainous regions of Armenia, Georgia, Mountainous-Karabakh and Ossetia, characterised by relatively high moisture supply and low temperatures during the vegetation period. The two-rowed hulled spring barleys are able to utilise late precipitation after long periods of soil drought. They have a high foliage density. They are susceptible to soil and air drought, but resistant to spring frosts. They have a low vernalisation requirement and a low reaction to daylength. Vavilov noted particular resistance properties, namely, the occurrence of forms with dominant genes for resistance to loose smut. Local barleys with good malting properties have been found in Georgia (Menabde, 1938).

Mountainous-Caucasian Group

This ecotype includes six-rowed spring barleys, mainly from the high altitudes of the Caucasian Ridge, which were first described by Regel (1910). This area is characterised by a short vegetation period, sufficient moisture and moderate temperatures during vegetation. The plant height is medium; the nodding spike with a fragile (easily breaking) spike neck becomes strongly anthocyan-coloured before ripening. The plants are generally early-ripening, susceptible to lodging, show a moderate drought resistance and good tolerance to late (spring) frosts, and are susceptible to fungal diseases. Their vernalisation requirement is very low, and their response to daylength neutral. They do not require high temperatures during ripening.

West-Caucasian Group

This group is dominated by hulled, six-rowed spring barleys cultivated in the western coastal area of Georgia, which is characterised by a humid warm temperate climate. The plants are very tall, with long, nodding spikes, a thick strong culm, a high foliage density, and a strong waxy cover on culm and leaves. The plants are late-ripening, resistant to lodging, prefer moist growing conditions, and are susceptible to drought. The vernalisation requirement and the reaction to daylength is medium. Some forms show resistance to loose smut and leaf rust.

Mediterranean Centre

Vavilov (1957) included the entire Mediterranean coastal area in this centre, covering North Africa (Egypt, Algeria, Tunisia), Greece with its islands, Spain, Italy, and western and south-western parts of Asia Minor. Trofimovskaya (1972) also included Syria, Jordan and Israel in this centre, whereas Vavilov (1957) had only included the "inner oases of Syria, Morocco and Algeria". Later, the term "Ancient Mediterranean" was coined, which also embraces the Arabian Peninsula and even India (cf. Filatenko *et al.*, 1998). According to Vavilov's theory, Mediterranean barleys originated from Near Eastern types. This is supported by archaeological findings (see also Chapter 2).

The barleys here are characterised by a rigid spike and a strong waxy cover that occurs especially in periods of drought. They show a complex resistance to fungal diseases (Vavilov, 1967). Accessions resistant to virulent races of leaf rust, powdery mildew and loose smut were found (Krivchenko *et al.,* 1972; Shchelko and Traynina, 1974; Shchelko *et al.,* 1976; Terentyeva, 1985; Grigoryev *et al.,* 1989; Khokhlova, 1990). Single samples from Southern Mediterranean (Morocco, Algeria, Tunisia) show resistance to powdery mildew, leaf rust and BYDV, but are susceptible to loose smut. Mediterranean barleys usually have a good lodging resistance.

Greek-Macedonian Group

This group is confined to the valleys of Thessalia, Macedonia, and the European part of Turkey, areas characterised by dry and hot summers with no precipitation from July to September. Barley is sown in late autumn and in spring (October and April) when rainfall occurs (Bakhteev, 1953). These hulled, mostly six-rowed spring and semi-winter barleys are medium to tall, with erect spikes, and have a high density of foliage with a strong waxy cover. The plants are early- to medium-ripening, with a low tendency to lodging, and susceptible to smut and *Drechslera*. They are tolerant to air drought, but sensitive to soil drought, and require high temperatures during grain filling. The vernalisation requirements are low in spring barleys and medium in semi-winter forms; the reaction to daylength is low.

Apennine Group

Sicily and Sardinia are characterised by typical Mediterranean conditions with less than 400mm of erratic rainfall, mild winters, light and poor soil, high temperature and drought during the grain filling period. The Apennine environment is characterised by lower temperatures during winter and more regular (more than 400mm) rainfall (Cattivelli *et al.*, 1994). Six-rowed forms predominate in this area. Hulled two-rowed spring barleys can be easily distinguished from Western European two-rowed barleys by their lower height, their rough, almost erect spike, stiffer straw and higher drought resistance; some taller two-rowed forms belong to the Western European group but have a higher drought resistance than these. The hulled six-rowed spring and semi-winter barleys are distinguished by their higher density of spikes and foliage, but traditional material with very dense spikes has been almost completely displaced by less dense forms (Hammer *et al.*, 1985). Some forms from Sicily and Sardinia have a particularly stiff straw. Naked two-rowed spring barleys differ from naked barleys of Asia Minor, their straw is softer, the plants are taller and their foliage has a higher density. The naked six-rowed barleys have stiff straw, large caryopses, a high yielding ability, and good drought resistance (Bakhteev, 1953). The reaction to vernalisation is low to medium, the daylength response is neutral. Barley has been traditionally used for bread making (Attene *et al.*, 1996) and, more recently, also for pasta (Cattivelli *et al.*, 1994).

Pyrenean Group

Hulled two- and six-rowed spring barleys, and hulled six-rowed semi-winter barleys are grown in Central and Eastern regions of Spain, and in Portugal. This region is characterised by a warm temperate climate with sufficient precipitation and hot dry summers. Barley is sown in autumn. This unique group differs from all other Mediterranean groups. The medium- to late-ripening plants are tall with large, dense spikes, dense foliage and thick straw, and a very strong response to vernalisation, but indifferent to daylength. They require high temperatures during ripening. Many forms are resistant to leaf rust.

Egyptian Group

The Egyptian Group includes extremely early hulled and naked six-rowed spring barleys from irrigated regions in Upper and Lower Egypt, and from Cyprus. The generally short plants with a high density of foliage are tolerant to air drought. They require high temperatures during ripening. The vernalisation requirement is medium, the reaction to daylength low. They show a high resistance to net blotch, bacterial diseases, leaf rust and powdery mildew (Bakhteev, 1953; Lukyanova *et al.*, 1990; Khalifa, 1996). Material from Cyprus is morphologically closer to the Syrian-Palestinian type, but the plants are taller, and the spike is softer.

North African Group

The North African Group consists of hulled, mostly six-rowed spring and semi-winter barleys and naked six-rowed spring barleys from coastal regions of Morocco, Algeria, Tunisia and Libya, including the pre-mountains of the Atlas. This area is characterised by abundant precipitation during autumn and winter, and less in spring; the summers are hot and dry. The barleys of Libyan desert oases show a connection to Ethiopia, e.g., naked forms occur (Hammer *et al.*,

1988). Barley is usually sown in autumn. The vernalisation requirement is medium in spring barleys, and high in semi-winter forms, the reaction to daylength is high. Many forms are resistant to several fungal diseases, but susceptible to frit fly. North African barleys are susceptible to net blotch and tend to lodging. The hulled barleys are rigid, with rough spike and awns, have large caryopses, and are resistant to heat and air drought. The naked barleys are softer and more susceptible to diseases and insect damage than the hulled ones.

Arabian Group

The main growing areas are mountainous regions of Arabia and Yemen, with a short vegetation period and damaging drought during the ripening period. This group includes two- and six-rowed hulled and naked forms. Together with the Egyptian and Indian barleys, the Arabian ones are the most early-ripening forms. The vernalisation requirement and the reaction to daylength are low. Their large caryopses suggest a close relationship to Ethiopian and Indian barleys (Bakhteev, 1953); they are extraordinarily drought-resistant and need high temperatures during ripening.

Indian Group

The barleys of the north-western dry areas of India and Pakistan, including Punjab, Sindh, and Kashmir, consist of two- and six-rowed hulled and naked spring and winter types. This region is characterised by monsoon climate. Semi-winter and winter forms are concentrated in Kashmir. The early-ripening plants have a low density of foliage. Some naked forms with short stiff straw are resistant to lodging. The vernalisation requirement is low; these barleys are almost daylength-insensitive. Most accessions are drought-resistant, but susceptible to net blotch, leaf rust and bacterial diseases. Pakistani barleys are highly susceptible to leaf rust.

Middle Asian Centre

The local dry-farming barleys from this region, which includes Afghanistan, Tajikistan, Uzbekistan and the western Tien-Shan area, are characterised by heat and drought tolerance. They are diverse in their morphological and adaptive characteristics. Due to the lack of periodical epiphytotics of powdery mildew and rusts, most barleys from this gene centre did not develop resistance to these diseases (Emmerikh, 1961). They are susceptible to lodging, but show higher drought tolerance than Mediterranean barleys. Some forms show frost resistance.

Most barleys from Middle Asia belong to the fodder type, characterised by high grain protein content. However, single six-rowed samples from Afghanistan with good malting quality have been found. Their enzymatic activity is higher than that of Western European malting cultivars, although they contain 2-3% less malt extract (Ilyina, 1972).

Naked barleys are frequent in mountainous regions. Among them, forms with good bread-baking properties were identified (Emmerikh, 1961).

Iranian Group

The naked, six-rowed barleys of the Iranian mountain ranges are sown in autumn and early spring. This area is characterised by rainfall in March and April, the summer is hot and dry. The lodging-resistant plants are short with thin straw. Medium-ripening forms predominate. These barleys are susceptible to drought, and need high temperatures before ripening. They do not show any vernalisation requirement; heading is earlier under long-day than under short-day conditions. They are very susceptible to rusts and bacterioses, and moderately susceptible to loose smut and leaf stripe.

Iranian-Turkestanian Group

This group consists of six- and two-rowed hulled spring barleys and hulled six-rowed semi-winter and winter forms. It is confined to the lowland and pre-mountain regions of Turkmenistan, Uzbekistan, Tajikistan, Kazakhstan, Kyrgyzstan, North-West China (Xinjiang region), Afghanistan and Iran. The precipitation is insufficient during the vegetation period, the summers are hot and dry and the winters are relatively mild. Barley is sown in autumn or early spring. The local barleys are generally drought-resistant, susceptible to lodging, and highly susceptible to fungal diseases. Six-rowed, hulled forms with small grains, a strong waxy cover on the spikes and a low density of foliage predominate. The vernalisation requirement and the reaction to daylength are low. In the lowlands of Kyrgyzstan and Uzbekistan, hulled two-rowed and, as admixture, six-rowed forms with black glumes can be found. The plants are generally tall with narrow-glumed caryopses of medium size. Landraces of Kazakhstan are usually mixtures of different forms. The dry-farming ecotype of the Turanian lowland has a rigid spike, and dry-farming barleys from Tajikistan are characterised by a high density of foliage and thick straw. Barleys from Turkmenistan have thin straw, a lax spike, and small caryopses. Forms with high frost tolerance are reported (Duganova, 1952). Within this region, the largest caryopses occur in Tajikistan, the smallest ones in Kazakhstan.

Pamir-Badakhshan Group

Six-rowed hulled and naked spring barleys predominate in the high-mountainous regions of Uzbekistan, Tajikistan, Turkmenistan, Pamir, Chitral, Badakhshan and Ladakh. This area is characterised by a strong winter with little snow and a short, dry vegetation period. The plants are short, with high density of foliage, with very broad, soft, nodding long leaves, and with thick lodging straw. In the Pamir and Badakhshan mountain ranges, medium- and late-ripening forms predominate. The vernalisation requirement and the reaction to daylength are generally medium. The barleys require soil moisture and are very susceptible to frit fly and fungal diseases, especially leaf and yellow rust and net blotch. They are tolerant to air drought, but not to heat; and they do not require high temperatures during ripening.

East Asiatic Centre

This centre includes China, Korea, Japan, the Tibetan plateau, Nepal and some adjacent areas. Vavilov first noted the diversity of endemic forms, which was not known from other centres. The barleys of Japan, China, Korea and India are distinctly different from the Occidental barleys (Takahashi, 1955). Barley first migrated to Japan from China, partly via Korea, and Japanese and Korean barleys have many traits in common with Chinese ecotypes, but have also developed specific characteristics. The diversity of six-rowed barleys is concentrated here, including awned and hooded naked winter forms. East Asiatic barleys are generally characterised by short straw (dwarfness), and genes for dwarfness occur only in the East Asiatic gene centre. However, tall and medium plants also occur (Trofimovskaya, 1972). They have dense, short spikes, and small, almost spherical caryopses. They are awnless or having short awns.

Dwarf and semi-dwarf forms from Japan and China have a low winter hardiness. East Asiatic barleys are characterised by early maturity (Trofimovskaya, 1972) and high temperature requirement during ripening.

Barley forms with naked, spherical grains are used for human consumption. In Korea and Japan, a beverage ("barley tea") is made from roasted barley. These countries have also produced many BYMV-resistant forms (Takahashi *et al.,* 1973).

Chinese Group

This group is confined to the humid regions of Central-Southern and Eastern China, characterised by a monsoon climate and hot summers; the winters are much more severe than in Japan. Barley is mainly sown in autumn, after the rice harvest. Landraces and cultivars bred in China are mostly hulled and naked six-rowed spring, semi-winter and winter forms. Short-awned, awnless, awnletted and hooded forms occur. Kernel colours include green, black and violet. They are similar to Japanese barleys, but differ from them with regard to taller plant height, narrower leaves, and thinner straw. Short-awned forms are endemic to Inner and Eastern China. Chinese barleys generally ripen earlier than other ecological groups. They show moderate resistance to spring drought, and moderate winter hardiness under mild winter conditions. Spring forms have a low vernalisation requirement; in winter barleys, it is very high. The reaction to daylength is medium. Morphologically, Chinese barleys are also close to those from mountainous Badakhshan (Pamir) and (especially western) Mongolia, but less drought-tolerant. Single accessions from China have good malting properties, combining a low protein content with high malt extract.

Mongolian-Tibetan Group

Barley found its way from China to the north via Mongolia, where it adapted to specific conditions, leading to the formation of the Mongolian ecotype. The main areas of this group are the oases of Inner and Outer Mongolia and the Tibetan plateau, generally characterised by a harsh continental climate, the winter temperatures in some regions reaching -45 °C; the summer is hot and dry, with large temperature differences between day and night; night frosts after hot days are common. With the exception of the south-eastern Tibetan plateau, precipitation is insufficient (Bakhteev, 1953). These mostly long-awned, six-rowed naked and hulled spring barleys are similar to the Chinese group, but differ in their taller plant height, lax spikes, the occurrence of hooded forms, susceptibility to diseases, and large leaves. Hulled forms predominate in Northern and Central Mongolia, whereas the frequency of naked forms increases towards south-west (Khoroshaylov, 1959). At high altitudes (3,000-4,000 m a.s.l.) of the Sichuan province, purple lemma colour is common, whereas black is rare. Purple, black and brown kernels are predominant (Takeda *et al.,* 2000). The plants have a moderate waxy cover on the spike, a high density of foliage and thick straw, but are susceptible to lodging. The spike neck is thick but quite brittle. Mongolian-Tibetan barleys range from early- to late-ripening; some of them are tolerant to high salinity. Under spring drought conditions, a certain drought-resistance was observed (Lukyanova and Kozhushko, 1969). The vernalisation requirement is medium, but the reaction to daylength strong. They easily stand low temperatures. Barleys of this group have a high protein content (average 15.6%).

Tibetan Winter Group

This group consists of an isolated ecotype of winter barleys from high altitudes in the Tibetan plateau, cultivated up to 4,500 m a.s.l., which is exceptional in barley. It is represented by highly diverse populations of six-rowed hulled and naked barleys with short, predominantly dense spikes and small to medium-sized caryopses. The plants are tall, extraordinarily prostrate, with thick straw, and long, very broad leaves, moderately resistant to lodging, with low frost resistance.

Nepalese Group

Earlier compilations did not include a Nepalese agro-ecological group, due to the low number of accessions in the Vavilov Institute's collection. Barleys from Nepal are predominantly six-rowed, naked (less frequently hulled) spring forms. The caryopses show a wide range of colours and other characteristics, awnless, awnletted and hooded forms occur (Witcombe, 1975). With respect to qualitative characters, barleys from Nepal show a greater diversity than those from all other Himalayan regions combined (Murphy and Witcombe, 1981). Hulled barleys are predominantly distributed in the lowlands and naked ones in the highlands (Takahashi, 1987; Takeda *et al.,* 2000). In a cluster analysis of agro-morphological characters of 1,118 spring barley accessions grouped by 25 countries, Nepalese barleys appeared to be most distinct from all other regions (Cross, 1994). Barleys of the Occidental and Oriental types may have hybridised around Central Nepal (Takahashi, 1959).

Korean Group

A Korean group was not delimited by Russian researchers. Most Korean barleys are six-rowed, some with naked grains (Kim *et al.,* 1987). Occasionally intermedium-types and a few two-rowed forms occur. Some South Korean barleys have laterally awnless six-rowed, dense spikes with short awns, uzu semi-dwarf types known as Oriental barleys (Takahashi, 1955), which are rare in North Korea. They are adapted to water-bed cultivation. Most South Korean barleys, but only half of the North Korean ones are winter forms. Early- and medium-ripening accessions are found. They are susceptible to lodging, and highly susceptible to powdery mildew, but resistant to BYMV. Local winter barleys with high resistance to BYDV were identified (Habekuß and Lehmann, 1991; Habekuß, 1995).

Japanese Group

This group is distributed over all Japanese islands and in Taiwan, an area generally characterised by monsoon climate without dry periods. The mean annual precipitation ranges from 1,000 to 3,000 mm, except for Hokkaido (750-1,000 mm). Barley is usually sown in autumn. There are hulled and naked six- and two-rowed spring, semi-winter and winter forms. The plants are short or dwarf, with thick straw and reduced length of the upper internodes, and not lodging (Khramysheva, 1971; Gudkova, 1989), with a strong waxy cover on culm and leaves. The leaves are short and broad. The spikes are short-awned or awnless and very dense, with small spherical caryopses, some with broad glumes (Bakhteev, 1953). The two-rowed barleys, and some taller six-rowed forms, are obviously not endemic, but most likely result from crosses between six-rowed Japanese barleys and two-rowed hulled European, Middle-Asian and Transcaucasian forms (Bakhteev, 1953). Under conditions of heavy rain, Japanese barleys show low resistance to root-crown lodging. In spring barleys, the vernalisation requirement is medium, but high or very high in winter forms. The reaction to daylength is medium. They are susceptible to drought, have a low winter-hardiness, and their susceptibility to powdery mildew, leaf rust and loose smut is medium to high. Rare forms with waxy, glutinous starch have been found, with up to 94.3% amylopectin (Trofimovskaya, 1972). The endemic semi-dwarf genotype called 'uzu', with short fleshy leaves forming a sharp angle with the culm, is popular as food barley in the western part of Japan. Single samples show good bread-making qualities (Trofimovskaya, 1972). High malt extractability has been found in the cultivars 'Hatakaze', 'Chusei Golden Melon', and 'Krin Choku 1' (Sergeeva, 1972).

European-Siberian Centre

The European-Siberian gene centre, first delimited by Zhukovskiy (1968), was defined for barley by Trofimovskaya (1972). Barley reached this region from the Near East (Sinskaya, 1969). Due to the long history of cultivation (see also Chapter 2), the intense agriculture and breeding (see also Chapter 3), and the variation in natural and climatic conditions, a number of specific ecotypes evolved (Orlov, 1936). The variation ranges from forms adapted to the moist climate of Western Europe to drought-resistant forms in Southern Ukraine, Northern Caucasus, Southeast Russia, Western and Eastern Siberia. Barley increased its tolerance to soil acidity in North and North-West Europe. The 'Hana' barley from Moravia, due to its technological properties, was the origin of the malting barleys currently cultivated worldwide. Material from this region has an optimal ratio of quantitative productivity characteristics, such as number of grains, TGW, productive tillering, and density of culms per square unit, allowing for high yields under conditions of intensive agriculture.

Northern Group

Early-ripening six-rowed hulled spring barleys dominate in Northern Scandinavia and the Northern European part of Russia. The older varieties are tall, with lax rectangular-rhombic spikes, nodding before ripening, and with thin awns. They are lodging and heavily susceptible to fungal diseases. Breeding progress drastically changed the plant architecture of Scandinavian barleys; recent cultivars are shorter. They do not require high temperatures during ripening and are adapted to periods of strong temperature reduction. The vernalisation requirement is low or absent, whereas the reaction to daylength is very strong. Barleys of this group, especially those from Finland, are highly susceptible to smuts and powdery mildew.

North European Naked Group

A small group of six-rowed naked spring barleys found among landraces of the European part of former USSR has obviously been introduced from Middle Asia and exposed to adaptation and selection (Bakhteev, 1953; Lukyanova *et al.,* 1990). These barleys have a long, lax spike; the soft caryopsis tends to pre-harvest sprouting. The plants are tall, with dense foliage, long and broad leaves and thick straw. They are susceptible to lodging and diseases, show medium to early ripening, and are moderately cold-resistant. They show a medium vernalisation requirement, but above medium reaction to daylength.

West European Group

This large group consists of very productive two-rowed hulled spring barleys grown in forest and forest-steppe areas of the European part of Russia, Germany, the Czech Republic, the UK, the Netherlands, Belgium, France, Austria, and Southern Scandinavia, where advanced breeding programmes have been established for more than a century. Older landraces from this region are characterised by tall plants with a particularly long upper internode and a high foliage density, lax and nodding spikes, large caryopses, and susceptibility to lodging. A smaller group of lodging-resistant forms with dense spikes forms an exception. The vernalisation requirement is above medium, the reaction to daylength generally strong. There are both high-quality forage (Russia, Netherlands, Southern Scandinavia) and low-protein malting barleys (Germany, Czech

Republic, the UK). These forms are generally not drought-resistant but can stand late frosts. Two-rowed winter forms were obtained by crossing two-rowed spring with six-rowed winter forms, e.g., in Austria, former Czechoslovakia, Germany and (former) Yugoslavia (cf. Chapter 3).

Modern cultivars from this region, delimited as a separate group (Lukyanova *et al.,* 1990) have short plants, suited for sowing in dense stands with well-balanced characters of ear productivity: 23-30 grains per spike, TGW above 40 g. Medium- and late-ripening cultivars, mainly of malting type, with good lodging resistance are predominant.

Central European Group

This group is formed by six-rowed hulled winter barleys of Central European countries and the Balkan region, often with small caryopses, which are also cultivated widely in the US. Within this group, lowland and pre-mountainous types can be distinguished. The former are tall, with dense foliage and long leaves, with quite large spikes and low drought resistance, while the latter have shorter plants, a smaller spike, a lower foliage density and shorter leaves, and thin but strong straw. The spike is nodding at ripening. These barleys do not require high temperatures during ripening, and can stand temperature reductions in spring. The vernalisation requirement is high and the reaction to daylength is low.

West-Ukrainian Group

The Transcarpathian and Ternopol regions comprise mainly six-rowed hulled winter and spring barleys with small caryopses. In the spring barleys, thin-hulled forms and large caryopses are most common. These cultivars are susceptible to lodging, and include medium- and early-ripening forms with moderate frost resistance and susceptibility to fungal diseases. The predominating winter forms of this group are tolerant to deep snow cover. The vernalisation requirement is medium to high. Spring barleys react very strongly to daylength, while winter barleys vary in their daylength reaction.

Steppe Group

These hulled six- and two-rowed barleys are cultivated in steppe and forest-steppe regions of the former USSR (dry steppes of Kazakhstan, Western Siberia, Volga region, Northern Caucasus, Crimea, Ukraine and Moldova). These cultivars were once widely used in breeding for dry regions. The plants are tall with an erect spike, and susceptible to lodging. The group includes early- and late-ripening forms. Forms with resistance to late-spring frost predominate. The vernalisation requirement is low to medium, whereas the reaction to daylength is above medium. This is the most widely spread type in the former USSR; it combines resistance to drought with good grain quality. Some Crimean barleys are also able to survive the winter in Southern areas if sown in autumn (Bakhteev, 1953).

North Caucasian Group

This group includes tall six-rowed hulled winter barleys with long, lax spikes, small caryopses and a high density of foliage, from the Northern lowlands and pre-mountains of the main Caucasus Ridge. They are mainly medium-early ripening and show good winter hardiness and moderate drought resistance. They have a high vernalisation requirement, and above medium response to daylength. North Caucasian barleys are moderately resistant to leaf rust and loose smut, but highly susceptible to powdery mildew.

West Siberian Group

These two- and six-rowed hulled barleys are cultivated in the regions of Omsk, Kemerovo and Novosibirsk. They differ from the barleys of the European part of Russia and of East Siberia by a longer period between seedling emergence and heading, thus resulting in a higher resistance to spring drought. The vernalisation requirement is low, the reaction to daylength medium. They are tolerant to stem-concealed pests and root rots, but susceptible to leaf rust, powdery mildew and loose smut.

East Siberian Group

These six-rowed hulled spring barleys with small, oblong caryopses are cultivated in the regions of Krasnoyarsk, Irkutsk, Yakutia, and Buryatia. The continental climate with little snow, short summers and large daily and seasonal temperature amplitudes led to the formation of unique, early-ripening forms adapted to early-spring drought and spring frosts. The lax, rhombic spikes are nodding at ripening. Under Northern conditions they ripen quickly at relatively low temperatures. The vernalisation requirement is low, the reaction to daylength very strong. The plants are susceptible to smuts, but their earliness allows them to escape frit fly attack.

Far-Eastern Group

In the region covering the Russian Far East (Primorsk and Khabarovsk regions, Sakhalin island) and North-East China, the vegetation period is long and characterised by abundant water supply (Bakhteev, 1953; Surin and Lyakhova, 1993). Six-rowed hulled spring forms predominate, accompanied by two-rowed hulled and naked barleys. The very tall plants have long, lax spikes, nodding at ripening, and dense foliage with long and broad leaves. The group includes medium- and late-ripening forms. They generally show good resistance to spring frosts; drought-resistant forms are lacking. The vernalisation requirement is low to medium; the reaction to daylength is above medium. The plants are highly infected by root rots. The plant habit and the softness of the spike resemble North Russian barleys, whereas the thick culm and broad leaves are closer to Japanese forms.

Ethiopian Centre

Harlan was the first to scientifically collect barley in Ethiopia in 1923 (cf. Vavilov, 1997). In 1927, Vavilov explored different agricultural regions of Ethiopia and compiled a unique collection reflecting the extraordinary botanical diversity of cultivated barleys. This comprises 38 botanical varieties, of which 16 are endemic. A larger number of endemic two-rowed botanical varieties exist than in any other country of the world. The highest diversity of forms with reduced lateral spikelets, i.e., Mansfeld's (1950) convar. *deficiens*, and with broad glumes was found around 2,540 m a.s.l., while at 2,700 m hulled forms with black or yellow kernels and broad glumes predominate (Orlov, 1936). Finally, at the altitude of 3,000 m, naked two- and six-rowed forms with various spike colours have their highest diversity. Asfaw (1990) notes 21 varieties in the provinces Jibat and Mecha. Measured by diversity indices, the genetic diversity increases from lower to medium altitude (2,400-3,000 m) and decreases thereafter (Engels, 1994). Ethiopian barleys include forms with narrow and broad glumes, smooth and serrated awns, hulled and naked caryopses, lax and dense spikes, and a variety of grain colours. The large morphological diversity was also described by Giessen *et al.* (1956), Zhukovskiy (1964), Harlan (1969), and Trofimovskaya (1972). A high diversity with respect to ear emergence, ripening, and yield has been reported by Lakew *et al.* (1995, 1997).

Orlov (1936) and Bakhteev (1953) noted a high yielding ability and large grain weight, and good lodging resistance. A high protein and lysine content was found (Orlov, 1929; Munck *et al.*, 1971; Lehmann *et al.*, 1978; Lukina, 1981). The unique 'Hiproly' was derived from an Ethiopian barley belonging to var. *neogenes*.

Barleys from this region are related to those from the Near East, which is within the distribution area of *H. vulgare* ssp. *spontaneum* (Nilan, 1964).

Vavilov (1965) paid particular attention to the disease resistance of Ethiopian barleys. Numerous evaluation reports on resistance to fungal (cf. Moseman, 1971; Lehmann *et al.*, 1976; Walther and Lehmann, 1980; Lakew *et al.*, 1995) and viral diseases (Proeseler and Lehmann, 1986) exist. A large diversity was found in resistance to loose smut, yellow and leaf rust, powdery mildew, root rots, leaf stripe, and frit fly, both under strong natural epiphytotic conditions and under artificial inoculation (Rigina, 1966; Krivchenko, 1984; Sidorov *et al.*, 1987; Khokhlova, 1990; Surin and Lyakhova, 1993; Terentyeva, 2001).

Ethiopian Group

The barleys from the highlands of Ethiopia and Eritrea, the Ethiopian Centre, form a single group, which includes two- and six-rowed hulled and naked spring barleys grown between 1,600 and 3,000 m. The precipitation is concentrated in two rainy seasons, namely, the large (July-September) and the small one (February). Barley is sown mainly in the beginning of the main rainy season, in July, and ripens during the dry period (Bakhteev, 1953). The plant height ranges from medium to tall. The spikes show violet tinges before ripening, except in six-rowed naked forms. The rachis is strong, and the spike is erect. The plants do not require high temperatures during ripening. The vernalisation requirement is low; forms with strong daylength response predominate. Large-grained forms combining resistance to loose smut with high protein and lysine contents have been found.

New World Centre

The development of barleys of the Americas and Oceania (Australia and New Zealand) is a relatively recent occurrence compared to the evolution in the ancient growing regions (see also Chapter 2).

Barley was introduced to America from Spain and Spanish colonies immediately after the discovery, and from England and The Netherlands in the 16th century (Orlov, 1936). The barleys introduced by Spanish and Portuguese immigrants adapted well and maintained their resistance to different rust species. Six-rowed barleys prevail in America. In the Eastern United States, six-rowed cultivars originating from continental Europe are better adapted than the two-rowed forms from England (Wiebe, 1979). Local cultivars from the steppe regions of Northern Caucasus and Southern Ukraine, characterised by drought resistance, contributed to the North American barley genepool (Lukyanova *et al.*, 1990). The awnless var. *dundar* from Asia Minor was used in many complex crosses, and cultivars from Western Europe, Turkey, China and North Africa were introduced. Cultivars grown in Alaska were created by crossing Siberian early-ripening forms (East-Siberian type of barley) with Ethiopian forms (Orlov, 1936).

Manchurian Group

This group consists of six-rowed hulled and smooth-awned malting spring barleys most widely grown in the humid section of the Upper Mississippi Valley, south of the Great Lakes, and in Canada (Åberg and Wiebe, 1946). Their pedigrees obviously contain Manchurian local material.

These barleys have a long, lax, nodding spike and a medium-size caryopsis. They are medium-ripening and drought-resistant. Modern Canadian and US malting cultivars of this group are characterised by resistance to lodging. The vernalisation requirement is low, the reaction to daylength medium (Lukyanova *et al.,* 1990). Compared to the European ecotype, they have lower starch contents, but higher enzymatic activity, and they are more tolerant to drought and high temperatures.

'Coast' Group

This group consists of six-rowed barleys with dense spikes and large caryopses cultivated in the Pacific coastal zone of the US. They are sown as winter barley in areas of the south-west characterised by a mild climate, such as California. They are also suited to spring sowing in areas of limited rainfall in the intermountain region (Åberg and Wiebe, 1946). They were developed from material of North African origin. The short to medium-height plants are moderately resistant to lodging, early-ripening, drought-resistant, and can stand high temperatures during the grain-filling period. The vernalisation requirement is low, the reaction to daylength medium (Lukyanova *et al.,* 1990). They show a good resistance to fungal diseases, especially to leaf rust.

'Tennessee Winter' Group

This group is mainly grown in the US winter-barley area located south and east of a curved line running from New York City through Kansas City and Western Texas. It is also grown on isolated farms in states outside this area, and occasionally in Canada (Åberg and Wiebe, 1946). This group comprises six-rowed, awned barleys of medium plant height. They are medium-ripening, and the lax spikes tend to nodding. These cultivars are among the most winter-hardy ones grown in the US. When sown in spring, they usually fail to head. This group supposedly originated from North-Caucasian ecotypes and landraces from Balkan countries with which they share morphological and physiological characteristics (Wiebe and Reid, 1961; Trofimovskaya, 1972). In Northern regions of North America characterised by a strong winter, barley is sown only in spring (Åberg and Wiebe, 1946).

South American barleys

The local forms and cultivars from South America (Bolivia, Peru, Ecuador, Colombia, Argentina, Chile) form a distinct ecotype characterised by a high foliage density. The period from seedling to ear emergence is relatively short, while the second period from ear emergence to waxy ripeness stage is quite prolonged. They are generally characterised by a relatively low number of large seeds per spike. The vernalisation requirement is medium, the reaction to daylength above medium (Gorelik *et al.,* 1985). Barleys are predominantly sown in autumn, and are late-ripening (Voronkin, 1985). In Bolivia and Ecuador, barley is the only grain fodder crop, and is grown up to altitudes of 3,800-4,000 m a.s.l. where they are used as green fodder in years when the grains do not ripen (Pukhalski *et al.,* 1973).

Landraces from Peru are grown in mountainous regions (up to 4,000 m a.s.l.) of the central and southern areas (Budin and Shmaraev, 1971). They show a wide morphological diversity, including six-rowed hulled and naked forms. They have lax spikes and large caryopses, are susceptible to lodging, medium-ripening, and highly susceptible to powdery mildew and leaf stripe, but also highly resistant to net blotch under artificial inoculation (Lukyanova *et al.,* 1996; Afanasenko *et al.,* 2000). Landraces resistant to loose smut were identified (Wolfe, 1976;

Lukyanova *et al.,* 1996). In Peru and Chile, only drought-resistant forms can be grown, since the annual precipitation does not exceed 350 mm (Pukhalski *et al.,* 1973).

Australian Group

Trofimovskaya (1972) delimited an Australian agro-ecological group, which includes two- and six-rowed hulled and naked forms. Barley was introduced at the end of the 18th century (Bakhteev, 1953), but was quickly adapted to the ecological conditions: "The hybrid varieties developed in Australia are of extraordinary interest, since they combine the most important characteristics acquired through intense and repeated hybridisation and selection from very distinct agro-ecological groups" (Vavilov, 1957). In the early period, old English late-ripening and moisture-dependent cultivars were introduced (Brezhnev and Shmaraev, 1974). The pedigrees of many contemporary Australian cultivars also include introductions from the US, Germany, Denmark and Czechoslovakia. 'Abyssinian' types from Ethiopia were grown commercially for forage. Australian cultivars are generally characterised by early ripening and susceptibility to lodging (Terentyeva *et al.*, 2000). The reaction to daylength is low to insensitive (Koshkin, pers. comm.). They are highly susceptible to powdery mildew. Net blotch is an important disease in West and South Australia (Platz *et al.,* 1996).

Conclusions and outlook

Over centuries and millennia, cultivated barley has adapted to different environmental conditions and different human utilisation requirements. The genepool of cultivated barley has thus acquired a very large morphological and adaptive diversity. The investigations of a large world collection of barley, carried out by Russian researchers and scientists in other countries, provide the basis for an ecogeographical classification. The Russian study reflects the situation in landraces collected in the first half of the 20th century (Bakhteev, 1953).

Recent diversity studies using molecular markers neutral to adaptation led to groupings, which reflect the geographical origin of the accessions. These studies also support the agro-ecological classification of cultivated barley. Strelchenko *et al.* (1998, 1999), using RAPD analysis, subdivided cultivated barley into three groups: (1) Occidental barleys, including accessions from European countries, the Mediterranean, Ethiopia and the New World; (2) Oriental barleys, including accessions from Central and East Asia; (3) naked barleys of Central Asia and the Caucasus region. Cross *et al.* (1992) and Cross (1994) found an Oriental-Occidental separation of barleys according to some morphological characteristics.

The characteristics of agro-ecological groups are based on the study of hundreds or thousands of accessions from the world collection. Various characters show a concentration in certain regions. This classification provides an orientation or entry point for studying the world resources of barley in more detail in search of interesting traits, such as drought resistance, naked grains, or resistance to lodging. It helps to identify regions with potentially interesting material for breeding or growing, to specify conditions to achieve this more efficiently and quickly, and allows the efficient utilisation of such material in breeding programmes.

Acknowledgements

The authors are grateful to Roland von Bothmer (Alnarp, Sweden), Theo van Hintum (Wageningen, The Netherlands) and Shozo Yasuda (Kurashiki, Japan) for helpful discussions and critically reading the text. Their valuable comments have improved the manuscript considerably. Jan Engels (Rome, Italy) commented on Ethiopian barley, Michele Stanca

(Fiorenzuola, Italy) on the Apennine group, Michael Mackay (Tamworth, Australia) on Australian barley.

References

Åberg, E., 1944. Studies of crop production in the United States 1940 to 1943. Almquist and Wiksells, Uppsala, 86 p.

Åberg, E., and G.A. Wiebe, 1946. Classification of barley varieties grown in the United States and Canada in 1945. Washington, USDA Agric. Bull. 907: 4-9.

Afanasenko, O.S., I.A. Terentieva and I.N. Makarova, 2000. Landraces from Peru – new sources of resistance to net blotch of barley. In: Barley Genetics VIII. Proc. 8th Int. Barley Genet. Symp., Adelaide, Australia, pp. 71-72.

Agadzhanyan, G.Kh., 1928. Zonal distribution of species and varieties of wheat, barley and rye in Armenia. Manuscript, VIR, 12 p. (In Russian).

Asfaw, Z., 1990. An ethnobotanical study of barley in the central highlands of Ethiopia. Biol. Zbl. 109: 51-62.

Attene, G., S. Ceccarelli and R. Papa, 1996. The barley (*Hordeum vulgare* L.) of Sardinia, Italy. Genet. Res. Crop Evol. 43: 385-393.

Bakhteev, F.Kh., 1953. Problems of ecology, phylogeny and breeding of barley (*Hordeum* L. section *Crithe* Döll). Moscow, Leningrad, Izd-vo AN SSSR, 218 p. (In Russian).

Bakhteev, F.Kh., 1960. Systematics of cultivated barleys. Izd-vo AN SSSR. (In Russian).

Bogdanova, G.M., I.A. Terentyeva, A.P. Khokhlova and N.N. Ivanova, 2001. The breeding value of naked barleys. In: Proc. Internat. Conf. "Genetic Resources of Cultivated Plants. Problems of mobilisation, inventarisation, preservation and study of the genepools of the most important agricultural crops for priority tasks in breeding", St. Petersburg, pp. 215-217. (In Russian).

Brezhnev, D.D., and G.E. Shmaraev, 1974. Plant production in Australia. Barley. Moscow, Kolos, pp. 87-94. (In Russian).

Budin, K.Z., and G.E. Shmaraev, 1971. A visit to the centres of provenance and varietal diversity of cultivated plants (Peru). Tr. prikl. bot. gen. i sel. 45(2): 238-271.

Cattivelli, L., G. Delogu, V. Terzi and A.M. Stanca, 1994. Progress in barley breeding. In: G.A. Slafer (ed), Genetic Improvement of Field Crops. Marcel Dekker, New York, pp. 95-181.

Cowan, P.R., 1936. Varieties of barley grown in Canada. Can. Soc. Tech. Agric. Rev. 11: 111-114.

Cross, R.J., 1994. Geographical trends within a diverse spring barley collection as identified by agro/morphological and electrophoretic data. Theor. Appl. Genet. 88: 597-603.

Cross, R.J., A.G. Fautrier and D.L. McNeil, 1992. IBPGR morphological descriptors, their relevance in determining patterns within a diverse spring barley germplasm collection. Theor. Appl. Genet. 85: 489-495.

Duganova, L.A., 1952. Local barleys of Middle Asia and their improvement. Diss. abstract, VIR, Leningrad. 16 p. (In Russian).

Emmerikh, F.D., 1961. Barleys of mountainous regions of different geographical origin, their biological characteristics and importance for breeding. Diss. abstract, Leningrad. 20 p. (In Russian).

Engels, J.M.M., 1994. Genetic diversity in Ethiopian barley in relation to altitude. Genet. Res. Crop Evol. 41: 67-73.

Filatenko, A.A., A. Diederichsen and K. Hammer, 1998. Vavilov's theories of crop domestication in the Ancient Mediterranean Area. In: A.B. Damania, J. Valkoun, G. Willcox and C.O. Qualset (eds), The Origins of Agriculture and Crop Domestication. ICARDA, Aleppo, Syria, pp. 9-24.

Flaksberger, C.A., 1935. Wheat. In: E.V. Wulff (ed), Flora of Cultivated Plants. Vol. 1. State Agric. Publ. Co., Moscow & Leningrad. (In Russian).

Gandilyan, P.A., 1980. Key to the determination of wheat, *Aegilops*, rye and barley. Izd-vo AN Armyanskoy SSR, Erevan. (In Russian).

Giessen, J.E., W. Hoffmann and R. Schottenloher, 1956. Die Gersten Äthiopiens und Erythräas. Z. Pflanzenzücht. 35: 377-440.

Gorelik, K.S., G.P. Voronkin, B.D. Karpenko and L.D. Pronina, 1985. The barleys of Argentina and their utilisation in breeding. Tr. prikl. bot., genet. i sel. 73: 100-105. (In Russian).

Grigoryev, M.V., M.V. Lukyanova, A.A. Sidorov, E.D. Emmerikh, O.S. Afanasenko, I.Yu. Kushnirenko, I.A. Terentyeva and N.D. Nikitina, 1989. Catalogue of world collection of VIR. Barley (Characterisation of cultivars for tolerance to root rots and loose smut). Leningrad, no. 517. 120 p. (In Russian).

Gudkova, G.N., 1989. Breeding value of Japanese barleys. Tr. prikl. bot., gen. i sel. 129: 52-55. (In Russian).

Habekuß, A., 1995. Results of ten-year selection of winter barley for tolerance to barley yellow dwarf virus. In: Int. Symp. 75 Years of Phytopathological and Resistance Research at Aschersleben. Ber. Bundesanst. Züchtungsforsch. Kulturpflanzen, pp. 99-103.

Habekuß, A., and C.O. Lehmann, 1991. Investigations of the Gatersleben winter barley collection for resistance to barley yellow dwarf virus. In: Barley Genetics VI. Proc. 6th Int. Barley Genet. Symp., Munksgaard Publ., Copenhagen, pp. 619-621.

Hammer, K., C.O. Lehmann and P. Perrino, 1985. Die in den Jahren 1980, 1981 und 1982 in Süditalien gesammelten Getreide-Landsorten. Kulturpflanze 33: 237-267.

Hammer, K., C.O. Lehmann and P. Perrino, 1988. A check-list of the Libyan cultivated plants, including an inventory of the germplasm collected in the years 1981, 1982 and 1983. Kulturpflanze 36: 475-527.

Harlan, H.V., 1914. Some distinctions in our cultivated barleys with reference to their use in plant breeding. U.S. Dept. Agric. Bull. 137, 38 p.

Harlan, H.V., 1918. The identification of varieties of barley. U.S. Dept. Agric. Bull. 622, 32 p.

Harlan, J. R., 1969. Ethiopia: a centre of diversity. Econ. Bot. 23: 309-314.

Ilyina, N.V., 1972. Study of agro-technical characters of local varieties of spring barley from the non-chernozem zone of the European part of USSR from the world collection of VIR. Diss. abstract cand. agric. sci., Leningrad. 25 p. (In Russian).

Khalifa, M.A., 1996. Germplasm collection and evaluation of barley in Egypt. In: Barley Genetics VII. Proc. 7th Int. Barley Genet. Symp., Saskatchewan, Canada, vol. 1, pp. 188-190.

Khokhlova, A.P., 1990. Resistance of barley to loose smut, and the gene complex controlling this character. Diss. abstract cand. agric. sci., Leningrad. 17 p. (In Russian).

Khoroshaylov, N.G., 1959. Field crops of the Mongolian People's Republic. Byul. VIR no. 7, pp. 37-41. (In Russian).

Khramysheva, L.I., 1971. Dwarf forms of winter barley from Japan in the pre-mountainous zone of the Krasnodar region. Byul. VIR, Leningrad, No. 18, pp. 20-22. (In Russian).

Kim, H.-S., K. Hammer, U.-X. Han, P. Hanelt and H.-S. Pak, 1987. Missions for the collection of plant genetic resources in the Democratic People's Republic Korea 1986. Kulturpflanze 35: 355-365.

Krivchenko, V.I., 1984. Resistance of grain cereals to the causal agents of ear diseases. Moscow, Kolos. 25 p. (In Russian).

Krivchenko, V.I., L.G. Shchelko and S.I. Traynina, 1972. Catalogue. Characterisation of resistance of barley to loose smut, powdery mildew and dwarf leaf rust. Leningrad, no. 97. 44 p. (In Russian).

Lakew, B., Y. Semeane and F. Alemayehu, 1995. Evaluation of Ethiopian barley landraces for disease and agronomic characters. Rachis 14: 1-2, 21-25.

Lakew, B., Y. Semeane, F. Alemayehu, H. Gebre, S. Grando, J.A.G. van Leur and S. Ceccarelli, 1997. Exploiting the diversity of barley landraces in Ethiopia. Genet. Res. Crop Evol. 44: 109-116.

Lehmann, C.O., I. Nover and F. Scholz, 1976. The Gatersleben barley collection and its evaluation. In: Barley Genetics III. Proc. 3rd Int. Barley Genet. Symp., Thiemig, München, pp. 64-69.

Lehmann, C.O., A. Rudolph, K. Hammer, A. Meister, K. Müntz and F. Scholz, 1978. Eiweißuntersuchungen am Getreide- und Leguminosen-Sortiment Gatersleben. Teil 2: Gehalt an Rohprotein und Lysin von Gersten (*Hordeum vulgare* L. s.l.). Kulturpflanze 26: 163-181.

Lukina, N.I., 1981. Study of barleys from the west European and Ethiopian ecological-geographical groups with respect to protein and lysine contents, with the aim to reveal promising forms for breeding. Diss. cand. agric. sci., Leningrad. 166 p. (In Russian).

Lukyanova, M.V., O.S. Afanasenko, I.A. Terentyeva, A.P. Khokhlova, L.R. Tyulina, G.M. Bogdanova and I.G. Makarova, 1996. Sources of resistance to net blotch, species of loose and covered smut, and dwarf leaf rust. Catalogue of the world collection of VIR, no. 684. Leningrad. 46 p. (In Russian).

Lukyanova, M.V., and N.N. Kozhushko, 1969. Drought resistant forms of barley from the Mongolian-Tibetan ecological group. Tr. prikl. bot., gen. i sel., 39: 209-221. (In Russian).

Lukyanova, M.V., A.Ya. Trofimovskaya, G.N. Gudkova, I.A. Terentyeva and N.P. Yarosh, 1990. Flora of cultivated plants, vol. 2, part 2, Barley. Leningrad, Agropromizdat. (In Russian).

Mansfeld, R., 1950. Das morphologische System der Saatgerste (*Hordeum vulgare* L. s.l.). Züchter 20: 8-24.

Maxted, N., P. Mabuza and S. Kell, 2000. Ecogeographic techniques and conservation: a case study for the legume genus *Vigna* in Africa. 7th MAFF Workshop on Genetic Resources, pp. 63-90.

Maxted, N., M.W. van Slageren and J.R. Rihan, 1995. Ecogeographic surveys. In: L. Guarino, V. Ramanatha Rao and R. Reid (eds), Collecting Plant Genetic Diversity, CAB International, Wallingford, U.K., pp. 255-285.

Menabde, V.L., 1938. The barleys of Georgia. Tr. Tbilisskogo botan. in-ta. 6, pp. 115-246. (In Russian).

Minasyan, A.K., 1961. The barleys of Armenia. Erevan. (In Russian).

Moseman, J.G., 1971. Co-evolution of host resistance and pathogen virulence. In: Barley Genetics II. Proc. 2nd Int. Barley Genet. Symp., Pullman, WA, USA, pp. 450-456.

Munck, L.K., E. Karlsson and A. Hagberg, 1971. Selection and characterization of a high protein, high-lysine variety from the world barley collection. In: Barley Genetics II. Proc. 2nd Int. Barley Genet. Symp., Pullman, WA, USA, pp. 544-558.

Murphy, P.J., and J.R. Witcombe, 1981. Variation in Himalayan barley and the concept of centres of diversity. In: Barley Genetics IV. Proc. 4th Int. Barley Genet. Symp., Edinburgh Univ. Press, Edinburgh, U.K., pp. 26-36.

Nilan, R.A., 1964. The cytology and genetics of barley 1951-1962. Monographic supplement No. 3, Research Studies Washington State University, vol. 32, no. 1.

Orlov, A.A., 1929. The barleys of Ethiopia and Eritrea. Tr. prikl. bot., gen. i sel., Leningrad, 20: 283-345. (In Russian).

Orlov, A.A., 1932. The barleys of Arabia. Leningrad. (In Russian).

Orlov, A.A., 1936. Barley. Flora of Cultivated Plants, vol. 2, pp. 97-332. (In Russian).

Platz, G.J., R.G. Rees and V.J. Galea, 1996. Virulence of net blotch in Australia. In: Barley Genetics VII. Proc. 8th Int. Barley Genet. Symp., vol. 2. Saskatoon, Canada, pp. 761-763.

Proeseler, G., and C.O. Lehmann, 1986. Resistenzeigenschaften im Gersten- und Weizensortiment Gatersleben. 25. Prüfung von Gersten auf ihr Verhalten gegenüber dem Gerstengelbmosaik-Virus (barley yellow mosaic virus). Kulturpflanze 34: 241-248.

Pukhalski, A.V., A.G. Zykin, N.K. Lemeshev and V.D. Nagorny, 1973. Plant production in Bolivia and Ecuador. Moscow.90 p. (In Russian).

Regel, R.E., 1910. The most important barley varieties and races of Russia. Tr. Byuro prikl. bot. 3: 229-257. (In Russian).

Rigina, S.I., 1966. Study of resistance of barley to infectious diseases. Diss. cand. agric. sci., Leningrad. 196 p. (In Russian).

Sergeeva, A.T., 1972. Study of agro-technological characters of barley from some Asian countries in the world collection of VIR. Diss. cand. agric. sci., Leningrad. 168 p. (In Russian).

Shchelko, L.G., V.I. Krivchenko, A.I. Tantsyura, V.A. Vershinina, V.V. Timoshenko and D.P. Pshinnik, 1976. Catalogue of the world collection of VIR. Samples of barley with characterisation of resistance to loose smut, powdery mildew and dwarf leaf rust. Leningrad, no. 177. 48 p. (In Russian).

Shchelko, L.G., and S.I Traynina, 1974. Immunological characteristics of barley of different origin. Tr. prikl. bot., genet. i sel. 53: 113-126. (In Russian).

Sidorov, A.A., M.F. Grigor'ev and M.V. Lukyanova, 1987. Resistance of barley to root rots. Byul. VIR 169: 59-63. (In Russian).

Sinskaya, E.N., 1969. Historical geography of the cultivated flora. Leningrad, Kolos. (In Russian).

Stoletova, E.A., 1929-1930. Field and garden crops of Armenia. Tr. prikl. bot., gen. i sel. 23: 376. (In Russian)

Strelchenko, P.P., N.K. Gubareva, O.N. Kovalyova and A. Graner, 1998. Geographical and breeding trends within Eurasian cultivated barley germplasm revealed by molecular markers. In: Plant Genetic Resources: Characterization and Evaluation. New Approaches for Improved Use of Plant Genetic Resources, Res. Council Secretariat MAFF, and Natl. Inst. Agrobiol. Resour., Tsukuba, Japan, pp. 115-132.

Strelchenko, P., O. Kovalyova and K. Okuno, 1999. Genetic differentiation and geographical distribution of barley germplasm based on RAPD markers. Genet. Res. Crop Evol. 46: 193-205.

Surin, N.A., and N.E. Lyakhova, 1993. Barley breeding in Siberia. Novosibirsk, pp. 33-62, 107-219. (In Russian).

Takahashi, R., 1955. The origin and evolution of cultivated barley. In: M. Demerec (ed), Advances in Genetics 7, Acad. Press, New York, pp. 227-266.

Takahashi, R., 1959. Phylogenetic differentiation of the barleys in the regions around Himalayas and Turkey. Jap. J. Breed. 9: 67. (In Japanese).

Takahashi, R., 1987. Genetic features of East Asian Barleys. In: Barley Genetics V. Proc. 5th Int. Barley Genet. Symp., Okayama, Japan, pp. 7-20.

Takahashi, R., I. Hayashi, T. Inoue, I. Moriya and C. Hirao, 1973. Studies on resistance to yellow mosaic disease in barley. I. Tests for varietal reactions and genetic analysis of resistance to disease. Ber. Ohara Inst. Landw. Biol. Okayama Univ. 16: 1-17.

Takahashi, R., S. Yasuda, J. Hayashi, T. Fukuyama, I. Moriya and T. Konishi, 1983. Catalogue of Barley Germplasm preserved in Okayama University 1983. Inst. Agric. and Biol. Sci., Okayama Univ., Kurashiki, Japan, 217 p.

Takeda, K., H. Yoshino, K. Kato, K. Sato, H. Tsujimoto and H. Tsuyuzaki. 2000. Wheat and Barley. In: Genetic assay and study of crop germplasm in and around China (2nd). Report. Res. Inst. Biores. Okayama Univ., pp. 37-71.

Terentyeva, I.A., 1985. Immunological resources of spring barley cultivars. In: Technology of grain and seed production in the Non-chernozem zone of Ural, Tr. Uralskogo NIISKh Sverdlovsk 44: 124-127. (In Russian).

Terentyeva, I.A., 2001. Evaluation of the world collection of spring barley for resistance to powdery mildew. In: Proc. Internat. Conf. "Genetic Resources of Cultivated Plants. Problems of mobilisation, inventarisation, preservation and study of the genepools of the most important agricultural crops for priority tasks in breeding", St. Petersburg, pp. 439-441. (In Russian).

Terentyeva, I.A., A.P. Khokhlova and O.N. Kovaleva, 2000. Australian spring barleys in Northwest Russia: sources of earliness, resistance to loose smut, drought tolerance. In: Barley Genetics VIII. Proc. 7th Int. Barley Genet. Symp., Adelaide, Australia, pp. 295-296.

Trofimovskaya, A.Ya., 1969. Breeding characters of barley in the gene centres detected by N.I. Vavilov. In: N.I. Vavilov and agricultural science. Moscow, Kolos, pp. 287-308. (In Russian).

Trofimovskaya, A.Ya., 1972. Barley (evolution, classification, breeding). Leningrad, Kolos. (In Russian).

Vavilov, N.I., 1926. Centres of origin of cultivated plants. Tr. prikl. botan., gen. i sel. 16(2): 3-248. (In Russian).

Vavilov, N.I., 1940a. The new systematics of cultivated plants. In: J. Huxley (ed), The New Systematics, Oxford Univ. Press, Oxford, pp. 549-566.

Vavilov, N.I., 1940b. The theory of origin of cultivated plants after Darwin. Sov. Nauka No. 2: 55-75. (In Russian).

Vavilov, N.I., 1957. World resources of varieties of small grains, grain legumes and flax, and their use in breeding. Moscow, Leningrad: Izd-vo AN SSSR. (In Russian).

Vavilov, N.I., 1965. Problems of origin, geography, genetics and breeding of plants, plant production and agronomy. In: Tr. Izbr, vol. 5. Moscow, Leningrad. (In Russian).
Vavilov, N.I., 1967. Selected works in two volumes. Leningrad, Nauka, vol. I, pp. 88-203, 248-327. (In Russian).
Vavilov, N.I., 1997. Five continents. International Plant Genetic Resources Institute, Rome, Italy.
Voronkin, G.P., 1985. The barleys of Latin American countries and their breeding value under conditions of the Krasnodar region. Tr. prikl. bot., genet. i sel. 73: 112-118.
Walther, U., and C.O. Lehmann, 1980. Resistenzeigenschaften im Gersten- und Weizensortiment Gatersleben. 24. Prüfung von Sommer- und Wintergersten auf ihr Verhalten gegenüber Zwergrost (*Puccinia hordei* Otth). Kulturpflanze 28: 227-238.
Wiebe, G.A., 1979. Introduction of barley into the New World. Sci. Educ. Admin., Beltsville, pp. 2-9.
Wiebe, G.A., and D.A. Reid, 1961. Classification of barley varieties grown in the United States and Canada in 1958. USDA Techn. Bull. 1224: 6-8.
Witcombe, J.R., 1975. Wheat and barley from two Himalayan regions. Euphytica 24: 431-434.
Wolfe, R.I., 1976. The response of some Peruvian cultivars to loose smut of barley (*Ustilago nuda*). Can. J. Plant Sci. 56: 430-431.
Zhukovskiy, P.M., 1933. Agricultural Turkey. N.I. Vavilov (ed), Moscow, Leningrad. (In Russian).
Zhukovskiy, P.M., 1964. Cultivated plants and their relatives. Kolos, Leningrad. (In Russian).
Zhukovskiy, P.M., 1968. New cradles of origin and gene centres of cultivated plants, and narrow-endemic microcentres of related species. Bot. zhurn. 53: 430-460. (In Russian).

Diversity in Barley *(Hordeum Vulgare)*
Roland von Bothmer et al (Editors).

Chapter 5

Diversity of barley mutants

Udda Lundqvist[a] and Jerome D. Franckowiak[b]

[a]SvalöfWeibull AB, SE-268 81 Svalöv, Sweden
[b]Department of Plant Sciences, North Dakota State University, Fargo, ND, 58105, USA

Introduction

Mutations, heritable changes in the DNA, afford opportunities to study developmental patterns and biochemical pathways in living organisms. Collections of visual and biochemical mutations represent only a small proportion of the possible natural and induced changes in DNA sequences. For a mutation to endure, DNA changes have to survive the array of repair mechanisms (Friedberg *et al.*, 1995), they must be present in reproductive tissue, and they must be transmitted to subsequent generations. Diversity among mutants is largely limited by the ability of researchers to isolate, preserve, and examine them. The array of mutants in barley, *Hordeum vulgare* L. ssp. *vulgare*, is unusual because morphological and biochemical variants have been isolated under many different climatic conditions and from various genetic backgrounds. Hence, barley mutants can provide insights about the genes and biochemical pathways that accommodate adaptation of barley to specific combinations of environmental stress.

Collections of landraces and modern cultivars contain genes and gene combinations that favour survival of barley as a domesticated crop. Populations of wild barley (*H. vulgare* ssp. *spontaneum*) contain considerable genetic diversity because only a small portion was sampled during the process of domestication (Harlan, 1976; Nevo, 1992; Lev-Yadun *et al.*, 2000).

Sources of hereditary variation

Heritable morphological and physiological changes, spontaneous mutations, have been identified in untreated barley populations. The frequency of spontaneous mutations, estimated primarily from the occurrence of chlorophyll-deficient seedlings in control material, varies from .0004 to .001 mutations per spike (Nilan, 1964). Results of spontaneous mutations include visual changes in morphological structures, modification of chlorophyll and anthocyanin pigmentation, variation in developmental and maturation patterns, reactions to pathogens and insects, and responses to climatic and edaphic factors (Smith, 1951; Nilan, 1964).

Research on induced mutations to improve barley was initiated in the 1920s. The first treatments used to induce mutations in barley were X-rays (Stadler, 1928), followed by neutrons, and later by various chemical mutagens. Since then barley has been used to investigate the

---

Lundqvist, U., and J.D. Franckowiak, 2003. Diversity of barley mutants. In: R. von Bothmer, Th. van Hintum, H. Knüpffer and K. Sato (eds), Diversity in Barley (*Hordeum vulgare*), pp. 77-96. Elsevier Science B.V., Amsterdam, The Netherlands.

mutation process and mutagen action. The most important physical and chemical mutagen treatments used are summarised in Table 5.1 (Lundqvist, 1992).

Alteration in the mutation spectrum, i.e., the relative proportions of various mutant types, is informative with regard to specific mutagen reactions. Distinct categories were established initially as a practical system for classification of mutants. Chlorophyll defects are the most common mutations and the first indication of treatment success. Studies in Sweden identified three major categories of mutants: (1) albina or albino seedlings; (2) viridis or light green lethals; and (3) less common chlorophyll defects including xantha, two-coloured, tigrina, striata, maculata, and zonata (Gustafsson, 1940). Irradiation treatments produced a higher frequency of albina mutants than treatments with chemicals. Chemicals induced a higher proportion of viridis mutants, but the proportion differences depended on which chemical was used. Some chemical treatments produced high rates of rare chlorophyll mutations. Different treatment procedures involving the same mutagen also contributed to the variation in the mutation spectra (Gustafsson, 1941, 1947; Ehrenberg *et al.,* 1959; von Wettstein *et al.,* 1959; Nilan, 1964; Lundqvist, 1992).

Mutagen specificity has been reported among various groups of viable mutants. Different treatments demonstrated that some surface wax loci show markedly different mutation reactions:

1. Large differences were found between the effects of chemicals and ionising radiation treatments, particularly densely ionising radiation;
2. No differences in effect were observed among various types of organic chemicals;
3. Sodium azide induced a significantly different array of mutants than organic chemicals;
4. The result of sodium azide treatment was very different from that of X-rays and even more so from that of neutrons;
5. Sulphonate treatments alone produced a mutant spectrum similar to combined sulphonate treatments, but different from sparsely ionising radiation alone;
6. Sparsely ionising X-rays and densely ionising neutrons had different effects on DNA;
7. Both spontaneous and induced mutants were found at some loci and had similar phenotypic effects (Lundqvist and Lundqvist, 1988 and references therein).

The erectoides or dense spike mutants were the first of the viable mutants induced by irradiation and the most commonly induced morphological changes. Densely ionising radiation induced more erectoides mutants than sparsely ionising radiation. Mutants at the *ert-a* locus produced no additional mutants after treatment with densely ionising radiation. All mutants at the

Table 5.1. Different types of mutagens used to induce genetic changes in barley.

A. Irradiations
a. X-rays and gamma-rays (sparsely ionising radiation)
b. Neutrons, electrons, and protons (densely ionising radiation)
c. Alpha-rays from radon, beta-rays from phosphorus 32 and sulphur 35
B. Chemical compounds
a. Colchicine, mustard gas substances
b. Epoxides: ethylene oxide, glycidol
c. Epimines: ethyleneimine
d. Purine and purine derivates: 8-ethoxycaffeine, nebularine
e. Various alkylating and oxidising compounds: iodine, dioxan, diethyl sulphate
f. Alkanesulphonic esters:
(i) Monofunctional: different sulphonates, i.e., ethyl methanesulphonate
(ii) Di- and trifunctional
g. N-alkyl-N-nitrosamides
h. Sodium azide (inorganic)

Table 5.2. A summarised grouping of mutations in barley.

Developmental stage	Different types of mutants
Reproduction	A. Floral bud development
	Heading time
	Delayed heading
	B. Spike morphology
	Spike length
	Rachis internode length
	Spikelets per rachis node
	Spike malformation
	Rachis bracts
	C. Awn characteristics
	Awn length
	Awn shape
	Barbs on the awn
	D. Spikelet and floret morphology
	Spikelet changes
	Floret changes
Gametic and zygotic formation	A. Meiosis and gamete formation
	B. Fertilisation and zygote development
Kernel development and distribution	A. Kernel formation
	B. Maturation pattern
	C. Seed distribution
Seedling developmental mutants	A. Germination and seedling vigour
	B. Chlorophyll development
	Lethal chlorophyll mutants
	Viable chlorophyll mutants
Vegetative growth	A. Root development
	B. Leaf development
	Leaf size and shape
	Absence of leaf organs
	Leaf pubescence
	Leaf blade necrosis
	C. Tillering and plant vigour
	D. Culm elongation
	E. Surface waxes
	F. Anthocyanin pigmentation
Environmental stress responses	A. Climatic factors
	B. Edaphic factors
	C. Host-pathogen reactions
	D. Yield and yield components
Nutritional quality factors	A. Starches and lipids
	B. Protein quality and quantity
	C. Malt quality
	D. Biochemical mutants

ert-b locus occurred after treatment only in 'Gull' (Gustafsson, 1941; Hagberg *et al.*, 1958; von Wettstein *et al.*, 1959). A similar phenomenon was reported for other loci, i.e., some breviaristatum or short awn loci (Kucera *et al.*, 1975).

Naming and grouping of barley mutants

Barley genes were first described by von Tschermak (1901). As more genes were reported, several different methods for assigning gene symbols were employed (Smith, 1951; Nilan, 1964). The first genes were assigned single-letter codes, and later two-letter codes were used. At an even later stage, phenotypically similar mutants were grouped together and assigned the same basic symbol plus a number and three-letter codes were recommended. Nomenclature rules for assignment of locus names and gene symbols to barley mutants were approved by the International Barley Genetic Symposium and were published in the Barley Genetics Newsletter (Franckowiak *et al.*, 1997). As an example, the symbol *msg1* is assigned to the male sterile genetic 1 locus and *msg1.ca* is a specific mutation at the *msg1* locus located in barley chromosome 1HL.

Information on the types and number of barley mutants and assigned gene symbols has been reviewed frequently with different strategies for grouping of mutants (Smith, 1951; Nilan, 1964; Hockett and Nilan, 1985; Søgaard and von Wettstein-Knowles, 1987; Davis *et al.*, 1997). However, we have chosen a system for grouping morphological mutants based on the life cycle of the barley plant (Table 5.2). The phytomer concept of plant development is used to organise stages of plant development (Bossinger *et al.*, 1992). The model assumes that formation of the barley plant occurs by adding repeats of a basic structural unit or phytomer. Pleiotropic effects are observed because some mutations modify the same component in both vegetative and reproductive phytomers. Because information about each mutant was limited when it was placed in a specific phenotypic group, the assigned gene symbol may not reflect current knowledge of that particular mutant. In this presentation only the most important features are described. The last groupings of 'Environmental stress responses' and 'Nutritional quality factors' listed in Table 5.2 are only briefly mentioned as they are reported in special chapters of this book.

Descriptions of different mutant groups

I. Reproductive mutants

A. Floral bud development

Major changes in the physiology of grass plants occur as the terminal bud of each tiller is converted into a reproductive bud. The timing of grain maturation is an important agronomic trait because of its relationships to seasonal changes in moisture and temperature, disease hazards, farm activities, crop rotations, and yield. The genes that control initiation of reproductive phases permit production of barley in an extremely wide array of environments. Many studies on early maturity were conducted in the 1920's because early maturity crops are important in regions with short vegetation periods (Smith, 1951).

Heading time

Many maturity genes were isolated as spontaneous or induced mutants in barley and its wild relative ssp. *spontaneum* (Nilan, 1964). Three spring growth habit (*sgh*) loci are associated with vernalisation requirement and winter growth habit. Wild barley and most cultivars grown under short-day conditions respond strongly to a photoperiod that is conditioned by the *Eam1* (*Ppd-H1*) gene. Most spring barley cultivars grown in temperate climates have a recessive allele at this locus. An early maturing, photoperiod-insensitive gene, the *mat-a* or *eam8* (*Ppd-H2*) gene, was found among the Swedish collection of praematurum mutants and released as 'Mari' (Hagberg, 1961). Plants with this gene matured 7 to 10 days earlier than midseason cultivars under long-day conditions and three or more weeks earlier under short-day conditions (Davis *et al.*, 1997). Extremely early maturation was accompanied by reduced culm length, fewer elongated internodes, and fewer kernels per spike. Early heading cultivars often have long upper culm internodes and short basal internodes (Gustafsson *et al.*, 1960). More than one thousand praematurum mutants have been studied and associated with 13 different loci (Gustafsson and Lundqvist, 1976; Lundqvist, 1992; Davis *et al.*, 1997).

The photo- and thermoperiod-insensitive mutants at the *eam8* locus mature earlier than most other cultivars in tests world-wide (Gustafsson *et al.*, 1971; Gallagher *et al.*, 1991). Extremely early maturation can also be produced by the *Eam1* gene under long-day conditions and by the *Eam1* gene plus mutants at the *eam7*, *eam9*, or *eam10* loci under short-day conditions. Modern cultivars for East Asia often contain the *Eam1* and *eam9* genes. Removal of the *Eam1* gene drastically reduced the effects of the *eam7* gene, but *eam9* and *eam10* genes do not interact strongly with the *Eam1* gene under short-day conditions. The *eam7* and *eam10* genes are expressed under long-day conditions, but *eam9* is not (Davis *et al.*, 1997).

Delayed heading

Some gigas (*gig*) mutants increase the size of vegetative organs and delay heading. The densinodosum (*den*) and many-noded dwarf (*mnd*) mutants delay heading by increasing the number of leaves produced prior to floral bud formation. Spike primordia can be aborted or delayed in the viviparoides (*viv*) mutants. A number of minor genes delay heading and increase kernels per spike (Nilan, 1964; Bossinger *et al.*, 1992; Davis *et al.*, 1997).

B. Spike morphology

More mutants for altered spike morphology have been isolated in barley than for any other structural characteristic. The traits modified include spike length, rachis internode length, subnodal bracts, fertile spikelets per rachis node, florets per spikelet, and pubescence on rachis margins and glumes. The rattail spike (*rtt*) mutant produces a highly underdeveloped, completely sterile spike.

Spike length

The number of fertile rachis internodes that develop per spike is dependent on environmental and nutritional as well as genetic factors. Early maturity genes are associated with fewer fertile spikelets per spike. Environmental stress can cause more sterility in the top of the spike in some genotypes than in others. The lesser internode number (*lin1*) gene reduces the number of fertile rachis nodes in each spike (Nilan, 1964; Tsuchiya and Haus, 1984).

Reduced rachis internode length

Shortening of the rachis internodes produces dense or compact spikes. Accessions with a semidominant dense spike trait were called zeocriton or 'little barley' (Smith, 1951). Swedish researchers induced a large number of erectoides (*ert*) mutants (Figure 5.1A) and allelism tests found 29 erectoides loci in only a portion of the collection. Many of the dense spike (*dsp*) mutants have been isolated and studied by other researchers including the *dsp1.a* gene, present in some Japanese cultivars. To facilitate allelism testing, this group of mutants was subdivided into

Figure 5.1. Mutations in barley. A: erectoides-c.1 (*ert-c.1*) spike to show the fusiform shape with shorter rachis internodes at the base of the spike to the left compared with normal 'Bowman'. B: elongated outer glume (*eog1.a*) spike to show increased outer glumes size and awnlength of the spikelets to the left compared with normal 'Bowman'. C: chlorina seedling 13 (*fch13.v*) to the left compared with normal 'Bowman'. D: single necrotic leaf spot (*nec3.e*) leaf blades developed with large tan blotches having distinct margins to the left compared with normal 'Bowman'.

length), pyramidatum (*Pyr*, spikes with shorter rachis internodes at the base of the spike), and dense spike (recessive genes with uniform shortening of rachis internodes) mutants. The *ert-m* and *ert-v* mutants have compact spikes plus variable rachis internode lengths. Failure of some upper culm internodes to elongate is also a characteristic of *ert-m* mutants. The opposite spikelet (*ops*) mutants have variable rachis internode length and irregular spikes (Nilan, 1964; Davis *et al.*, 1997).

Increased rachis internode length

A number of mutants have one or more elongated rachis internodes compared with parental strains. A large collection of laxatum (*lax*) mutants has been assembled, but only a few are associated with distinct pleiotropic effects (Larsson, 1985). The *lax-a* mutants have the lodicules converted into two extra anthers. The *lax-c* mutants have a broad junction between the lemma and its awn. Plants with accordion-like (*acr*) spikes have very long rachis internodes and may and its awn. Plants with accordion-like (*acr*) spikes have very long rachis internodes and may show partial trapping of the spike in the leaf sheath. Mutants at the basal rachis internode (*lbi*) locus have an elongated basal rachis internode, the first internode above the collar being less than 1 cm to more than 20 cm long, depending on genetic background. The accordion basal rachis internode (*abr*) mutants have very long basal rachis internodes plus a slight elongation of other lower rachis internodes (Davis *et al.*, 1997).

Spikelets per rachis node

The number of fertile spikelets per rachis node is one of the most constant and easily recognised characteristics of barley. Three fertile spikelets are formed at each rachis node in six-rowed barley, and only the central spikelet at each rachis node will set seed in two-rowed barley. The *vrs1.a* allele is present in most six-rowed barley and the common allele in two-rowed barley is *Vrs1.b*. The *Vrs1.t* allele (deficiens) is characterised by the absence or extreme size reduction of the lateral spikelets. The *vrs1.c* allele in 'Nudihaxtoni' and 'Bozu' produces a six-rowed spike with small, fertile lateral spikelets that have very short awns. The hexastichon (*hex-v*) mutants induced in two-rowed cultivars are alleles at the *vrs1* locus, and the fertile lateral spikelets have awns that are about ¾ of normal length. Intermedium spike (*Int-d*) mutants at the *vrs1* locus have lateral spikelets that are slightly larger than those of two- by six-rowed hybrids. The 'labile' (*lab*) mutant interacts with the *vrs1.a* gene to produce irregular spikes, with variable abortion of spikelets. Specific alleles at the intermedium spike (*int-c*) locus interact with *vrs1* alleles to produce spike phenotypes that are typical of two- and six-rowed barley. Mutants at other intermedium loci have spikes that are intermediate between those of two- and six-rowed barley. Mutants at the *vrs4* (*mul1*) and *int-h* loci produce more than three spikelets at each rachis node. Plants with the *int-m* mutants have spikes that are less than ¼ of normal length (Gustafsson and Lundqvist, 1980; Lundqvist, 1992; Davis *et al.*, 1997).

Spike malformation

The multiflorus (*mul*) mutants are characterised by variable expression of supernumerary florets and occasionally nodal branches in the spike. Rachilla malformation in lateral spikelets of *mul2* plants is most pronounced near the middle of the spike. Extra florets develop under favourable environmental conditions. The branched ear (*bre*) mutants are more difficult to study because their expression is greatly influenced by environmental conditions and background genotype (Smith, 1951; Davis *et al.*, 1997).

Compositum (*com*) mutants are characterised by the failure of the rachilla to abort in some spikelets located in the basal half of the spike. The curly lateral (*crl*) mutant causes malformation of lateral spikelets and development of wavy, twisted awns. The small lateral spikelet (*sls*) gene is found in six-rowed cultivars, but it is expressed only in two-rowed spikes as a progressive decrease in size of sterile lateral spikelets near the spike tip. Other genes for small lateral spikelets exist in two-rowed barley, but they are not associated with an alteration in size of the lateral spikelets (Davis *et al.*, 1997).

Rachis bracts

The barley spike rarely has vestigial leaf-like bracts on the rachis; however, they are observed in several mutants. The leafy bract (*Lfb*) and semidwarf 2 (*sdw2*) mutants form a leaf-like bract below the spike at the collar. The *Lfb* mutant forms a large, leaf-like bract in six-rowed lines. In the third outer glume (*trd*) and bracteatum (*bra*) mutants, a bract develops below the two empty glumes of the central spikelet. The bract below the lowest spikelet is the largest and may embrace nearly one-half the spike. The basal rachis internodes are elongated and the upper culm internodes are affected in some mutants. A glume-like bract may arise at the outer glumes of some central spikelets in the subnodal bract (*snb*) mutant. The extra floret (*flo*) mutants produce extra bracts on the abaxial side of the rachis at the base of a few central spikelets (Davis *et al.*, 1997).

C. Awn characteristics

The size and photosynthetic efficiency of awns play a role in determining grain size as a yield factor in barley improvement. Since changes in awn length and shape are easy to observe, a large number of awn variants have been isolated. Much smaller awns are present on the outer glumes of the spikelet and two awns can occasionally develop from the palea.

Awn length

About 700 short awn (*lks*) and breviaristatum (*ari*) mutants have been analysed genetically. Awn lengths vary from awnless to slightly shorter than those of the parental cultivar. Some genes control differential development of awns on central and lateral spikelets. Study of awn mutants is difficult in certain genetic backgrounds because many cultivars derived from Oriental barleys have awns that are about ¾ of normal length. About half of the breviaristatum mutants were localised to 19 loci, but Kucera *et al.* (1975) reported that 13 other *ari* mutants had unique phenotypes.

The awn length mutants contain at least two different phenotypic groups: (1) those with changes only in length and brittleness of the awn; and (2) those with changes affecting plant height, leaf size, reproductive organs, and kernel size and shape. When the awnless (*Lks1*) gene is present, little or no development of the appendage on the lemma occurs. Heterozygotes may be awnless or awnletted depending upon the source of the *Lks1* gene and the genetic background. Each source of the *Lks1* gene is associated with a specific spike type because the *Lks1* locus is closely linked to the *vrs1* locus in chromosome 2HL. The *Lks1* gene from 'Engleawnless' does not recombine with the two-rowed spike (*Vrs1.b*) gene. The *lks2* and *lks5* (*ari-c*) mutants reduce awn length, but have little effect on other traits. The *ari-k* mutants have dehiscent awns (Davis *et al.*, 1997).

Several breviaristatum mutants have proven to be alleles at other loci. The dwarfing mutant in 'Golden Promise' (*Gpert* or *ari-e.GP*) is allelic to *ari-e* and was associated with salt tolerance, a lower accumulation of sodium ions (Forster, 1994). The *ari-d* mutants are allelic to the unbranched style 4 (*ubs4*) mutant in which the stigma has very few branches and seed set is greatly reduced. The *ari-i* and *ari-l* mutants are allelic to the first and second brachytic (*brh*) mutants. Some *ari* mutants have globe-shaped grain (Kucera *et al.*, 1975; Davis *et al.*, 1997).

Awn shape

The lemma awn can be changed into a trifurcate appendage, called a hood, by the *Kap* gene. Just above the lemma apex a structure develops that may contain a deformed floret with two triangular leaf-like projections called lemma wings. This supernumerary floret may contain stamens with fertile pollen grains and occasionally bears a kernel. Ectopic expression of the *Kap1.a* allele forms the extra floret and is associated with a 305-base pair duplication in intron 4 of the *Knox3* sequence (Müller *et al.*, 1995). The hooded gene is present in several cultivars, but probably represents one spontaneous event occurring early in the domestication of barley. The subjacent hood (*sbk*) mutant forms a central cup-like cavity in the lemma appendage that is situated just below the top of the lemma. No rudimentary floral organs exist within the hood-like cavity. Plants with a calcaroides (*cal*) mutant develop a semi-hooded phenotype as a consequence of bending and malformation of the awn near its base. The short crooked awn (*sca*) mutant produces a very short awn with the tip curved outward. Plants expressing the triple awned lemma (*trp*) gene can have three awns on the lemma, but expression is highly restricted in certain genetic backgrounds (Davis *et al.*, 1997).

Barbs on the awn

The barbs on the awn decrease the desirability of barley as a livestock feed. The smooth-awned mutants (*raw1* and *raw2*) reduce the number and size of barbs and have no apparent effect on seed size. Although the *raw1.a* gene is more effective and is present in many cultivars, it also reduces the number of stigma branches and has a very low female fertility leading to reduced seed set (Davis *et al.*, 1997).

D. Spikelet and floret morphology

Spikelet changes

Plants with outer glumes wider than 1mm commonly have a mutant allele at the elongated outer glume (*eog*) locus (Figure 5.1B). When awns on the outer glumes are longer than the lemma, a semidominant allele is generally present at the long glume awn (*Lga*) locus. The lemma becomes a leaf-like structure when the leafy lemma (*lel*) mutant is present. The barbs are absent from the upper part of the lateral lemma nerves in plants with a recessive allele at the toothed lemma (*gth*) locus. A few hairs, 1 to 2 mm in length and mixed with the barbs on the lateral nerves of the lemma, are present when the hairs on lemma nerves (*Hln*) gene is expressed. The palea elongates to form two awns in plants with the awned palea (*adp*) mutant. The pistils are often transformed into leafy structures and female fertility is low. Plants with the short rachilla hair (*srh*) gene have short, branched multicellular hairs on the rachilla instead of long unicellular hairs. This characteristic is used to identify different cultivars both in the field and in threshed grain. Presence of the *srh* gene is highly correlated with the length of hairs on the glumes and rachis edges. Several genes restrict development of pubescence on the glumes (*Pbg*) and rachis margin (*Pbr*) (Smith, 1951; Nilan, 1964; Davis *et al.*, 1997).

Floret changes

Lodicules in the floret of the laxatum-a (*lax-a*) mutants are transformed into two extra anthers. Lodicule size and opening of the floret is controlled by a number of genetic factors. The anthers of the multi-ovary (*mov5*) mutants are converted into pistils and the lodicules become half-curled greenish leaflets. The number of stigma branches is greatly reduced by mutants at the unbranched style (*ubs4*), the gigas 1 (*gig1*), and the eceriferum *cer-yd* and *cer-yh* loci. These stigmas of mutants have very few short branches that limit pollen reception and seed set. Abortion of the developing anther is observed in several of the male sterile genetic (*msg*) mutants. The filament of the stamen is greatly thickened in the *tfm* mutant and plants form little or no selfed seeds (Davis *et al.*, 1997, and references therein).

II. Gametic and zygotic mutants

A. Meiosis and gamete formation

Smooth functioning of biochemical events during meiosis and gamete formation is critical to survival of annual, self-pollinated plants such as barley. Reduced fertility mutants are easy to observe in barley, but researchers have chosen mainly to study those that drastically reduce fertility and gamete transmission. Failure of the anthers to develop normally is observed in some male sterile genetic (*msg*) and in the cytoplasmic male sterile maternal (*msm*) mutants. The collection of male sterile mutants contains 650 entries (Hockett, 1988) from which 49 loci have been identified. Although most *msg* mutants show normal female fertility, reduced female fertility is associated with the *msg40* and *msg41* mutants. The pollen grain aperture fails to open in the *msg6* and *msg39* mutants even though pollen grain development is normal. The level of self-fertility in *msg37* mutants is temperature-sensitive. Desynapsis (*des*) mutants affect condensation and synapsis of chromosomes during meiosis and show reduced fertility of both male and female gametes. The high lysine 3 (*lys3*) locus is linked to gametophyte factor 3 (*Gam3*) for low male gamete transmission (Yamae and Konishi, 1993). Several other pollen lethal (*let*) or haplo-viable mutants have been described in barley (Hockett and Nilan, 1985; Davis *et al.*, 1997).

B. Fertilisation and zygote development

A number of genes alter fertilisation and seed development in barley. The floral sensitivity (*fls*) gene causes abortion of spikes following emasculation and hand pollination. The haploid initiator (*hap*) mutant causes failure of double fertilisation in 15 to 40% of the selfed seeds (Hagberg and Hagberg, 1981). The triploid inducer (*tri*) generates some triploid progeny. The shrunken endosperm genetic (*seg*) mutants cause partially shrivelled seeds associated with the maternal genotype. Complete or partial shrivelling of seed is controlled by the zygote genotype in the defective endosperm xenia (*dex*) and shrunken endosperm xenia (*sex*) mutants, respectively. The high lysine (*lys*) mutants increase the percentage of lysine in the kernel, but also cause partial shrivelling of the endosperm (Hockett and Nilan, 1985; Davis *et al.*, 1997).

III. Kernel development and distribution

A. Kernel formation

Retention of the lemma and palea during threshing is important in determining how barley is used for human food and livestock feed. The naked caryopsis (*nud*) gene prevents adhesion of the lemma and palea to the caryopsis and permits free threshing at maturity. The seminudoides (*smn*) and narrow leaf dwarf (*nld*) mutants have partial covering of the caryopsis. Poor hull attachment during threshing and grain handling is associated with mutants at the *cer-zv*, *cer-yl* and *cer-ym* loci and a number of minor genes. The globosum (*glo*) mutants form rounded or globe-shaped kernels by reducing kernel length. The granum (*gra-a*) plants produce small, thin kernels. Some gigas and other leaf width mutants have a pleiotropic effect on the size and shape of kernels (Nilan, 1964; Davis *et al.*, 1997).

B. Maturation pattern

At least two pigments are responsible for kernel colour in barley: (1) anthocyanin that is red or violet in acid, and blue in alkaline solution; and (2) melanin-like pigments associated with grey or black coloration. These pigments may occur in the hull, pericarp, aleurone layer, and occasionally deeper in the endosperm. Complementary genes for purple (red) lemma and pericarp (*Pre*) control development of a deep red or purple coloration of the hull and pericarp during the soft dough stage of grain fill. The colour fades as the grain matures. Many of the anthocyanin-less (*ant*) mutants prevent pigmentation development in kernels. The blue aleurone colour is caused by the accumulation of anthocyanin pigments in aleurone cells as lumps inside the aleurone granules. The non-blue aleurone xenia (*blx*) mutants prevent development of blue pigments in the aleurone. Expression of the blue colour varies from dark blue to off-white depending on environmental conditions and the presence of a gene for intense blue aleurone (*ibl*). The *blx4.d* gene produces a pink aleurone colour; this is brick red when the *ibl* gene is expressed. The black lemma and pericarp (*Blp*) locus controls the black pigments that develop in the lemma and pericarp slightly before maturation. Alleles at the *Blp* locus determine the intensity of black pigmentation and its distribution to the spikes, awns, peduncles, and upper leaves (Smith, 1951; Davis *et al.*, 1997).

Maturation of the barley plant and its kernels is influenced by many environmental and genetic factors. However, a few mutants cause abnormal or premature senescence of plants. The dusky (*dsk*) mutant causes dwarfing and an olive-grey colour as plants mature. Since the anthers remain green at anthesis, the *dsk* gene may control breakdown of chlorophyll. Differential responses to biotic and abiotic stress can cause premature ripening of some barley cultivars, but the prematurely ripe (*pmr*) mutant causes senescence of plants about two weeks after anthesis. The necrotic leaf margin (*nlm*) mutant causes the margins of the leaf blade to lose chlorophyll at or shortly after anthesis (Davis *et al.*, 1997).

C. Seed distribution

Cultivated barley has two closely linked genes for non-brittle rachis (*Btr*) that prevent disarticulation of rachis segments at maturity (see Chapter 2). Cultivated barley differs from wild barley with respect to several threshability genes that permit easy removal of the awns. In addition, some cultivars have dehiscent awns that break near the tip of the kernel shortly before maturation. Variable shattering of the rachis after maturity has been associated with several tough *vs.* weak rachis loci. Clean threshing in many two-rowed cultivars is associated with weak attachment of the lemma and palea to the rachis node. Strong hull attachment to the rachis improves threshing in hulless barley (Smith, 1951; Nilan, 1964; Davis *et al.*, 1997).

IV. Seedling developmental mutants

A. Germination and seedling vigour

The post-harvest dormancy (*phd*) genes control an important attribute of barley cultivars. Romagosa *et al.* (1999) identified four loci associated with dormancy in the 'Steptoe'/'Morex' mapping population. Thousands of mutants with drastically reduced seedling vigour have been observed after treatment with mutagenic agents, but only a few are described as genetic stocks. The dominant dwarf 2 (*Dwf2*) mutant is insensitive to gibberellic acid (GA) application (Ivandic *et al.*, 1999). The GA insensitive (*gai*) and GA sensitive (*gal*) mutants differ in their response to exogenously applied GA_3 (Börner *et al.*, 1999). Induced dwarf mutants were classified as GA-responsive dwarf (*grd*), GA sensitivity (*gse*), and elongation (*elo*) mutants. Constitutive production of GA_3 occurs in the slender (*sln*) seedlings due to a mutant in a negative regulator of GA-synthesis (Chandler and Robertsson, 1999). Seedlings with the angustifolium-b (*Fol-b*) mutant have thread-like leaves. Plants with the semi-minute (*min1*) and enhancer of minute (*en-min*) genes have very short and thick leaf blades and are less than 5 cm tall. The presence of mixaploid nuclei in the leaf tissue is caused by incomplete formation of cell walls during mitosis (Davis *et al.*, 1997).

B. Chlorophyll development

Numerous chlorophyll mutants have been induced and studied by barley researchers because chlorophyll deficiencies are easily detected in seedlings. Inheritance studies have shown that recessive, monofactorial nuclear genes control most chlorophyll deficient mutants. Some chlorophyll mutants are maternally inherited and a few show aberrant segregation ratios. Chlorophyll mutants have been used to investigate submicroscopic plastid structures and the biochemical steps in the development of the chloroplast (Nilan, 1964; Hockett and Nilan, 1985; von Wettstein *et al.*, 1995). The chlorophyll mutants can be divided into two main categories: lethal and viable chlorophyll mutants.

Lethal chlorophyll mutants

Lethal chlorophyll mutants must be maintained in heterozygous stocks because seedlings die at a two- or three-leaf stage. Albino and albina seedling (*abo*, *alb*) mutants are white and have no carotinoids or chlorophylls. The expression of the albina trait in some seedlings can be modified, i.e., the *alb-c* mutant produces normal chlorophyll after feeding with certain amino acids. Xantha seedling (*xnt*, *xan*) mutants have a yellow colour because carotinoids, but not chlorophylls, are produced. Several *xan* loci are structural genes for chlorophyll synthesis. The viridis (*vir*) mutants are characterised by a uniform light-green or yellowish-green colour in the seedling leaves. Some *vir* mutants survive to maturity in favourable environments, and others were selected for high chlorophyll content and high fluorescence. Striata (*str*) mutants are characterised by vertical and white/yellow stripes, and tigrina (*tig*) mutants have alternating light and dark regions. In continuous light or continuous darkness, the seedling leaf colour of some *tig* mutants is uniform. Two-coloured chlorophyll mutants are fairly common, the seedling leaves having either green tips and a white or yellow base (viridoalbina, viridoxantha), or white or yellow tips and green bases (alboviridis, xanthviridis). Mutants with white tips and yellow bases (alboxantha) occur after artificially induced treatments. Virescent seedling (*yvs*) mutants often have a green tip similar to viridoalbina mutants (Gustafsson, 1940; Simpson and von Wettstein, 1991; Davis *et al.,* 1997).

Viable chlorophyll mutants

Most mutants in this group are viable, but plant vigour is often reduced, development is delayed, and seed yield is often low. The expression in many of them is more distinct when grown at low temperatures. Chlorina seedling (*fch*) (Figure 5.1C) and light green (*lgn*) mutants have yellowish green, light or pale green seedling colour that may or may not persist until maturity. Seedlings of the grandpa (*gpa*) mutant may display a pattern of transverse or alternating white and green bands, while adult plants frequently have striped upper leaves and white spikes. The flavum (*fla-a*) mutant produces chlorina seedlings but later a distinct yellowing of the lemma and awns occurs; the spikes and nodes of the yellow head 1 (*yhd*) gene appear ivory in colour, but the lemma tips and awns are green. The albino lemma (*alm*) and orange lemma (*rob*) genes produce the same chlorophyll development pattern except the ivory coloured tissues are white or orange. In addition, two main changes in chlorophyll pattern can be distinguished: (1) vertical white or yellow streaks of variable width and number in the leaf blades, i.e., white streak (*wst*), yellow streak (*yst*), variegated (*var*) and midseason stripe (*mss*) mutants. Among the last group, mutants develop chlorophyll deficient sectors after heading in the upper culm and spike. (2) mutant types with a typical development of pale green to yellow, or white bands or necrotic blotches partially across seedling leaves, i.e., zebra stripe (*zeb*) and mottled leaf (*mtt*) mutants (Davis *et al.*, 1997).

V. Vegetative growth

A. Root development

Hockett and Nilan (1985) reviewed studies on heritable differences in the root system of barley. Few simply inherited genes have been identified that modify growth of the seminal and adventitious root systems. The few roots (*fer*) mutant decreases the number of seminal roots to one or two from the normal five to seven. Several of the curly (*cur*) mutants cause coiling of the roots in addition to coiling of the leaves and other culms. When the subcrown internode length (*Sil*) gene is expressed, the first culm internode does not elongate, and placement of the crown is determined by planting depth. Differential responses of barley cultivars to aluminium toxicity, boron toxicity, drought, salinity, and water-saturated soils are mediated by the root system and have been associated with quantitative trait loci (QTLs). Tolerance to low pH (*Pht*) and aluminium tolerance (*Alp*) have been associated with chromosome 4H (Davis *et al.*, 1997; Tang *et al.*, 2000).

B. Leaf development

This group of characteristics is concerned with the blade, leaf sheath, auricle and ligule. Genetic factors modify leaf characteristics including width, length, size, orientation, appendages, and pubescence.

Leaf size and shape

Leaf blade width and leaf blade angle are heritable traits that are influenced by environmental and genetic factors (Hockett and Nilan, 1985). A few mutants such as broad leaf 1 (*blf1*) and gigas (*gig*) produce large leaf blades that are up to twice the normal width. In contrast, the angustifolium-a (*fol-a*) and narrow leaf dwarf (*nld*) mutants have small, narrow leaf blades. The brachytic (*brh*) mutants often have relatively short leaf blades. The leaf blade margins of the revoluted leaf (*rvl*) mutant tend to roll outwards forming a flat surface. The scirpoides leaf (*scl*) and scirpoides (*sci*) mutants, and mutants at the breviaristatum-g (*ari-g*) locus are characterised by narrow involuted or inward-folding leaf blades. Leaf blades of the curled leaf dwarf (*clh*) are erect and folded inwards. The curly (*cur*), curly dwarf (*cud*), and lazy dwarf (*lzd*) mutants have leaf blades that are coiled or twisted to some degree (Davis *et al.*, 1997).

Absence of leaf organs

The liguleless (*lig*) mutants are characterised by the absence of the ligule and auricles, and the leaf blades are erect along the stem. The exligulum (*eli*) mutants have no auricles, rudimentary ligules, and semicompact spikes with enlarged glumes near the tip. Upper leaf blades are aborted or greatly decreased in size by the leafless (*lfl*) mutant (Davis *et al.*, 1997).

Leaf pubescence

The pubescent leaf blade (*Pub*) locus controls development of short hairs or trichomes, 1 to 2 mm, on the upper and lower surface of the leaf blade. The hairs tend to be aligned along the smaller leaf veins and are more difficult to observe on older leaves. Cultivars expressing the *Pub* gene have widely scattered hairs, while some ssp. *spontaneum* accessions have numerous hairs. The semi-dominant hairy sheath (*Hsh*) gene controls development of hairs on the leaf sheath of lower leaves. The large trichome (*Ltc*) gene produces wide hairs that are four times the normal size and give the leaf blade a distinct roughness (Sato and Takeda, 1993; Davis *et al.*, 1997).

Leaf blade necrosis

The leaf blade and sometimes the leaf sheath will form necrotic or discoloured spots and blotches even when a pathogen is not present. Several necrotic leaf spot (*nec*) genes (Figure 5.1D) control formation of small, dark blotches on the leaf blade and sheath. The flecks are also observed on the awns when *nec1* is expressed. The maculosus (*mac*) mutants form small, nearly black flecks on the leaf blade after heading. Several *nec* mutants and the concentric leaf spot (*cls*) genes form large irregular blotches on the leaves. Numerous, large spots in some *nec* mutants can cause the plant to die prematurely (Davis *et al.*, 1997).

C. Tillering and plant vigour

The number and size of tillers is greatly influenced by environmental factors and interplant competition and generally has a quantitative inheritance pattern. However, several simply inherited mutants decrease or increase the number of tillers per plant. Most tillering mutants have other pleiotropic effects on plant development. The lazy (*lzd*) mutant produces prostrate tillers with no apparent geotropism response (Takeda, unpublished). The uniculm 2 (*cul2*) gene produces plants with a single culm or stem, which is increased in diameter and length and heads earlier than normal sibs. Swedish researchers isolated several uniculm mutants that have no secondary tillers or a few malformed ones. The low number of tillers (*lnt, int-l.81*) and absent lower laterals (*als*) mutants reduce the number of culms per plant to two to four, which form soon after seedling emergence. The culms are large, thick, and stiff similar to those of the *cul2* mutant. The corn stock (*cst*) reduces plant vigour and the number of tillers to one or two in six-rowed plants. Plants with the densinodosum (*den.6*), angustifolium (*fol-a*), granum (*gra*), and many-noded dwarf (*mnd1* and *mnd4*) mutants have numerous thin tillers with narrow leaves and short internodes (Davis *et al.*, 1997).

D. Culm elongation

Rapid elongation of the apical dome and an increase in apex diameter are observed when the shoot apex enters the reproductive stage. Elongation of culm internodes is dependent on environmental factors such as temperature and many genetic factors. The brachytic (*brh*) mutants generally have short culms, spikes, awns, and kernels. Plants of the uzu dwarf or semi-brachytic (*uzu*) mutant show temperature sensitive height reductions of less than 15% in cool environments and over 75% in warm ones (Takahashi and Yamamoto, 1951). Many semidwarf (*sdw*) mutants appear to affect only plant height. The semidwarf 1 (*sdw1*) gene, which is widely used to reduce plant height in barley cultivars, delays maturity and reduces yield in some environments (Hellewell *et al.*, 2000). These observations involve both temperature and genetic factors, such as the short culm (*hcm*) gene. Some of the genes that reduce rachis internode length also reduce plant height or have positive interactions with other plant height genes. The winding dwarf (*wnd*) mutant causes coiling of the upper portion of peduncle, but reduced plant height was observed only when the *dsp1* gene is present. Slender dwarf (*sld*) mutants reduce plant height and the size of leaves, culms, and spikes. The single internode dwarf (*sid*) mutants are characterised by elongation of only the peduncle and maybe one or two additional internodes. Plants with a fragile stem (*fst*) mutant have very fragile, easily broken leaves, stems and spikes, and show a slight to drastic reduction in plant height. Expression of the vascular bundle number (*Vbn*) gene is associated with improving lodging resistance in barley (Davis *et al.*, 1997).

E. Surface waxes

Presence of the wax coating and its composition reduces evaporation of water from the plant and helps protect it against pathogens. Surface wax mutants are some of the most common mutant types induced in barley. Most surface wax mutants, the eceriferum (*cer*), glossy sheath (*gsh*), and glossy leaf (*glf*) loci, affect the presence and type of epicuticular waxes on the leaf blades and sheaths, culms, and spikes. When the wax coating is completely absent, various organs appear as a bright, glossy green colour. When mutants have a reduced surface wax coating, the glossy colour is not as bright a green (Lundqvist *et al.*, 1968; Lundqvist, 1992).

At least 1,900 different surface wax mutants have been isolated as induced or spontaneous mutants in barley. Allelism tests localised 1,631 mutants to 79 *cer*, *gsh*, and *glf* gene loci. The five phenotypic classes 'spike and leaf sheath', 'partial', 'spike', 'leaf blade', and 'spike, leaf sheath and leaf blade' contain 8, 19, 23, 25, and 4 loci, respectively (Lundqvist and Lundqvist, 1988). Biochemical and ultrastructural examination of the composition of the epidermal wax coating identified alkanes, primary alcohols, free fatty acids, aldehydes, beta-diketones, hydroxy-beta-diketones, esters and hydrocarbons. Both the occurrence and amount of different wax constituents and the development and shape of the wax crystals were observed to be organ-specific (von Wettstein-Knowles, 1992). Some *cer* mutants exhibit morphological defects as reduced plant vigour and height, reduced awn length, low seed set, narrow leaf blades, and chlorotic or necrotic leaves. Some *cer* and *gsh* mutants and the yellow node (*Ynd*) gene show a specific pattern of wax development on the nodes of the culm. Surface wax mutants that develop an excessive amount of wax on the spikes have been isolated (Davis *et al.*, 1997).

F. Anthocyanin pigments

The barley plant has various patterns of reddish purple coloration in the leaf blade and sheath, culm, auricles, kernel, awn, and at the base of the plant. The colour patterns are produced by variable distribution of anthocyanin pigment. They are greatly influenced by environmental conditions and are more pronounced after periods of cool, wet weather. The development of reddish purple sheath and stem pigments as the barley plant matures is a characteristic used in cultivar identification. Failure of this pigment pattern to develop in some cultivars is controlled by two red stem genes. Alleles at the anthocyanin-less 1 (*ant1*) locus greatly restrict pigment development, while alleles at the anthocyanin-less 2 (*ant2*) locus prevent development of pigments. Pigment development at the base of seedlings or in the auricles of the flag leaf can help identify *ant* mutants in environments where stem pigmentation does not develop. A series of *ant* mutants were isolated in different cultivars by Scandinavian researchers. When the plants with the *ant13.13* allele at the *ant13* locus were demonstrated to be free of anthocyanin and proanthocyanidin, which is involved in haze formation in beer, many other programmes induced mutants with no anthocyanidin, proanthocyanidin and/or catechin. Allelism tests have identified a hundred mutants at 30 different *ant* loci (Davis *et al.*, 1997; Jende-Strid *et al.*, 1999).

Many of the *ant* mutants restrict or prevent anthocyanin pigmentation of all plant parts. However, several other loci that control pigmentation of specific tissues have been identified. The coloured leaf tip (*Clt*) loci affect purple pigmentation of the tips of leaf blades. The purple auricle (*Pau*) gene is characterised by strong purple pigmentation of the auricles. The purple veined lemma (*Pvc*) gene can control purple pigmentation of the lemma late during grain fill. Mutants at the red lemma and pericarp (*Pre*) loci prevent development of purple kernels. As plant tissues die or mature, purple pigments fade (Smith, 1951; Nilan, 1964; Davis *et al.*, 1997).

Environmental stress responses

Barley is grown in many environments where climatic conditions are marginal for growth of other annual crops. Cultivars have been selected that respond well to stresses imposed by specific combinations of climate and edaphic factors. Estimates of heritability and combining ability have demonstrated that the responses have a genetic basis and can be measured at various stages of plant growth. Recently, molecular mapping studies have associated some responses with specific regions of the barley genome. For example, relatively few loci control winter growth habit, but winter hardiness is often location specific and has been associated with many interacting traits and genetic factors (Hockett and Nilan, 1985). Many genes for pest resistance have been identified and mapped in barley. Inheritance patterns vary from monogenic dominant resistance to multigenic additive resistance. Some resistant genes are effective only for specific physiological races of the pest (see also Chapter 8).

Nutritional quality factors

The grain differs among cultivars with regard to starch content, grain protein level, lipid content, amino acid balance, and enzymatic activity. Many desirable morphological and biochemical traits have relatively high heritability estimates in barley (Hockett and Nilan, 1985). A gene from the cultivar 'Karl' decreases grain protein level. The high amylose (*amo*) mutant increases the percentage of amylose starch to almost 50%, while the waxy endosperm (*wax*) mutants decrease the amylose content to almost zero and increases the β-glucan content. The high lysine (*lys*) mutants increase the lysine levels in endosperm proteins and nutritional value. Structural genes for alpha- and beta-amylase have been mapped and sequenced. The loci controlling many other biochemical processes have been studied and isozyme variants have been identified at many of these loci (Hockett and Nilan, 1985; Søgaard and von Wettstein-Knowles, 1987; Kleinhofs, 2000) (see Chapters 7 and 9).

Conclusions and outlook

The barley plant is a very complex organism but is one of the best investigated crop plants, and our knowledge has increased considerably by using many modern techniques of biotechnology. Many different marker genes provide us with a detailed understanding of the genetic composition of the barley genome, especially when combined with increased chromosome and genetic analyses of linkage and biochemical studies of the DNA constitution and the amino acid composition. The collections of barley mutants with their defined source of mutant genes will form a major input for future gene mapping and be immensely valuable for molecular genetic analyses of cloned mutant genes. The advantage of using mutant genes is that we already know something of the organs of the plants where these genes are expressed because of the mutant phenotype. They have also contributed significantly to the construction of genetic maps.

Many genetic studies have been performed to determine the number of gene loci involved in the different morphological groups and characteristics. Due to the low chromosome number, 2n=2x=14, and the annual life cycle of barley, chromosome maps for the morphological marker genes could be established quite soon. Interchange breakpoints have been used to establish the independence of seven genetic linkage groups and associate them with the individual barley chromosomes, 1H - 7H. Many morphological markers, pest resistance genes, isozyme loci, and molecular markers have been mapped. Since the number of mapped loci and quantitative traits increases each year, linkage maps for the seven barley chromosomes need to be updated frequently. The distribution of the different gene loci in the seven chromosomes is shown in

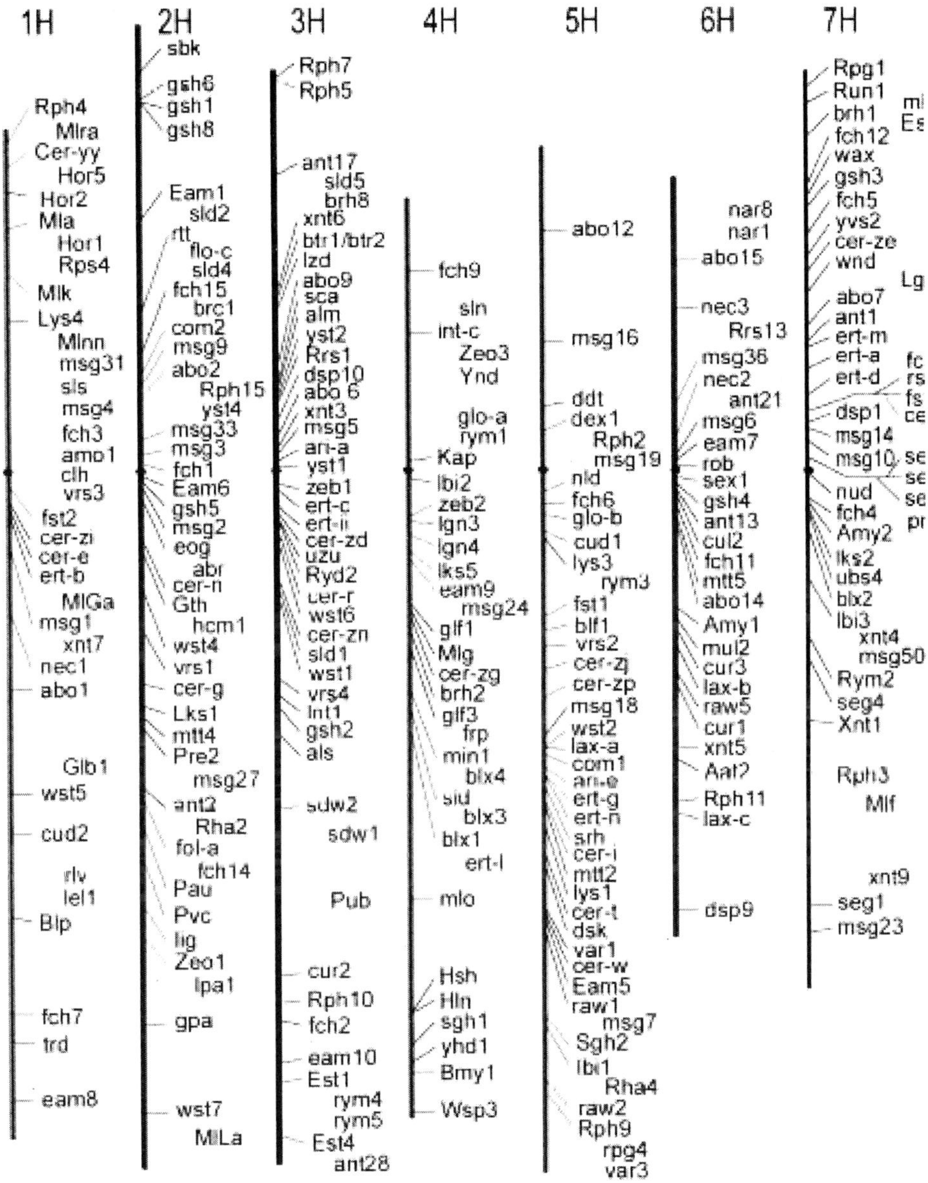

Figure 5.2. Barley chromosome maps showing estimated locations of genes controlling the expression of morphological markers, disease resistance, and isozymes.

Figure 5.2. Recent research work using translocation lines has provided a physical map of the gene spaces in all seven chromosomes. The gene spaces are separated by segments with a large amount of repetitive DNA sequences and low recombination frequencies. Gene spaces amount to 12% of the genome (Künzel *et al.*, 2000). During the last decade, many different molecular marker gene maps have been established and should be integrated. Current information on the integrated molecular, morphological, and physiological marker maps are presented as BIN maps at http://barleygenomics.wsu.edu (Søgaard and von Wettstein-Knowles, 1987; Davis *et al.*, 1997; Kleinhofs, 2000).

Collections of hundreds of different morphological and physiological mutants, either spontaneously or artificially induced, have been brought together from all over the world, over a period of several decades. They originate from different parts of the world, occurring in different genetic backgrounds and isolated in different environments and climatic conditions. During the last decade, major efforts have been made to transfer the different marker genes into a common genetic background, to make them useful and easier available to plant breeders and researchers worldwide. Several large collections show the broad diversity of the different mutant types existing in barley today. Many mutant accessions, cultivars and breeding lines are kept in different local and national collections of barley germplasm but with a high degree of duplication. Some of them deserve to be listed here: National Small Grain Collection, Aberdeen ID, USA, the center for all Barley Genetic Stocks; USDA/ARS/NPS collection, Beltsville MD, USA, mainly the center for resistance gene stocks; Vavilov Institute, St. Petersburg, Russia; The Nordic Gene Bank, Alnarp, Sweden, taking care of the nearly 10,000 induced mutant lines and Scandinavian barley accessions; Institute of Plant Genetics and Crop Plant Research (IPK), Gatersleben, Germany; the Barley Germplasm Center at Okayama, Kurashiki, Japan, the center for all induced and genetically analysed Japanese barley genes; the National Institute for Agrobiological Resources, Tsukuba, Japan; and the Plant Genetic Resources Centre (PGRC/E), Ethiopia.

There will always be a large demand for a broad diversity, and genetically characterised mutants and genebank collections are a gold mine for molecular genetics and barley breeding. They will serve as the basic material for all kinds of barley research.

References

Börner, A., V. Korzun, S. Malyshev and V. Ivandic, 1999. Molecular mapping of two dwarfing genes differing in their GA response on chromosome 2H of barley. Theor. Appl. Genet. 99: 670-675.

Bossinger, G., U. Lundqvist, W. Rohde and F. Salamini, 1992. Genetics of plant development in barley. In: Barley Genetics VI. Proc. 6th Int. Barley Genet. Symp., Helsinborg, Sweden, pp. 989-1022.

Chandler, P.M., and M. Robertsson, 1999. Gibberellin dose-responses curves and the characterization of dwarf mutants of barley. Plant Physiol. 120: 623-632.

Davis, M.P., J.D. Franckowiak, T. Konishi and U. Lundqvist (eds), 1997. Special Issue, Barley Genet. Newsl. 26, 533 p.

Ehrenberg, L., Å. Gustafsson and U. Lundqvist, 1959. The mutagenic effects of ionising radiations and reactive ethylene derivates in barley. Hereditas 45: 351-368.

Forster, B.P., H. Pakniyat, M. Macaulay, W. Matheson, M.S. Phillips, W.T.B. Thomas and W. Powell, 1994. Variation in the leaf sodium content of the *Hordeum vulgare* (barley) cultivar Maytorpe and its derived mutant cv. Golden Promise. Heredity 73: 249-253.

Franckowiak, J.D., U. Lundqvist and T. Konishi, 1997. New and revised names for barley genes. Barley Genet. Newsl. 26: 4-8.

Friedberg, E.C., G.C. Walker and W. Siede, 1995. DNA Repair and Mutagenesis. ASM Press, Washington DC.

Gallagher, L.W., K.M. Soliman and H. Vivar, 1991. Interactions among loci conferring photoperiod insensitivity for heading time in spring barley. Crop Sci. 31: 256-261.
Gustafsson, Å., 1940. The mutation system of the chlorophyll apparatus. Lunds Universitets Årsskrift, N.F. Avd. 2. Bd. 36: 1-40.
Gustafsson, Å., 1941. Mutation experiments in barley. Hereditas 27: 225-242.
Gustafsson, Å., 1947. Mutations in agricultural plants. Hereditas 33: 1-100.
Gustafsson, Å., A. Hagberg and U. Lundqvist, 1960. The induction of early mutants in Bonus barley. Hereditas 46: 675-699.
Gustafsson, Å., A. Hagberg, G. Persson and K. Wiklund, 1971. Induced mutations and barley improvement. Theor. Appl. Genet. 41: 239-248.
Gustafsson, Å., and U. Lundqvist, 1976. Controlled environment and short-day tolerance in barley mutants. In: Induced mutations in cross-breeding. Proc. of an advisory Group. Vienna, 1975. IAEA Vienna, pp. 45-53.
Gustafsson Å., and U. Lundqvist, 1980. Hexastichon and intermedium mutants in barley. Hereditas 92: 229-236.
Hagberg, A., 1961. Svalöfs original Mari-korn. "Aktuellt från Svalöf", Allmänna Svenska Aktiebolaget, pp. 13-16.
Hagberg, A., Å. Gustafsson and L. Ehrenberg, 1958. Sparsely contra densely ionising radiation and the origin of erectoid mutants in barley. Hereditas 44: 523-530.
Hagberg, G., and A. Hagberg, 1981. Haploidy initiator gene in barley. In: Barley Genetics IV. Proc. 4th Int. Barley Genet. Symp., Edinburgh, UK, pp. 686-689.
Harlan, J.R., 1976. Barley. In: N.W. Simmonds (ed), Evolution of Crop Plants. Longman, London. pp. 93-98.
Hellewell, K.B., D.C. Rasmusson and M. Gallo-Meagher, 2000. Enhancing yield of semidwarf barley. Crop Sci. 40: 352-358.
Hockett, E.A., 1988. New mutants in the genetic male sterile barley collection. Barley Genet. Newsl. 18: 70-73.
Hockett, E.A., and R.A. Nilan, 1985. Genetics. In: D.C. Rasmusson (ed), Barley, Agronomy Monograph No. 26. American Society of Agronomy, Madison, WI, pp. 187-230.
Ivandic, V., R.A. Malyshev, V. Korzun, A. Graner and A. Börner, 1999. Comparative mapping of gibberellic acid-insensitive dwarfing gene (*Dwf2*) on chromosome 4HS in barley. Theor. Appl. Genet. 98: 728-731.
Jende-Strid, B., U. Lundqvist and J.D. Franckowiak, 1999. Descriptions of barley genetic stocks for 1998. Barley Genet. Newsl. 29: 80-123.
Kleinhofs, A., 2000. Integrating molecular and morphological/physiological marker maps. Barley Genet. Newsl. 30: 84-86.
Kucera, J., U. Lundqvist and Å. Gustafsson, 1975. Inheritance of breviaristatum mutants in barley. Hereditas 80: 263-278.
Künzel, G., L. Korzun and A. Meister, 2000. Cytologically integrated physical restriction fragment length polymorphism maps for the barley genome based on translocation breakpoints. Genetics 154: 397-412.
Larsson, H.E.B., 1985. Morphological analysis of *laxatum* barley mutants. Hereditas 103: 239-253.
Lev-Yadun, S., A. Gopher and S. Abbo, 2000. The cradle of agriculture. Science 288: 1602-1603.
Lundqvist, U., 1992. Mutation Research in Barley. Dissertation, Dept. of Plant Breeding Research. The Swedish University of Agricultural Sciences, 283 p.
Lundqvist, U., and A. Lundqvist, 1988. Mutagen specificity in barley for 1580 *eceriferum* mutants localized to 79 loci. Hereditas 108: 1-12.
Lundqvist, U., P. von Wettstein-Knowles and D. von Wettstein, 1968. Induction of *eceriferum* mutants in barley by ionising radiations and chemical mutagens. II. Hereditas 59: 473-504.
Müller, K.J., N. Romano, O. Gerstner, F. Gracia-Maroto, C. Pozzi, F. Salamini and W. Rohde, 1995. The barley *Hooded* mutation caused by a duplication in a homeobox gene intron. Nature 374: 727-730.

Nevo, E., 1992. Origin, evolution, population genetics and resources for breeding of wild barley, *Hordeum spontaneum*, in the Fertile Crescent. In: P.R. Shewry (ed), Barley, Genetics, Biochemistry, Molecular Biology and Biotechnology. CAB International, Wallingford, U.K., pp. 19-43.

Nilan, R.A., 1964. The cytology and genetics of barley, 1951-1962. Monogr. Suppl. 3, Res. Stud. Vol. 32, No. 1. Washington State Univ. Press, Pullman.

Romagosa, I., F. Han, J.A Clancy and S.E. Ullrich, 1999. Individual locus effects on dormancy during seed development and after ripening in barley. Crop Sci. 39: 74-79.

Sato, K., and K. Takeda, 1993. Genetic analysis of large trichome on barley leaf blade. Barley Genet. Newsl. 22: 50-52.

Simpson, D.J., and D. von Wettstein, 1991. Coordinator's report: Nuclear genes affecting the chloroplast. Stock list of mutants kept at the Carlsberg Laboratory. Barley Genet. Newsl. 21: 102-108.

Smith, L., 1951. Cytology and genetics of barley. Bot. Rev. 17: 1-51, 133-202, 285-355.

Søgaard, B., and P. von Wettstein-Knowles, 1987. Barley: genes and chromosomes. Carlsberg Res. Comm. 52: 123-196.

Stadler, L.J., 1928. The rate of induced mutation in relation to dormancy, temperature, and dosage. Anatomical Record 41, 97 p.

Takahashi, R., and J. Yamamoto, 1951. Studies on the classification and geographical distribution of the Japanese barley varieties. III. On the linkage relation and the origin of the "uzu" or semi-brachytic character in barley. Ber. Ohara Inst. Landw. Forsch. 9: 399-410.

Tang, Y., M.E. Sorrells, L.V. Kochian and D.F. Garvin, 2000. Identification of RFLP markers linked to the barley aluminum tolerance gene *Alp*. Crop Sci. 40: 778-782.

Tschermak, E. von, 1901. Über Züchtung neuer Getreiderassen mittelst künstlicher Kreuzung. Kritisch-historische Betrachtungen. Z. landw. Versuchswesen Oesterreichs 4: 1029-1060.

Tsuchiya, T., and T.E. Haus, 1984. Stock description of lesser internode (*lin*). Barley Genet. Newsl. 14: 91.

Wettstein, D. von, S. Gough and C.G. Kannangara, 1995. Chlorophyll biosynthesis. Plant Cell 7: 1039-1057.

Wettstein, D. von, Å. Gustafsson and L. Ehrenberg, 1959. Mutationsforschung und Züchtung. Arbeitsgemeinschaft für Forschung des Landes Nordrhein-Westfalen. Köln und Opladen, pp. 7-50.

Wettstein-Knowles, P. von, 1992. Molecular genetics of lipid synthesis in barley. In: Barley Genetics VI. Proc. 6th Int. Barley Genet. Symp., Helsinborg, Sweden, pp. 753-771.

Yamae, G., and T. Konishi, 1993. Distorted segregation of the *lys3a* mutant in barley kernels. Barley Genet. Newsl. 22: 68-70.

Diversity in Barley *(Hordeum Vulgare)*
Roland von Bothmer et al (Editors).

Chapter 6

Cytogenetic diversity

Shin Taketa[a], Ib Linde-Laursen[b] and Gottfried Künzel[c]

[a]Department of Life Sciences, Faculty of Agriculture, Kagawa University, Ikenobe, Kagawa 761-0795, Japan
[b]Botanical Section, Department of Ecology, The Royal Veterinary and Agricultural University, Rolighedsvej 21, DK-1958 Frederiksberg C (Copenhagen), Denmark
[c]Institute of Plant Genetics and Crop Plant Research, Corrensstrasse 3, D-06466 Gatersleben, Germany

Introduction

Barley is one of the crop plants most suitable for cytogenetic studies because it is a diploid self-fertilizer with a low number (2n=2x=14) of relatively large chromosomes. The cytogenetics of *Hordeum vulgare* L. ssp. *vulgare* (cultivated barley) and its wild progenitor, ssp. *spontaneum* (C. Koch) Thell. has been reviewed extensively (e.g., Smith, 1951; Nilan, 1964; Burnham, 1966; Ramage, 1985, 1991; Hagberg and Hagberg, 1991b; Tsuchiya, 1991; Linde-Laursen *et al.*, 1997). However, these reviews focus upon cytogenetics *per se* and little attention has been paid to diversity. Because of the labor-intensive nature of cytogenetic techniques, cytogenetic diversity has been studied in a limited number of accessions. Moreover, comparison of results is sometimes difficult to carry out because of modifications of techniques applied in different laboratories. Despite these limitations, the present chapter attempts to review cytogenetic diversity of barley as revealed by conventional, banding and molecular techniques. In addition, the power of a collection of translocation stocks for the construction of physically integrated linkage maps of high resolution for the barley genome is demonstrated.

General karyotype variation

Mitosis

Hordeum vulgare has a common 'standard' karyotype comprising seven pairs of relatively large chromosomes. The 2C nuclear DNA content is ~11 pg with little variation (~4%) between entries as measured by Feulgen microdensitometry (Bennett and Smith, 1976). However, DAPI-staining and flow cytometry produce wider variation (~18%) ranging from 8 to 9.4 pg (Kankanpää *et al.*, 1996). The karyotype is symmetrical and has a quotient (ratio) longest/shortest chromosome of ~1.3 (Fukui and Kakeda, 1990; Jensen and Linde-Laursen, 1992). The seven chromosomes of the genome are more or less metacentric (Table 6.1). Five non-satellited chromosomes (1H, 2H, 3H, 4H, 7H) are so similar in length and arm ratios (Table

---

Taketa, S., I. Linde-Laursen and G. Künzel, 2003. Cytogenetic diversity. In: R. von Bothmer, Th. van Hintum, H. Knüpffer and K. Sato (eds), Diversity in Barley (*Hordeum vulgare*), pp. 97-119. Elsevier Science B.V., Amsterdam, The Netherlands.

Table 6.1. Relative lengths of individual chromosome arms measured in milliGeNomes (mGN) (Pedersen, 1995).

Chromosome	Fukui and Kakeda (1990)	Jensen and Linde-Laursen (1992)	Marthe and Künzel (1994)	Mean	Arm ratio L/S
1HS	52	51	50	51	1.4
1HL	73	70	72	72	
2HS	69	65	71	68	1.2
2HL	84	82	84	84	
3HS	65	61	66	64	1.3
3HL	83	82	82	82	
4HS	64	67	66	66	1.2
4HL	79	77	77	78	
5HS[1]	60	63	59	61	1.5
satellite	18	18	16	17	
5HL	93	92	90	92	
6HS[1]	67	70	65	67	1.0
satellite	23	26	24	24	
6HL	70	67	70	69	
7HS	73	76	75	75	1.0
7HL	71	74	73	73	

[1] including satellite

6.1) that identification based on these characters is less reliable (Jensen and Linde-Laursen, 1992; Fukui and Kakeda, 1994). Furthermore, the chromosomes undergo dynamic changes during metaphase (Kakeda and Fukui, 1994). Two chromosomes carry satellites. Chromosome 5H has a smaller and 6H a larger satellite on the short arm. Besides centromeric (primary) and nucleolar (secondary) constrictions, one or more small 'tertiary' constrictions may occasionally be observed (cf. Nilan, 1964).

The published data on the lengths of the somatic chromosomes vary widely. Most variation is certainly caused by different levels of chromosome contraction. Kakeda and Fukui (1994) thus reported that total chromosome length in 250 haploid cells ranged between 42 and 96 μm at somatic metaphase, giving mean chromosome lengths of between 6.0 and 13.7 μm. However, identification of mutants having long (Burnham, 1946) and short chromosomes (Moh and Nilan, 1954) demonstrates genetic variation for the character.

Before the introduction of chromosome banding techniques (Giemsa C- and N-banding), barley chromosomes could not be unequivocally identified (Linde-Laursen, 1975, 1981; Vosa, 1976; Noda and Kasha, 1978; Islam, 1980; Singh and Tsuchiya, 1982a; Künzel, 1987). These techniques reveal between one and four small to large (>10% arm length) bands per arm through visualization of constitutive heterochromatin. In spite of minor differences in size and visualization frequency of some bands, each of the seven different banding patterns produced identifies one and the same chromosome in all lines. This verifies the existence of one common basic banding pattern specific to *H. vulgare.* Consistently observed C-bands are foremostly located on one or both sides of the centromeres, at juxtacentromeric positions on one or both arms and at the nucleolar constrictions. Few bands are located distally. Telomeric bands, when observed, are very small. N-banding patterns are rather similar to C-banding patterns, but generally do not include bands at the nucleolar constrictions, at the telomeres and at a few distal intercalary sites presenting C-bands. On the other hand, N-banding specifically produces bands on both sides of all centromeres (Linde-Laursen, 1981; Singh and Tsuchiya, 1982a, b; Marthe and Künzel, 1994). Modifications of the technique may result in minor differences (Kakeda *et al.*, 1991). Banding patterns deviating relatively little from C- and N-banding may be produced after *in situ* digestion with restriction endonucleases (Kamisugi *et al.*, 1992), by G-banding (Song *et al.*, 1994), by GAA-banding through *in situ* hybridization (see section Molecular cytogenetic variation) and through visualization of late replicating chromosomal regions (Uozu *et al.*, 1997).

Prior to the advent of the banding techniques, the seven barley chromosomes were best identified through their different effects on the morphology of trisomic lines (Tsuchiya, 1960). Therefore, the specific banding pattern of each chromosome and chromosome arm was established through the application of C-banding or N-banding to the respective trisomics and monotelotrisomics (Linde-Laursen, 1978b; Singh and Tsuchiya, 1982b).

The individual chromosomes, and linkage groups of barley were previously designated by small letters, or Roman or Arabic numerals (cf. Burnham and Hagberg, 1956). In the latter system each non-satellited chromosome was designated by a number between 1 and 5 according to their suggested decreasing length. The chromosome with the larger satellite was designated 6 and the chromosome with the smaller satellite 7. This system was in general use until the Triticeae numbering system was adopted at the 7th International Barley Genetics Symposium (Linde-Laursen *et al.*, 1997). In the Triticeae system the chromosomes are designated 1H to 7H in accordance with their homoeology with the wheat chromosomes, which corresponds to chromosomes 5, 2, 3, 4, 7, 6 and 1, respectively, in the system proposed by Burnham and Hagberg (1956). The capital letter H indicates the genome symbol for *Hordeum vulgare* and *H. bulbosum*. Apart from small duplications and/or deletions of particular markers and some apparently inverted segments, the barley chromosomes show collinearity with the wheat chromosomes (cf. Linde-Laursen *et al.*, 1997). Chromosome 2H is considered the longest, followed in length by 5H, 3H, 7H, 4H, 6H and 1H (cf. Table 6.1). Short arms are designated by the letter 'S' and long arms by the letter 'L'. In accordance with Singh and Tsuchiya (1982b), the physically longer arm of chromosome 7H is designated 7HS and the physically shorter arm 7HL (Linde-Laursen, 1978b; Fukui and Kakeda, 1990; Jensen and Linde-Laursen, 1992; Marthe and Künzel, 1994).

Various approaches have been applied for comparisons of chromosome lengths measured in different studies. Burnham and Hagberg (1956) assigned a relative length of 100 to chromosome 6H excluding the satellite. Jensen and Linde-Laursen (1992) defined total genome length at mitotic metaphase as "one GeNome" (GN) and gave arm lengths and distances of chromosomal markers from centromeres (position 0) in milliGeNomes (mGN), i.e., one thousandth of genome length (Table 6.1; see also Marthe and Künzel, 1994). Fraction length (FL) puts the length of each arm as measured from the centromere at 100 (e.g., Künzel *et al.*, 2000). Measurements of chromosomes correctly identified by C- or N-banding indicate no significant differences in the relative length of the same chromosome arm or satellite of various entries with the standard karyotype (Table 6.1).

Meiosis

Karyotypes prepared from studies of meiotic chromosomes resemble karyotypes of somatic chromosomes (cf. Ramage, 1985). The study of the early meiotic prophase stages up to diakinesis is not practicable. In pollen, as well as embryo sac mother cells at diakinesis and meiotic metaphase I (MI), barley normally forms seven closed bivalents with a distal chiasma in each arm. Occasionally one or two of the closed bivalents may be replaced by open bivalents.

The majority of these include satellited chromosomes (Stoinova, 1994). In Northern Europe mean chiasma frequencies lie around 14 with restricted, but significant variation among cultivars, subspecies and years (Gale and Rees, 1970; Nilsson and Pelger, 1991). In contrast, a study in Pakistan reports chiasma frequencies between 15 and 20 (Jahan *et al.*, 1992). A comparison of recombination rates in male- (anther culture) and female- (*H. bulbosum* method) derived doubled haploid populations has indicated an overall 18% higher recombination rate on the male side for every chromosome and most chromosome arms (Devaux *et al.*, 1995). In diploid barley the duration of male meiosis has been estimated to last ~39 hours at 20°C (with prophase stages lasting ~33 h, metaphase I ~1.6 h, anaphase I-telophase II ~5.2 h), followed by a tetrad stage of ~8.0 h with little variation between cultivars. In contrast, meiosis lasted ~31 hours only in an autotetraploid barley (Finch and Bennett, 1972). The duration of female meiosis, which occurs synchronously with male meiosis, was not significantly different from male meiosis (Bennett *et al.*, 1973). A few studies have used N-banding of bivalents at meiotic MI for chromosome identification. Xu and Snape (1988) and Xu and Kasha (1992) verified the homology of 6H and the satellited chromosome of *H. bulbosum* L., and Künzel and Marthe (1991) localized breakpoints to chromosome arms in translocations. The study and analysis of chromosomal rearrangements using banding techniques are covered extensively in the section Chromosome structural variation.

Polymorphism of banding patterns

Linde-Laursen (1991a) reported 68 C-banding pattern variants of barley chromosomes 1H through 7H in a comprehensive sample including cultivars, landraces, "*H. agriocrithon*" and ssp. *spontaneum.* Comparisons of banding patterns in cultivars and lines of varying geographic origin indicated less diversity among entries from the same area than among entries from different areas (Linde-Laursen, 1978a; Muravenko *et al.*, 1996). Modern cultivars were less polymorphic than older, exotic entries, but even in the modern ones, different combinations of variants generally identified the single entry. For instance, a pedigree of 60 related cultivars and lines presented 44 C-banded karyotypes, 15 of which appeared in more than one line (Linde-Laursen *et al.*, 1982). A few modern cultivars carried banding pattern variants introduced through the use of exotic breeding material.

Inheritance of bands and banding patterns

Generally, bands located at the same position in a chromosome show Mendelian single-gene segregation (Linde-Laursen, 1979, 1982; Kjær *et al.*, 1991), which opens up the possibility of using them as markers in genetic studies and of comparing cytological and genetic maps. For instance, a study including C-bands and marker genes located on the long arm of 3H showed absence of recombination and no genes in the proximal 30% of the arm, and a low recombination frequency (12%) in the middle part verifying that most recombination, and most segregating genes are found in distal parts (e.g., Linde-Laursen, 1979, 1982; Künzel, 1982; Künzel *et al.*, 2000, see also the sections Chromosome structural variation and Molecular cytogenetic variation). The findings are supported by observations on the 44 C-banded karyotypes identified in the pedigree of Northwest European barley. Only five of these presented chromosomes with rearranged banding patterns, all of which were referable to distal recombination (Linde-Laursen *et al.*, 1982). This indicates that the proximal parts of the chromosomes are inherited as blocks. Inheritance studies including identification of banding patterns of the barley chromosomes may render it possible to trace the fate of proximal chromosome regions (Linde-Laursen, 1982; Linde-Laursen *et al.*, 1982; Künzel *et al.*, 2000). For instance, a C-banding pattern variant including the innermost 30% of the long arm of 3H that was first identified in cultivars released around 1930, was present in all later marketed cross-bred offspring (Linde-Laursen *et al.*, 1982). Assuming a neutral effect of the constitutive heterochromatin located in the bands, the 'unconscious' selection has apparently been for linked traits of agronomic value (Kjær *et al.*, 1991).

Polymorphism of NORs

Barley chromosome pair 6H of the standard karyotype deviates from 5H because of its larger satellites. Except in a few cultivars (see later), it is further characterized by having larger $AgNO_3$-stained nucleolus organizing regions (NORs) and longer nucleolar constrictions with a greater nucleolus forming capacity. This is indicated by the formation of the larger two of a maximum of four nucleoli of 'standard' size (range from ~20 to ~200 μm^3) observed in somatic cells (e.g., Linde-Laursen, 1984a; Ramage, 1985). The mean volume of the nucleoli formed at the smaller NORs/shorter nucleolar constrictions of 5H constitutes about 45% of the mean volume of the larger ones (Linde-Laursen, 1984a). Position effects of NORs were verified by the decreased or failing activity of 5H NORs at interphase in translocation lines with 5H and 6H NORs located on opposite arms or the same arm of one chromosome (Nicoloff *et al.*, 1979; Anastassova-Kristeva *et al.*, 1980; Linde-Laursen, 1984a; Schubert and Künzel, 1990). The observed dominance of the NOR on 6H over that on 5H represented the first case of intraspecific 'nucleolar dominance' as opposed to the well-known nucleolar dominance observed in plant and animal interspecific hybrids (for review, see Rieger *et al.*, 1979). Interphases of the cultivar 'Wong' may contain one or two micronucleoli in addition to up to four nucleoli of standard size (Linde-Laursen, 1984a) indicating the presence of an additional chromosome pair with nucleolus forming capacity (cf. Ramage, 1985). The nucleolus forming activity giving rise to the micronucleoli is probably located on the short arm of chromosome 1H (see section Molecular cytogenetic variation). Although 'Wong' has the standard karyotype (Linde-Laursen, 1978a), this cultivar, like one of its parents, 'Orel', has the larger NORs and longer nucleolar constrictions on 5H not on 6H, suggesting, in this case, nucleolar dominance of 5H. In 'Wong' the suggested change in nucleolar dominance is probably connected with the observation that the mean volume of the smaller pair of nucleoli of standard size constitutes 20% of the volume of the larger pair only, i.e., less than half the volume estimated in a cultivar with the longer nucleolar constriction on 6H. In 'Orel' the relative sizes of the nucleoli corresponded to those found in the majority of cultivars. The observations indicate a correlation between length of nucleolar constriction and size of NOR and the nucleolus produced at it (Linde-Laursen, 1984a).

Chromosomal diversity in autotetraploid barley

Autotetraploid barley lines (2n=4x=28) are characterized by having a very high frequency of aneuploids with 2n=26 to 31 as well as a few hexaploids (2n=6x=42) or near-hexaploids (Rommel, 1961). Linde-Laursen (1984b) reported 40% aneuploids in C_2 and C_3 generations. In a sample of 28 plants with 2n=28, only one was aneuploid. It had lost one chromosome 3H, but gained a chromosome 6H. Among 34 aneuploid plants, half were hypoploids with 2n=26 (1 plant) and 27, and half hyperploids with 2n=29. Although 1H was lost comparatively often, there was no solid indication that a particular chromosome was preferentially lost or gained. A single hypohexaploid plant had lost chromosomes 1H and 7H.

Barley × alien species hybrids

The C-banded karyotype of barley deviates so markedly from those of other species of *Hordeum* (Linde-Laursen *et al.*, 1995) and related genera (*Triticum, Secale, Psathyrostachys*) that it has been possible to study the chromosomal composition and distribution in cells of interspecific hybrids and derivatives (e.g., Islam, 1980; Linde-Laursen and Jensen, 1984, 1991; Linde-Laursen and Bothmer, 1984, 1988, 1999; Jørgensen and Andersen, 1989; Linde-Laursen, 1991b; Taketa *et al.*, 1995). For instance, an offspring of a barley × *Psathyrostachys fragilis* (2n=2x=14) hybrid having the genomes of both parents was a haploid of barley produced through elimination of the *Psathyrostachys* chromosomes (Linde-Laursen and Bothmer, 1984). The uniparental elimination of the chromosomes of the alien parent corresponds to the elimination pattern in barley × *H. bulbosum* (2x) hybrids (e.g., Kasha and Kao, 1970). However, monosomic and double monosomic substitutions of homoeologous or probably homoeologous *H. bulbosum* chromosomes into barley have been reported (Pickering, 1992). Barley × rye hybrids harboring rye B-chromosomes, sometimes observed as isochromosomes, only show elimination of the B-chromosomes (Linde-Laursen, 1991b). Observations on the chromosome complements of hybrids between barley and polyploid species of *Hordeum* and *Triticum* demonstrate that elimination in these mostly affects barley chromosomes. Barley chromosomes are probably also duplicated more often (Islam, 1980; Linde-Laursen and Bothmer, 1988, 1999; Taketa *et al.*, 1995; Bothmer *et al.*, 1999). Combinations of certain genotypes produce euploid hybrids only, comprising the complete genomes of both parents, whereas other combinations also produce aneuploid hybrids and (poly)haploids through elimination of one to all barley chromosomes. The frequency of elimination can be modified by external influences (Pickering and Morgan, 1985; Linde-Laursen and Bothmer, 1988, 1999; Bothmer *et al.*, 1999). The complements of both euploid and aneuploid hybrids may have one or more chromosomes of one parent replaced by chromosomes of the other parent (Linde-Laursen and Bothmer, 1988, 1999; Taketa *et al.*, 1995). In a few plants with the (poly)haploid chromosome number, substitutions of three or four chromosomes of the alien parent by barley chromosomes suggested homoeology. Finch (1983) observed alternative species-specific elimination in a *H. marinum* × 'Tuleen 346' barley hybrid. The *H. marinum* genome was eliminated in embryos and the barley genome in endosperms. The difference might be ascribed to a changed balance between genomes: a 1:1 ratio in embryos and a 2:1 ratio in endosperms. However, in a barley × *H. bulbosum* (2x) hybrid, *H. bulbosum* chromosomes were eliminated from both tissues.

The identification of the barley chromosomes in aneuploid interspecific hybrids with elimination of one to six barley chromosomes has shown that elimination patterns differ among combinations. In 48 hybrids of the cross *H. lechleri* (2n=6x=42) accession H 504 × 'Wong', the barley chromosomes were preferentially eliminated in the order: 1H-4H-5H-3H-7H-2H-6H (Linde-Laursen and Bothmer, 1999). Other combinations presented other patterns, but likewise with a tendency that chromosome 1H was eliminated as the first or second one. This agrees with observations in plants regenerated from callus cultures of similar interspecific hybrids (Jørgensen and Andersen, 1989). However, in a *H. marinum* × barley hybrid, a translocated barley chromosome including the greater part of the long arm of 1H (Finch and Bennett, 1982) was eliminated last (Finch, 1983). In contrast to the H 504 × 'Wong' combination with late elimination of 6H, chromosomes 2H or 3H were eliminated as the last ones in other combinations (Linde-Laursen and Bothmer, 1999). Comparisons of the barley chromosome 'complements' present in aneuploid hybrids of various combinations suggested that the chromosomes are arranged in different specific orders within individual genomes, elimination beginning with one of the chromosomes located at either end of an order. For instance, the order was 1H-5H-2H-6H-7H-3H-4H in the combination *H. lechleri* H 504 × 'Wong', and 1H-5H-2H-4H-7H-6H-3H in the combination H 504 × 'Igri'. The suggestion of different chromosomal orders within barley genomes does not support the theory of one order ('natural karyotype') specific for each species (Bennett, 1984).

The interspecific differences among chromosome banding patterns of hybrids having barley as one parent further demonstrated that the barley chromosomes were located closer to the cell center than the alien chromosomes (Linde-Laursen and Bothmer, 1984; Linde-Laursen and Jensen, 1984, 1991) as also observed in other investigations (e.g., Bennett, 1984). These studies also verified that the centromeres of the centrally located barley chromosomes were more clearly expressed, and that the nucleolus organizing regions of the two satellited barley chromosomes exerted nucleolar dominance over those of the more distally located genome(s).

In aneuploid and euploid *H. lechleri* (2n=6x=42) × barley hybrids, the presence of the barley chromosomes generally resulted in higher chromosome pairing at MI than in trihaploids, contradicting previous reports of a depressing effect of these in interspecific hybrids (Bothmer *et al.*, 1999). However, 4H had a negative influence. Some combinations showed a large difference in pairing according to parental origins. The meiotic chromosomes were not identified, but pairing between barley and alien chromosomes was considered low.

Wheat-barley addition lines

N- and C-banding have been used to identify unmodified, telocentric and translocated barley chromosomes in wheat-barley addition and substitution lines derived from interspecific hybrids (Islam, 1980, 1983; Koba *et al.*, 1997; Linde-Laursen *et al.*, 1997; Taketa and Takeda, 1997, 2001; Islam and Shepherd, 2000; Taketa *et al.*, 2002). Furthermore, use of the banding techniques has established the absence of addition lines disomic for complete chromosome 1H. Identified lines have been used to demonstrate the homoeologous relationship of wheat and barley chromosomes, and the chromosomal location of barley genes (review by Islam and Shepherd, 1990; Linde-Laursen *et al.*, 1997). Recent progress in generating, isolating and defining deletion and wheat/barley translocation chromosomes added to common wheat resulted in a number of lines suitable for deletion mapping in barley (Schubert *et al.*, 1998; Shi and Endo, 1999, 2000). A first deletion-based physical mapping was performed for chromosome 7H (Serizawa *et al.*, 2001). This will support and further refine the physical mapping results obtained from PCR-mapping of translocation breakpoints (see section Chromosome structural variation).

Chromosome structural variation

This section deals with reciprocal translocations, para- and pericentric inversions, duplications, and terminal or intercalary deletions. Over various periods of history, it is mainly translocations, but also the other chromosomal reconstructs, that have provided unique tools for genetic, cytogenetic research as well as for breeding research in barley.

History

Translocations and inversions were at first described and analyzed with regard to their influence on meiotic chromosome pairing during the early 1940s (Smith, 1941). Partial sterility served as a phenotypic marker for translocation heterozygotes, which allowed the use of translocations in genetic linkage studies as markers without any microscopic control (for review, see Smith, 1951). Since at this time only few genetic markers were at hand, this was an important advance. There followed a period of prosperity in utilizing chromosome structural variants. In the 1950s, 1960s and 1970s, many results, especially from linkage studies with translocations, contributed significantly to barley genetics. This period ended in the 1980s with the advent of molecular markers, which were easier to handle, gave more reliable results and became rapidly available in ample quantity. Translocations widely lost their importance in genetic linkage studies. In the 1990s, a new strategy for physical mapping of plant chromosomes was devised by integrating translocation breakpoints as physical landmarks into barley molecular linkage maps (Sorokin *et al.*, 1994; Künzel *et al.*, 1995; Künzel and Korzun, 1996). This strategy combines classical cytogenetics (defined translocations) with microdissection of individual chromosomes, polymerase chain reaction (PCR) with chromosome-specific DNA and primers for 'sequence-tagged-sites', derived from genetically mapped DNA markers. It resulted in cytologically integrated physical linkage maps of high resolution (240 breakpoints included) for all barley chromosomes (Künzel *et al.*, 2000). Thus translocations became once more an important tool to address a genetic problem which otherwise was not resolvable (see *Translocations*).

Deletions

Microscopically detectable deletions are usually lethal for true diploids like barley. As an exception, a viable homozygous deletion for the 18S-25S rRNA gene family in chromosome 6H was reported by Gecheff *et al.* (1994). This is explainable because this multigene family is carried on two chromosomes (5H and 6H), which can obviously substitute for each other. Another viable barley plant carrying a heterozygous deficiency for 67% of the long arm of chromosome 7H was described by Hang and Satterfield (1997). If individual barley chromosomes deficient for whole arms (telocentrics) or smaller segments in one or both arms ('acrocentrics' or 'metacentrics') are added to the normal diploid complement, their missing genetic information is compensated for by the normal chromosomes. Eleven out of the 14 possible telotrisomics have been reported and used to locate the centromere positions within genetic linkage maps of all seven barley chromosomes. Several genes were assigned to physical chromosome segments by means of linkage studies with acrotrisomics (cf. Tsuchiya, 1991).

Duplications

A total of 64 lines containing duplicated chromosome segments were reported (Hagberg and Hagberg, 1991a; Hagberg, 1995). These lines are collected in the Nordic Gene Bank, Alnarp, Sweden. All of the duplications originated from free recombination in individuals heterozygous for different translocations involving the same two chromosomes (for details see Hagberg and Hagberg, 1991b). Fifty-one of the lines involve the satellited chromosomes 5H and 6H, and 13 lines involve chromosome 1H. However, duplications in these lines have not been confirmed by banding techniques. Since precise prediction of the cytological size and location of duplicated segments is possible provided suitable translocation lines with defined breakpoints are available, 'directed' duplications were produced for breeding purposes. Envisaged was, for instance, to combine closely linked loci for different disease resistances, or different alleles of certain loci controlling race-specific pathogen resistance, or to duplicate genes for enhancing enzymatic activity (cf. Ramage, 1991). However, such ambitious programes failed for several reasons: (1) The physical positions of breakpoints and their distance to the candidate genes were not known with sufficient precision; (2) Most of the duplication lines proved to be inferior to the parental lines in vigor and yield, due to harmful effects of the dose changes caused by duplicated segments. Two duplication lines were reported to be 'high-yielders' as compared to their parental translocation lines (cf. Hagberg and Hagberg, 1991b; Liljeroth *et al.*, 1994). These seem, however, to be rare exceptions.

Inversions

Inversions have not been exploited extensively in genetic, cytogenetic or breeding studies in barley. Nevertheless, first hints of suppressed meiotic recombination within the proximal regions of barley chromosomes came from observations on inverted chromosomes more than 30 years ago (Holm, 1960; Nilan *et al.*, 1968; Kreft, 1969). Only a small number of spontaneous (e.g., Konishi and Linde-Laursen, 1988) and induced (e.g., Nilan, 1964) inversions were reported in barley, in spite of notable efforts to find them (cf. Nilan *et al.*, 1968). Most inversions were recognized by the presence of bridges and fragments at anaphase I and II in heterozygotes, typical for paracentric inversions (cf. Ramage, 1985). A few pericentric inversions which include the centromere, and therefore do not result in aberrant meiotic stages were discovered by Giemsa banding. Gecheff (1989) found such an inversion, involving chromosome 6H in a multiply restructured karyotype with all chromosomes, to be morphologically identifiable. Gecheff (1996) also reported that six of 22 chromosomal rearrangements induced in a barley karyotype with morphologically distinguishable chromosomes were pericentric inversions. Three further pericentric inversions were found in translocation lines as additional rearrangements, involving chromosome 3H in line T1-6i (Linde-Laursen, 1983), chromosome 4H in line T1-5am, and chromosome 5H in line T5-6ap, respectively (Marthe and Künzel, 1994). Only once was inversion analysis used for genetic linkage studies by Ekberg (1974). Using the degree of sterility in heterozygotes of a paracentric inversion as an estimate for the rate of recombination within meiotic inversion loops, Säll *et al.* (1990) found remarkable variations in recombination frequency among different barley cultivars.

Translocations

Worldwide, ~1000 lines with single reciprocal translocations and ~280 stocks with karyotypes combining two to nine homozygous translocations were collected (Ramage, 1973, 1975; Künzel, 1992). Furthermore, 72 of these lines, homozygous for single or multiple translocations, are available as colchicine-induced tetraploids (Künzel, 1994). The entire translocation material is centered at the IPK Gatersleben and available for distribution upon request. Major parts of this material are also maintained in the Nordic Gene Bank at Alnarp, Sweden and the USDA-ARS National Small Grains Germplasm Research Facility at Aberdeen, Idaho, USA.

Most of the translocations have been induced by irradiation. In addition, a few of spontaneous origin were identified in both cultivated and wild barleys. Among 1,240 lines of cultivated barley, the same rearranged karyotype was found in four lines collected in different areas of Ethiopia. Among 120 accessions of ssp. *spontaneum* three carried translocations (Konishi and Linde-Laursen, 1988). Another translocated karyotype was identified in two lines from Tunisia (Kurauchi, 1997).

Translocations have been used extensively for genetic linkage studies, which contributed to associate the seven linkage groups with individual chromosomes, to assign new genes to linkage groups, and to determine relative positions of marker genes, translocation breakpoints and centromeres in linkage maps (see reviews by Nilan, 1964; Ramage, 1985).

Moreover, translocations were used for producing duplications, as outlined above. They were also an important source of trisomics in barley (Ramage, 1960). Among the different types of trisomics resulting from translocation heterozygotes, tertiary trisomics (a translocated chromosome is added to the normal diploid complement) were of special importance (Ramage, 1965). For a limited time, so-called 'balanced tertiary trisomics' (BTTs) based on genes for male sterility were used for commercial production of hybrid barley cultivars (for review, see Ramage, 1991). BTT stocks were also used to demonstrate marked differences between genetic and physical distances (Künzel, 1982). About 120 BTT stocks have been established (Künzel, 1993).

Multiple translocations have been constructed in which all seven chromosome pairs are morphologically identifiable without banding procedures (Tuleen, 1973; Künzel and Nicoloff, 1979; Finch and Bennett, 1982; Gecheff, 1989, 1996). Such stocks were used to study the intrachromosomal distribution pattern of sister chromatid exchanges (Schubert *et al.*, 1980) and to demonstrate the non-random distribution of mutagen-induced chromatid aberrations along chromosomes (e.g., Nicoloff *et al.*, 1975, 1981).

Recently, translocation lines with distinct length differences of interchanged chromosomes were successfully used to flow-sort chromosomes (Lysák *et al.*, 1999).

Despite this progress, all applications using translocations were hampered by the difficulty in determining the positions of breakpoints accurately. For more than 50 years many efforts have been directed at localizing these:

(1) Analyses of pairing configurations at MI in hybrids with translocation tester sets (Burnham, 1962; Hagberg, 1986);
(2) Analyses of meiotic pairing configurations at MI, in combination with fertility studies on hybrids from large sets of intercrosses between individuals homozygous for different translocations involving the same two chromosomes (cf. Kasha and Burnham, 1965b; Hagberg and Hagberg, 1968, 1969);
(3) Linkage studies with translocations and marker gene stocks (e.g., Kasha and Burnham, 1965a:, Persson, 1969a, b; Jensen, 1971; Tuleen, 1971);
(4) Somatic metaphase analysis considering modified length ratios of short and long arms and/or positional changes of cytogenetic markers, like nucleolar constrictions/satellites (e.g., Tjio and Hagberg 1951; Hagberg *et al.*, 1975, 1978; Hagberg, 1986), or Giemsa bands.

Especially by using banding techniques, it became feasible to identify the positions of many breakpoints in somatic and meiotic chromosomes (e.g., Finch and Bennett, 1982; Georgiev *et al.*, 1985; Künzel, 1987; Konishi and Linde-Laursen, 1988; Linde-Laursen, 1988; Kakeda and Yamagata, 1991; Xu and Kasha, 1992; Marthe and Künzel, 1994; Gecheff, 1996). Of 70 translocations analyzed, Linde-Laursen (1988) could localize breakpoints by changes in banding patterns and/or arm length ratios precisely in 16, and to smaller or larger chromosomal segments in 36 translocations. The breakpoints could not be identified in 18 symmetrical translocations because they were located distally to marker bands. Kakeda and Yamagata (1991) also failed to identify the breakpoints in 10 of 58 translocations studied. Few breaks have been located with certainty within a band (Kakeda and Yamagata, 1991; Künzel *et al.*, 2000), while many apparently occur at the borders between euchromatin and heterochromatin (Linde-Laursen, 1988; Gecheff, 1996). Breaks within centromeres producing 'semi-dicentric' chromosomes are rare (Linde-Laursen, 1988), whereas several breaks were found within nucleolar constrictions (Anastassova-Kristeva *et al.*, 1980; Konishi and Linde-Laursen, 1988; Linde-Laursen, 1988; Schubert and Künzel, 1990; Marthe and Künzel, 1994). A comparison of the chromosomal distribution of radiation-induced translocation breakpoints in first post-treatment mitoses versus those transmitted to viable progenies showed that the breakpoints were non-randomly distributed along barley chromosomes (Künzel *et al.*, 2001). In first post-treatment mitoses, centromeres and the heterochromatin-containing proximal segments tended to be more than randomly involved and terminal segments to be less than randomly involved in translocations. Contrary to this, small chromosomal regions in median and distal arm positions, characterized by high recombination rates and high gene density according to Künzel *et al.* (2000), were identified as preferred sites for the origination of viable translocations, probably due to deviations in chromatin organization. Apparently the position of a breakpoint has an influence on the rate of viability versus elimination of the carrier cells. Surprisingly, breakpoints within centromeres and heterochromatin-containing segments seem to be more harmful for survival than those induced in gene-rich regions. For localization of breakpoints by analysis of pairing configurations at MI in hybrids, Linde-Laursen (1988) and Marthe and Künzel (1994) proposed translocation tester sets having distally located breakpoints.

For the use of translocation breakpoints as physical landmarks to relate physical to genetic distances within molecular linkage maps (Sorokin *et al.*, 1994), initially the positions of many breakpoints could not be determined precisely enough even by Giemsa banding (Künzel and Korzun, 1996). Eventually, the PCR-mediated mapping of many breakpoints into different linkage groups increasingly refined their karyological positions. The physical range to which a breakpoint was originally assigned by chromosome measurement, gradually decreased as the more breakpoints were correctly mapped between markers of a defined genetic distance. This was due to the mutual determination of the respective genetic and physical positions. Using a software developed to process the corresponding data, the mean extension of segments to which the breakpoints were karyologically assigned (23.6 mGN) could be refined to 5.1 mGN for the 240 breakpoints of the 120 translocations which were integrated as physical landmarks into linkage maps of the seven barley chromosomes (for details, see also Künzel *et al.*, 2000). In this way a basis was established for precisely relating physical and genetic map distances. Altogether 138 physical subregions of the genome could be compared to their corresponding linkage segments. This high degree of resolution, together with the comprehensive results for all linkage groups, enabled the following realistic conclusions:

(1) Recombination is mainly confined to a few relatively small chromosomal areas separated by large segments of severely suppressed recombination;
(2) Most recombination occurs in gene-rich regions which correspond to only small chromosomal areas (Figure 6.1);
(3) The gene-density and recombination frequency seem to be similar for gene-rich regions in plant species, irrespective of their genome sizes;
(4) Gene-rich regions of all species should therefore be equally accessible to map-based cloning (Künzel *et al.*, 2000)

Images of the physical versus genetic maps for the seven barley chromosomes and idiograms of the 120 translocations used for the PCR-mediated mapping are available on-line at http://wheat.pw.usda.gov/ggpages/Barley_physical/.

Molecular cytogenetic variation

Since more than 75% of the barley genome comprises repetitive DNA sequences (Flavell *et al.*, 1977), their isolation, localization and characterization are important research topics in barley molecular cytogenetics. With the advent of the non-isotopic *in situ* hybridization technique, many DNA sequences, mostly repetitive ones, have been physically mapped to barley chromosomes. Several single or low copy sequences have also been localized, providing valuable chromosome landmarks for the integration of genetic and physical maps. Thus, *in situ* hybridization along with other molecular techniques considerably improved barley cytogenetics.

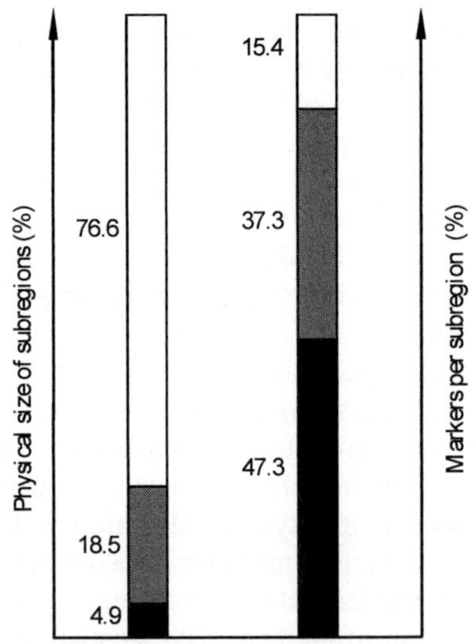

Figure 6.1. Distribution of recombination rates and markers within the barley genome according to Künzel *et al.* (2000). The megabase/centimorgan ratios (Mb/cM) were calculated on the basis of a DNA content of 5350 Mb/haploid barley genome (Bennett and Smith, 1976) and relative chromosome measurements of Marthe and Künzel (1994). Considering the entire genetic length of the 'Igri' x 'Franka'-map of 1214.2 cM (Graner, 1994), an average value of 4.4 Mb/cM is assumed for the barley genome. White represents regions of suppressed (>4.4 Mb/cM), grey of increased (1.0 – 4.4 Mb/cM) and black of strongly increased (≤1.0 Mb/cM) recombination. The proportional amount of these regions is shown on the left and the proportion of markers they contain on the right.

Physical mapping of repetitive DNA sequences

In barley the 18S-5.8S-25S (18S-25S) ribosomal RNA genes and intergenic spacers, together called the rDNA, was one of the first coding sequences localized by *in situ* hybridization. When the 18S-25S rDNA clone pTa71 of wheat, which contains the coding sequences for the 18S, 5.8S and 25S rRNA genes and the intergenic spacer sequences (Gerlach and Bedbrook, 1979) was used as a probe, two major and four minor 18S-25S rDNA sites were detected. The two major sites were located in the nucleolar organizer region (NOR) of the two satellited chromosomes 5H and 6H. Minor sites were found on the short arms of four other chromosomes, namely 1H, 2H, 4H and 7H (Leitch and Heslop-Harrison, 1992; Pedersen and Linde-Laursen, 1994). The locus on 1H has the strongest signal among the four minor sites and probably has nucleolus-forming activity because up to two micronuclei were detected by silver staining in 'Wong' (Linde-Laursen, 1984a). The weaker signals detected on 2H and 7H are of similar strength (Leitch and Heslop-Harrison, 1992). The locus on 4H has the weakest signal (Pedersen and Linde-Laursen, 1994). Although the activity of these three loci has not been confirmed by silver staining, Tsuchiya (1960) reported association of meiotic trivalents or univalents with the nucleolus in ssp. *spontaneum* lines trisomic for chromosomes 2H, 3H and 7H. The two major and four minor 18S-25S rDNA loci were observed in four barley cultivars including both two- and six-rowed types and winter and spring growth habits (Pedersen and Linde-Laursen, 1994) and one ssp. *spontaneum* accession (Taketa *et al.*, 1999). These observations suggest that these six 18S-25S rDNA loci are generally present in *H. vulgare*. In contrast, the 18S rDNA probe of mosquito origin detected only the two pairs of well-known major 18S-25S rDNA sites (Brown *et al.*, 1999).

Four 5S rDNA loci were detected on chromosomes 2H, 3H, 4H and 7H of cultivated barley when the 5S rDNA clone pTa794 of wheat, which includes a 120-bp coding sequence for the 5S rRNA gene and the intergenic spacer (Gerlach and Dyer, 1980) was used as a probe (Leitch and Heslop-Harrison, 1993; Pedersen *et al.*, 1995). The strengths of the signals were greatest on 2H and weakest on 4H. The signals on 3H and 7H had intermediate strength (Leitch and Heslop-Harrison, 1993). Brown *et al.* (1999) confirmed these four 5S rDNA loci using the 5S rDNA from *Matthiola incana*. Combination of both 5S and 18S-25S rDNA sites allows the identification of all chromosome arms. Fukui *et al.* (1994) used directly cloned 5S rDNA sequences of barley that includes coding and spacer regions as a probe. This detected up to six pairs of additional minor hybridization sites as well as the above-mentioned four loci. Japanese barley cultivars were polymorphic for some of the additional sites. Since these additional sites were observed only when the probe included the non-coding spacer regions, they may represent relics of active 5S rDNA loci or unrelated sequences that show considerable homology. Using clone pTa794 as a probe, Taketa *et al.* (1999) detected only three 5S rDNA loci in a ssp. *spontaneum* accession. The locus on 4H was absent in this accession. Since the 5S rDNA locus on 4H has the weakest signal among the four 5S rDNA loci, the 4H locus of this accession may be either deleted or reduced in copy number below the detection limit of *in situ* hybridization.

Several repetitive DNA sequences have been isolated and characterized in barley. A tandemly repeated DNA sequence, the HvRT sequence family (Belostotsky and Ananiev, 1990), is located within the subtelomeric region of all barley chromosome arms proximal to the telomere repeats and named telomere associated sequences (Kilian and Kleinhofs, 1992; Röder *et al.*, 1993). The signal intensity of this sequence family differs among chromosome arms and the signals in the long arms of chromosomes 2H and 5H are very small. The HvRT family was also detected within an interstitial region of the long arm of chromosome 3H and its signal intensity is polymorphic among barley cultivars (Röder *et al.*, 1993; Pedersen *et al.*, 1995).

The dpTa1 sequence family, originally isolated from wheat (Vershinin *et al.*, 1994), is another tandem repeat abundantly present in barley. The chromosomal distribution of this 340-bp sequence or sequences homologous to it have been reported for dpTa1 (Taketa *et al.*, 2000), pAs1 isolated from *Aegilops tauschii* (Brandes *et al.*, 1995), and pHvMWG2315 (Busch *et al.*, 1995) and pHv14A (Tsujimoto *et al.*, 1997), both isolated from cultivated barley. These four sequences share 80% or more sequence homology and show rather similar dispersed *in situ* hybridization patterns having about 40 hybridization sites both near the ends of most chromosome arms and at multiple intercalary sites. Because of the differences of clones, barley accessions and *in situ* hybridization techniques employed in each experiment, it is not certain whether or not the observed differences reflect polymorphism in the distribution pattern of dpTa1 homologous sequences. Sequences homologous to dpTa1 provide useful chromosomal landmarks especially at subtelomeric to telomeric positions, where C- and N-bands are scarce. The dpTa1 homologous sequences are more proximally located compared to the HvRT family (Brandes *et al.*, 1995; Tsujimoto *et al.*, 1997).

In barley, the GAA-rich sequences are repetitive DNA sequences useful for chromosome identification (Dennis *et al.*, 1980, Pedersen *et al.*, 1996). *In situ* hybridization of the polypurine motif $(GAA)_7$, or GAA-rich genomic DNA clones, produces a signal pattern similar to C- and N-banding patterns except for the N-positive and C-negative bands at the centromeric sites, the telomeric C-bands and the NORs (Pedersen and Linde-Laursen, 1994; Pedersen *et al.*, 1996).

The GAA bands are similar in their FL positions to the previously reported C- and N-bands, but there was a tendency toward higher FL values for most proximal bands (Pedersen *et al.*, 1996). *In situ* hybridization with the labeled GAA-satellite sequence may substitute for N-banding because an efficient *in situ* hybridization protocol has been established (Pedersen and Linde-Laursen, 1994).

The pSc119.2 sequence was among the first tandemly repeated DNA sequences to be isolated in rye (McIntyre *et al.*, 1990). Sequences homologous to this 118-bp repeat have been found widely distributed in Triticeae (e.g., Mukai *et al.*, 1993). However, Southern and *in situ* hybridization cannot detect its presence in barley and the diploid cytotype of *H. murinum*, although it is present in many other *Hordeum* species, including *H. bulbosum*, which is considered to carry the same H genome as barley (Gupta *et al.*, 1989; Taketa *et al.*, 2000). By PCR analysis, Vershinin *et al.* (1996) demonstrated the presence of this sequence in low copy number in barley.

Svitashev *et al.* (1994) described a dispersed distribution of five repetitive DNA sequences isolated from barley by Vershinin *et al.* (1990). The barley retrotransposable element *BARE-1* shows a dispersed distribution pattern along chromosomes except for pericentromeric regions, telomeres and NORs (Suoniemi *et al.*, 1996; Heslop-Harrison *et al.*, 1997). These DNA sequences do not appear to be promising as probes for studying cytogenetic diversity because of their uniformly dispersed hybridization patterns. Presting *et al.* (1998) reported the location of a *Ty3/gypsy*-like retroelement at all centromeric regions of barley. The sequence has not yet been tested for inter-varietal polymorphism.

Physical mapping of low- or single-copy sequences

As for other plants, *in situ* hybridization of low- or single-copy sequences has been reported to be difficult in barley and so far this can be achieved reliably with large, single-copy or multiple-copy probes only. Genes encoding B hordein (*Hor2*) (Lehfer *et al.*, 1993; Pedersen and Linde-Laursen, 1995), C hordein (*Hor1*) (Pedersen and Linde-Laursen, 1995), alpha-amylase 1 (Pedersen *et al.*, 1995; Butnaru *et al.*, 1998) and alpha-amylase 2 (Leitch and Heslop-Harrison, 1993; Pedersen *et al.*, 1995; Brown *et al.*, 1999) were localized. In these studies, no significant differences in the hybridization sites among different barley cultivars were noticed. Pedersen *et al.* (1995) localized five more genetically mapped genes and confirmed by comparison of physical and genetic distances a strong reduction of recombination in the proximal half of the chromosome arms. Physical mapping of the anonymous RFLP clone pMSU12 detected signals on chromosomes other than that assigned by genetic mapping (Pedersen *et al.*, 1995). Such a discrepancy may be caused by hybridization to relic genes or to sites that could not be genetically mapped due to lacking polymorphism. Using *in situ* hybridization, Pedersen *et al.* (1997) localized transgenes (8.5kb) introduced by microprojectile bombardment. The integration sites of transgenes varied among transformants with a tendency toward distal regions of the chromosomes. Bacterial artificial chromosome (BAC) clones with inserts varying from 20 to 150kb were successfully used as probes for *in situ* hybridization in the presence of excess amounts of a C_0t-1 fraction of barley DNA (Lapitan *et al.*, 1997).

Genomic in situ *hybridization*

Genomic *in situ* hybridization (GISH), which uses total genomic DNA as a probe, is a powerful tool for determining the origin of genomes or chromatin in hybrids. This technique successfully detected barley chromosomes in hybrids with wheat (Mukai and Gill, 1991; Schwarzacher *et al.*

1992). The technique also permitted discrimination between species-specific chromosomes in *H. vulgare* × *H. bulbosum* hybrids (Anamthawat-Jónsson *et al.*, 1993; Pickering *et al.*, 1997), although these species are considered to carry the same H genome. This suggests differences in the composition of repetitive DNA sequences between the two H genomes. In a barley line with leaf rust resistance, homozygous introgression of *H. bulbosum* chromatin to the long arm of chromosome 2H was detected by sequential GISH and *in situ* hybridization with a labeled $(CTT)_{10}$ oligonucleotide (Pickering *et al.*, 2000). Sequential N-banding and GISH were performed to select aberrant barley chromosomes in a genetic background of wheat with a gametocidal chromosome, which causes chromosomal breakages (Schubert *et al.*, 1998; Shi and Endo, 1999, 2000).

Conclusions and outlook

Cytogenetic information on barley has hitherto mainly been obtained from studies of one or a few accessions. To date, cytogenetic diversity appears to be best surveyed by Giemsa banding techniques. Thus, banding analysis of more than 100 accessions detected a considerable number of polymorphisms. Banding analyses have also revealed the approximate positions of breakpoints in ca. 200 reciprocal translocations. However, even with the use of banding techniques, the number of accessions studied has to be extended in order to obtain a more comprehensive coverage of cytogenetic diversity inherent to barley. Since very few of the up to 40 conspicuous C-/N-bands are located in distal chromosomal segments, variation in these regions can hardly be detected by banding analysis. This, together with the poor reproducibility of minor bands, hampers the efficiency of the banding techniques for evaluating cytogenetic diversity.

To study cytogenetic diversity more thoroughly, more informative and reproducible techniques are necessary. In future, *in situ* hybridization will probably be the technique of choice for that purpose, even though it is still time-consuming and costly. The obtainable information will increase the more DNA sequences suitable as hybridization probes become available, and the more efficient hybridization protocols are developed. The use of 'standardized' *in situ* hybridization protocols for evaluating and comparing entries of the Barley Core Collection (BCC, see also Chapter 13) may lead to a deeper understanding of cytogenetic diversity.

Improved techniques for *in situ* hybridization and signal detection will render it possible to localize many more DNA sequences, including small single-copy sequences, on the metaphase and interphase chromosomes. This may help to solve basic questions, for instance, to reveal the order of chromosomes within nuclei and to elucidate the interactions between chromosomes/genomes in interspecific hybrids. Chromosomal *in situ* suppression (CISS) hybridization for 'painting' of individual chromosomes within species-specific karyotypes, though not possible at present for plant genomes (cf. Fuchs *et al.*, 1996), may enable accurate physical positioning of translocation breakpoints.

High resolution physical maps of the barley genome have been constructed that integrate 240 translocation breakpoints and more than 300 molecular markers. In the future, these maps may be physically refined down to the molecular level. Once contigs of large-insert clones are available for extended stretches of barley DNA, they may be physically mapped on chromosomes by anchoring markers that define 'landmark' loci for subarm regions. This would greatly improve our understanding of barley genome organization and function.

References

Anamthawat-Jónsson, K., T. Schwarzacher and J.S. Heslop-Harrison, 1993. Behavior of parental genomes in the hybrid *Hordeum vulgare* × *H. bulbosum*. J. Hered. 84: 78-82.

Anastassova-Kristeva, M., R. Rieger, G. Künzel, H. Nicoloff and A. Hagberg, 1980. Further evidence on "nucleolar dominance" in barley translocation lines. Barley Genet. Newsl. 10: 3-6.

Belostotsky, D.A., and E.V. Ananiev, 1990. Characterization of relic DNA from barley genome. Theor. Appl. Genet. 80: 374-380.

Bennett, M.D., 1984. The genome, the natural karyotype, and biosystematics. In: W.F. Grant (ed), Plant Biosystematics. Academic Press, Toronto, pp. 41-66.

Bennett, M.D., R.A. Finch, J.B. Smith and M.K. Rao, 1973. The time and duration of female meiosis in wheat, rye and barley. Proc. Roy. Soc. Lond. B. 183: 301-319.

Bennett, M.D., and J.B. Smith, 1976. Nuclear DNA amounts in angiosperms. Phil. Trans. Roy. Soc. Lond. B Biol. Sci. 274: 227-274.

Bothmer, R. von, B. Salomon and I. Linde-Laursen, 1999. Chromosome pairing patterns in interspecific *Hordeum lechleri* × *H. vulgare* (cultivated barley) hybrids with 2n=21-29. Hereditas 131: 109-120.

Brandes, A., M.S. Röder and M.W. Ganal, 1995. Barley telomeres are associated with two different types of satellite DNA sequences. Chromosome Res. 3: 315-320.

Brown, S.E., J.L. Stephens, N.L.V. Lapitan and D.L. Knudson, 1999. FISH landmarks for barley chromosomes (*Hordeum vulgare* L.). Genome 42: 274-281.

Burnham, C.R., 1946. A gene for "long" chromosomes in barley. Genetics 31: 212-213.

Burnham, C.R., 1962. Discussions in Cytogenetics. Burgess Publ., Minneapolis, 375 p.

Burnham, C.R., 1966. Cytogenetics in plant improvement. In: Frey, K.J. (ed), Plant Breeding, Symp. Iowa State University, Ames, Iowa, pp. 139-187.

Burnham, C.R., and A. Hagberg, 1956. Cytogenetic notes on chromosomal interchanges in barley. Hereditas 42: 467-482.

Busch, W., R. Martin, R.G. Herrmann and U. Hohmann, 1995. Repeated DNA sequences isolated by microdissection. I. Karyotyping of barley (*Hordeum vulgare* L.). Genome 38: 1082-1090.

Butnaru, G., J. Chen, P. Goicoechea and J.P. Gustafson, 1998. *In situ* hybridization mapping of genes in *Hordeum vulgare* L. J. Hered. 89: 366-370.

Dennis, E.S., W.L. Gerlach and W.J. Peacock, 1980. Identical polypyrimidine-polypurine satellite DNAs in wheat and barley. Heredity 44: 349-366.

Devaux, P., A. Kilian and A. Kleinhofs, 1995. Comparative mapping of the barley genome with male and female recombination-derived, doubled haploid populations. Mol. Gen. Genet. 249: 600-608.

Ekberg, I., 1974. Cytogenetic studies of three paracentric inversions in barley. Hereditas 76: 1-30.

Finch, R.A., 1983. Tissue-specific elimination of alternative whole parental genomes in one barley hybrid. Chromosoma 88: 386-393.

Finch, R.A., and M.D. Bennett, 1972. The duration of meiosis in diploid and autotetraploid barley. Can. J. Genet. Cytol. 14: 507-515.

Finch, R.A., and M.D. Bennett, 1982. The karyotype of Tuleen 346 barley. Theor. Appl. Genet. 62: 53-58.

Flavell, R.B., J. Rimpau and D.B. Smith, 1977. Repeated sequence DNA relationships in four cereal genomes. Chromosoma 63: 205-222.

Fuchs, J., A. Houben, A. Brandes and I. Schubert, 1996. Chromosome 'painting' in plants - a feasible technique? Chromosoma 104: 315-320.

Fukui, K., and K. Kakeda, 1990. Quantitative karyotyping of barley chromosomes by image analysis methods. Genome 33: 450-458.

Fukui, K., and K. Kakeda, 1994. A critical assessment of karyotype analysis by imaging methods. Jap. J. Genet. 69: 537-544.

Fukui, K., Y. Kamisugi and F. Sakai, 1994. Physical mapping of 5S rDNA loci by direct-cloned biotinylated probes in barley chromosomes. Genome 37: 105-111.

Gale, M.D., and H. Rees, 1970. Genes controlling chiasma frequency in *Hordeum*. Heredity 25: 393-410.

Gecheff, K.I., 1989. Multiple reconstruction of barley karyotype resulting in complete cytological marking of the chromosome complement. Theor. Appl. Genet. 78: 683-688.
Gecheff, K.I., 1996. Production and identification of new structural chromosome mutations in barley (*Hordeum vulgare* L.). Theor. Appl. Genet. 92: 777-781.
Gecheff, K., T. Hvarleva, S. Georgiev, T. Wilkes and A. Karp, 1994. Cytological and molecular evidence of deletion of ribosomal RNA genes in chromosome 6 of barley (*Hordeum vulgare*). Genome 37: 419-425.
Georgiev, S., K. Gecheff, H. Nicoloff, G. Künzel and R. Rieger, 1985. Giemsa banding as a means of identification of reconstructed chromosomes in barley. Biol. Zbl. 104: 29-34.
Gerlach, W.L., and J.R. Bedbrook, 1979. Cloning and characterization of ribosomal RNA genes from wheat and barley. Nucleic Acids Res. 7: 1869-1885.
Gerlach, W.L., and T.A. Dyer, 1980. Sequence organization of the repeated units in the nucleus of wheat which contains 5S-rDNA genes. Nucleic Acids Res. 8: 4851-4865.
Graner, A., 1994. Igri × Franka map. on-line, http://wheat.pw.usda.gov/ggpages/maps.html#barley.
Gupta, P.K., G. Fedak, S.J. Molnar and R. Wheatcroft, 1989. Distribution of a *Secale cereale* DNA repeat sequence among 25 *Hordeum* species. Genome 32: 383-388.
Hagberg, A., 1986. Induced structural rearrangements. In: W. Horn, C.J. Jensen, W. Odenbach and O. Schieder (eds), Genetic Manipulation in Plant Breeding. Walter de Gruyter, Berlin, pp. 17-36.
Hagberg, A., 1995. Coordinator's report: Duplication of chromosome segments. Barley Genet. Newsl. 25: 114.
Hagberg, A., and G. Hagberg, 1968. Induced translocations in barley. Mutations in Plant Breeding II, Proc. Panel Vienna 1967, IAEA, Vienna, pp. 129-137.
Hagberg, A., and G. Hagberg, 1969. Use of induced structural rearrangements in plant breeding. Induced Mutations in Plants, Proc. Symp., Pullman 1969, IAEA and FAO, IAEA, Vienna, pp. 647-653.
Hagberg, A., and P. Hagberg, 1991a. Coordinator's report: Chromosome duplications. Stock list of lines to be included in the Nordic Gene Bank. Barley Genet. Newsl. 21: 130-136.
Hagberg, A., and P. Hagberg, 1991b. Production and analysis of chromosome duplications in barley. In: P. K. Gupta and T. Tsuchiya (eds), Chromosome Engineering in Plants: Genetics, Breeding, Evolution, Part A, Elsevier, Amsterdam-Oxford-New York-Tokyo, pp. 401-410.
Hagberg, G., L. Lehmann and P. Hagberg, 1975. Segmental interchanges in barley. I. Translocations involving chromosomes 5 and 6. Hereditas 80: 73-82.
Hagberg, A., L. Lehmann and P. Hagberg, 1978. Segmental interchanges in barley. II. Translocations involving chromosomes 6 and 7. Z. Pflanzenzücht. 81: 89-110.
Hang, A., and K. Satterfield, 1997. *In vitro* culture induced deficiency in barley. Cer. Res. Comm. 25: 21-26.
Heslop-Harrison, J.S. (Pat), A. Brandes, S. Taketa, T. Schmidt, A.V. Vershinin, E.G. Alkhimova, A. Kamm, R.L. Doudrick, T. Schwarzacher, A. Katsiotis, S. Kubis, A. Kumar, S.R. Pearce, A.J. Flavell and G.E. Harrison, 1997. The chromosomal distribution of Ty1-*copia* group retrotransposable elements in higher plants and their implications for genome evolution. Genetica 100: 197-204.
Holm, G., 1960. An inversion in barley. Hereditas 46: 274-278.
Islam, A.K.M.R., 1980. Identification of wheat-barley addition lines with N-banding of chromosomes. Chromosoma 76: 365-373.
Islam, A.K.M.R., 1983. Ditelosomic additions of barley chromosomes to wheat. In: S. Sakamoto (ed), Proc. 6th Int. Wheat Genet. Symp., Kyoto. Maruzen, Kyoto, pp. 233-238.
Islam, A.K.M.R., and K.W. Shepherd, 1990. Incorporation of barley chromosomes into wheat. In: Y.P.S. Bajaj (ed), Biotechnology in Agriculture and Forestry, Vol. 13. Wheat. Springer, Berlin, pp. 128-151.
Islam, A.K.M.R., and K.W. Shepherd, 2000. Isolation of a fertile wheat-barley addition line carrying the entire barley chromosome 1H. Euphytica 111: 145-149.
Jahan, B., A.A. Vahidy, and S.I. Ali, 1992. Chiasma frequency and chromosome associations in sixteen taxa of *Hordeum* species. Cytologia 57: 51-57.

Jensen, J., 1971. Mapping of 10 mutant genes for necrotic spotting in barley by means of translocations. In: Barley Genetics II. Proc. 2nd Int. Barley Genet. Symp., Pullman, WA, USA, pp. 213-219.

Jensen, J., and I. Linde-Laursen, 1992. Statistical evaluation of length measurements on barley chromosomes with a proposal for a new nomenclature for symbols and positions of cytological markers. Hereditas 117: 51-59.

Jørgensen, R.B., and B. Andersen, 1989. Karyotype analysis of regenerated plants from callus cultures of interspecific hybrids of cultivated barley (*Hordeum vulgare* L.). Theor. Appl. Genet. 77: 343-351.

Kakeda, K., and K. Fukui, 1994. Dynamic changes in the morphology of barley chromosomes during the mitotic metaphase stage. Jap. J. Genet. 69: 545-554.

Kakeda, K., K. Fukui and H. Yamagata, 1991. Heterochromatic differentiation in barley chromosomes revealed by C- and N-banding techniques. Theor. Appl. Genet. 81: 144-150.

Kakeda, K., and H. Yamagata, 1991. Localization of breakpoints in barley reciprocal translocations involving chromosome 4 by banding techniques. In: Barley Genetics VI. Proc. 6th Int. Barley Genet. Symp., Helsingborg, Sweden, pp. 280-282.

Kamisugi, Y., Y. Ikeda, M. Ohno, M. Minezawa and K. Fukui, 1992. *In situ* digestion of barley chromosomes with restriction endonucleases. Genome 35: 793-798.

Kankanpää, J., L. Mannonen and A.H. Schulman, 1996. The genome sizes of *Hordeum* species show considerable variation. Genome 39: 730-735.

Kasha, K.J., and C.R. Burnham, 1965a. The location of interchange breakpoints in barley. I. Linkage studies and map orientation. Can. J. Genet. Cytol. 7: 62-77.

Kasha, K.J., and C.R. Burnham, 1965b. The location of interchange breakpoints in barley. II. Chromosome pairing and the intercross method. Can. J. Genet. Cytol. 7: 620-632.

Kasha, K.J., and K.N. Kao, 1970. High frequency haploid production in barley (*Hordeum vulgare* L.). Nature 225: 874-876.

Kilian, A., and A. Kleinhofs, 1992. Cloning and mapping of telomere-associated sequences from *Hordeum vulgare* L. Mol. Gen. Genet. 235: 153-156.

Kjær, B., V. Haahr and J. Jensen, 1991. Associations between 23 quantitative traits and 10 genetic markers in a barley cross. Plant Breed. 106: 261-274.

Koba, T., S. Takumi and T. Shimada, 1997. Isolation, identification and characterization of disomic and translocated barley chromosome addition lines of common wheat. Euphytica 96: 289-296.

Konishi, T., and I. Linde-Laursen, 1988. Spontaneous chromosomal rearrangements in cultivated and wild barleys. Theor. Appl. Genet. 75: 237-243.

Kreft, I., 1969. Cytological studies on an inversion in barley. Hereditas 62: 14-24.

Künzel, G., 1982. Differences between genetic and physical centromere distances in the case of two genes for male sterility in barley. Theor. Appl. Genet. 64: 25-29.

Künzel, G., 1987. Giemsa banding of barley chromosomes in meiosis improves identification of chromosome arms involved in translocations. Barley Genet. Newsl. 17: 83-86.

Künzel, G., 1992. Coordinator's report: Translocations and balanced tertiary trisomics. Barley Genet. Newsl. 22: 80-102.

Künzel, G., 1993. Coordinator's report: Translocations and balanced tertiary trisomics. Barley Genet. Newsl. 23: 163-102.

Künzel, G., 1994. Coordinator's report: Translocations and balanced tertiary trisomics. Barley Genet. Newsl. 24: 153-155.

Künzel, G., K.I. Gecheff and I. Schubert, 2001. Intrachromosomal distribution patterns of radiation-induced interchange breakpoints in barley: First post-treatment mitosis versus viable offspring. Genome 44: 928-932.

Künzel, G., and L. Korzun, 1996. Physical mapping of cereal chromosomes, with special emphasis on barley. In: Barley Genetics VII. Proc. 7th Int. Barley Genet. Symp., Saskatchewan, Canada, pp. 197-206.

Künzel, G., L. Korzun and A. Meister, 2000. Cytologically integrated physical RFLP maps for the barley genome based on translocation breakpoints. Genetics 154: 397-412.

Künzel, G., and F. Marthe, 1991. Segmental interchanges in barley: State of analysis. In: Barley Genetics VI. Proc. 6th Int. Barley Genetics Symp., Helsingborg, Sweden, pp. 283-286.
Künzel, G., and H. Nicoloff, 1979. Further results on karyotype reconstruction in barley. Biol. Zentralbl. 98: 587-592.
Künzel, G., A. Sorokin and F. Marthe, 1995. Progress in relating genetic to physical distances of marker sequences on micro-isolated barley chromosomes. In: P.E. Brandham and M.D. Bennett (eds), Kew Chromosome Conf. IV, Royal Bot. Gardens, Kew, pp. 149-155.
Kurauchi, N., 1997. Detection of spontaneous reciprocal translocation lines in Tunisian barley. Jap. J. Tropic. Agric. 41: 264-267. (In Japanese with English summary).
Lapitan, N.L.V., S.E. Brown, W. Kennard, J.L. Stephens and D.L. Knudson, 1997. FISH physical mapping with barley BAC clones. Plant J. 11: 149-156.
Lehfer, H., W. Busch, R. Martin and R.G. Hermann, 1993. Localization of the B-hordein locus on barley chromosomes using fluorescence *in situ* hybridization. Chromosoma 102: 428-432.
Leitch, I.J., and J.S. Heslop-Harrison, 1992. Physical mapping of the 18S-5.8S-26S rRNA genes in barley by *in situ* hybridization. Genome 35: 1013-1018.
Leitch, I.J., and J.S. Heslop-Harrison, 1993. Physical mapping of four sites of 5S rDNA sequences and one site of the alpha-amylase-2 gene in barley (*Hordeum vulgare*). Genome 36: 517-523.
Liljeroth, E., P. Hagberg and A. Hagberg, 1994. Bacterial abundance on roots of a high-yielding duplication line and some related varieties. Hereditas 121: 197-198.
Linde-Laursen, I., 1975. Giemsa C-banding of the chromosomes of 'Emir' barley. Hereditas 81: 285-289.
Linde-Laursen, I., 1978a. Giemsa C-banding of barley chromosomes. I. Banding pattern polymorphism. Hereditas 88: 55-64.
Linde-Laursen, I., 1978b. Giemsa C-banding of barley chromosomes. II. Banding patterns of trisomics and telotrisomics. Hereditas 89: 37-41.
Linde-Laursen, I., 1979. Giemsa C-banding of barley chromosomes. III. Segregation and linkage of C-bands on chromosomes 3, 6 and 7. Hereditas 91: 73-77.
Linde-Laursen, I., 1981. Giemsa banding patterns of the chromosomes of cultivated and wild barleys. In: Barley Genetics IV. Proc. 4th Int. Barley Genet. Symp., Edinburgh, UK, pp. 786-795.
Linde-Laursen, I., 1982. Linkage map of the long arm of barley chromosome 3 using C-bands and marker genes. Heredity 49: 27-35.
Linde-Laursen, I., 1983. A pericentric inversion in chromosome 3 of a translocation line of barley. Barley Genet. Newsl. 13: 54-55.
Linde-Laursen, I., 1984a. Nucleolus organizer polymorphism in barley, *Hordeum vulgare* L. Hereditas 100: 33-43.
Linde-Laursen, I., 1984b. Giemsa C-banding of barley chromosomes. IV. Chromosomal constitution of autotetraploid barley. Hereditas 100: 219-223.
Linde-Laursen, I., 1988. Giemsa C-banding of barley chromosomes. V. Localization of breakpoints in 70 reciprocal translocations. Hereditas 108: 65-76.
Linde-Laursen, I., 1991a. Giemsa C-banded karyotypes of cultivated and wild barley. Barley Genet. Newsl. 20: 47-52.
Linde-Laursen, I., 1991b. Barley × rye hybrids eliminating accessory chromosomes of rye. Hereditas 115: 127-132.
Linde-Laursen, I., and R. von Bothmer, 1984. Somatic cell cytology of the chromosome-eliminating, intergeneric hybrid *Hordeum vulgare* × *Psathyrostachys fragilis*. Can. J. Genet. Cytol. 26: 436-444.
Linde-Laursen, I., and R. von Bothmer, 1988. Elimination and duplication of particular *Hordeum vulgare* chromosomes in aneuploid interspecific *Hordeum* hybrids. Theor. Appl. Genet. 76: 897-908.
Linde-Laursen, I., and R. von Bothmer, 1999. Orderly arrangement of the chromosomes within barley genomes of chromosome-eliminating *Hordeum lechleri* × barley hybrids. Genome 42: 225-236.
Linde-Laursen, I., R. von Bothmer and N. Jacobsen, 1995. Karyotype differentiation and evolution in the genus *Hordeum* (Poaceae). In: P.E. Brandham and M.D. Bennett (eds), Kew Chromosome Conference IV Roy. Bot. Gard., Kew, pp. 233-247.

Linde-Laursen, I., H. Doll and G. Nielsen, 1982. Giemsa C-banding patterns and some biochemical markers in a pedigree of European barley. Z. Pflanzenzücht. 88: 191-219.

Linde-Laursen, I., J.S. Heslop-Harrison, K.W. Shepherd and S. Taketa, 1997. The barley genome and its relationship with the wheat genomes. A survey with an internationally agreed recommendation for barley chromosome nomenclature. Hereditas 126: 1-16.

Linde-Laursen, I., and J. Jensen, 1984. Separate location of parental chromosomes in squashed metaphases of hybrids between *Hordeum vulgare* L. and four polyploid, alien species. Hereditas 100: 67-73.

Linde-Laursen, I., and J. Jensen, 1991. Genome and chromosome disposition at somatic metaphase in a *Hordeum* × *Psathyrostachys* hybrid. Heredity 66: 203-210.

Lysák, M.A., J. Cíhalíková, M. Kubaláková, H. Šimková, G. Künzel and J. Doležel, 1999. Flow karyotyping and sorting of mitotic chromosomes of barley (*Hordeum vulgare* L.). Chromosome Res. 7: 431-444.

Marthe, F., and G. Künzel, 1994. Localization of translocation breakpoints in somatic metaphase chromosomes of barley. Theor. Appl. Genet. 89: 240-248.

McIntyre, C.L., S. Pereira, L.B. Moran and R. Appels, 1990. New *Secale cereale* (rye) DNA derivatives for the detection of rye chromosome segments in wheat. Genome 33: 635-640.

Moh, C.C., and R.A. Nilan, 1954. "Short" chromosome - A mutant in barley induced by atomic bomb irradiation. Cytologia 19: 48-53.

Mukai, Y., and B.S. Gill, 1991. Detection of barley chromatin added to wheat by genomic *in situ* hybridization. Genome 34: 448-452.

Mukai, Y., Y. Nakahara and M. Yamamoto, 1993. Simultaneous discrimination of the three genomes in hexaploid wheat by multicolor fluorescence *in situ* hybridization. Genome 36: 489-494.

Muravenko, O.V., M.A. Nazarova, E.D. Badaeva and A.V. Zelenin, 1996. Intraspecific C-banding chromosome polymorphism of *Hordeum spontaneum*. In: A. Slinkard, G. Scoles and B. Rossnagel (eds), Proc. 5th Int. Oat Conf. and 7th Int. In: Barley Genetics Symp., Saskatoon 1996, Vol. 1. Univ. Extension Press, Saskatoon, pp. 363-365.

Nicoloff, H., M. Anastassova-Kristeva, R. Rieger and G. Künzel, 1979. 'Nucleolar dominance' as observed in barley translocation lines with specifically reconstructed SAT chromosomes. Theor. Appl. Genet. 55: 247-251.

Nicoloff, H., R. Rieger and G. Künzel, 1981. Effects of chromosome reconstruction on deletion clustering in *Hordeum vulgare*. Theor. Appl. Genet. 59: 225-229.

Nicoloff, H., R. Rieger, G. Künzel and A. Michaelis, 1975. Non-random intrachromosomal distribution of chromatid aberrations induced by alkylating agents in barley. Mutation Res. 6: 257-261.

Nilan, R.A., 1964. The cytology and genetics of barley 1951-1962. Monogr. Suppl. 3. Research Studies vol. 32, no. 1. Washington State Univ. Press, Pullman, 278 pp.

Nilan, R.A., J.B. Powell, B.V. Conger and C.E. Muir, 1968. Induction and utilization of inversions and mutations in barley. Mutations in Plant Breeding II, Proc. Panel Vienna 1967, IAEA, Vienna, pp. 193-203.

Nilsson, N.-O., and S. Pelger, 1991. The relationship between natural variation in chiasma frequencies and recombination frequencies in barley. Hereditas 115: 121-126.

Noda, K., and K.J. Kasha, 1978. A modified Giemsa C-banding technique for *Hordeum* species. Stain Technol. 53: 155-162.

Pedersen, C., 1995. The relationship between physical and genetic distances on the barley chromosomes studied by *in situ* hybridization. Ph.D. thesis. Royal Vet. and Agric. Univ., Copenhagen, 113 p. + 6 app.

Pedersen, C., H. Giese and I. Linde-Laursen, 1995. Towards an integration of the physical and the genetic chromosome maps of barley by *in situ* hybridization. Hereditas 123: 77-88.

Pedersen, C., and I. Linde-Laursen, 1994. Chromosomal locations of four minor rDNA loci and marker microsatellite sequence in barley. Chromosome Res. 2: 65-71.

Pedersen, C., and I. Linde-Laursen, 1995. The relationship between physical and genetic distances at the *Hor1* and *Hor2* loci of barley estimated by two-colour fluorescent *in situ* hybridization. Theor. Appl. Genet. 91: 941-946.

Pedersen, C., S.K. Rasmussen and I. Linde-Laursen, 1996. Genome and chromosome identification in cultivated barley and related species of the Triticeae (*Poaceae*) by *in situ* hybridization with the GAA-satellite sequence. Genome 39: 93-104.

Pedersen, C., J. Zimny, D. Becker, A. Jähne-Gärtner and H. Lörz, 1997. Localization of introduced genes on the chromosomes of transgenic barley, wheat and triticale by fluorescence *in situ* hybridization. Theor. Appl. Genet. 94: 749-757.

Persson, G., 1969a. An attempt to find suitable genetic markers for dense ear loci in barley I. Hereditas 62: 25-96.

Persson, G., 1969b. An attempt to find suitable genetic markers for dense ear loci in barley II. Hereditas 63: 1-28.

Pickering, R.A., 1992. Monosomic and double monosomic substitutions of *Hordeum bulbosum* L. chromosomes into *H. vulgare* L. Theor. Appl. Genet. 84: 466-472.

Pickering, R.A., A.M. Hill and R.G. Kynast, 1997. Characterization by RFLP analysis and genomic *in situ* hybridization of a recombinant and a monosomic substitution plant derived from *Hordeum vulgare* L. × *Hordeum bulbosum* L. cross. Genome 40: 195-200.

Pickering, R.A., S. Malyshev, G. Künzel, P.A. Johnson, V. Korzun, M. Menke and I. Schubert, 2000. Locating introgressions of *Hordeum bulbosum* chromatin within the *H. vulgare* genome. Theor. Appl. Genet. 100: 27-31.

Pickering, R.A., and P.W. Morgan, 1985. The influence of temperature on chromosome elimination during embryo development in crosses involving *Hordeum* spp., wheat (*Triticum aestivum* L.) and rye (*Secale cereale* L.). Theor. Appl. Genet. 70: 199-206.

Presting, G.G., L. Malysheva, J. Fuchs and I. Schubert, 1998. A *Ty3/gypsy* retrotransposon-like sequence localizes to the centromeric regions of cereal chromosomes. Plant J. 16: 721-728.

Ramage, R.T., 1960. Trisomics from interchange heterozygotes in barley. Agron. J. 52: 156-159.

Ramage, R.T., 1965. Balanced tertiary trisomics for use in hybrid seed production. Crop Sci. 5: 177-178.

Ramage, R.T., 1973. Reports from coordinators: Translocations and balanced tertiary trisomics. Barley Genet. Newsl. 3: 85-86.

Ramage, R.T., 1975. Coordinator's reports: Translocations and balanced tertiary trisomics. Barley Genet. Newsl. 5: 76-83.

Ramage, R.T., 1985. Cytogenetics. In: D.C. Rasmusson (ed), Barley. Agronomy Monogr. no. 26. ASA-CSSA-SSSA, Madison, Wisconsin, pp. 127-154.

Ramage, R.T., 1991. Chromosome manipulations in barley breeding. In: P.K. Gupta and T. Tsuchiya (eds), Chromosome Engineering in Plants: Genetics, Breeding, Evolution. Part A. Elsevier, Amsterdam, pp. 385-400.

Rieger, R., H. Nicoloff and M. Anastassova-Kristeva, 1979. 'Nucleolar dominance' in interspecific hybrids and translocation lines - a review. Biol. Zbl. 98: 385-398.

Röder, M.S., N.L.V. Lapitan, M.E. Sorrells and S.D. Tanksley, 1993. Genetic and physical mapping of barley chromosomes. Mol. Gen. Genet. 238: 294-303.

Rommel, M., 1961. Aneuploidy, seed set and sterility in artificially induced autotetraploid *Hordeum vulgare* L. Can. J. Genet. Cytol. 3: 272-282.

Säll, T., J. Flink and B.O. Bengtsson, 1990. Genetic control of recombination in barley. I. Variation in recombination frequency measured with inversion heterozygotes. Hereditas 112: 157-170.

Schubert, I., and G. Künzel, 1990. Position-dependent NOR activity in barley. Chromosoma 99: 352-359.

Schubert, I., G. Künzel, H. Bretschneider, R. Rieger and H. Nicoloff, 1980. Sister chromatid exchanges in barley. Theor. Appl. Genet. 56: 1-4.

Schubert, I., F. Shi, J. Fuchs and T.R. Endo, 1998. An efficient screening for terminal deletions and translocations of barley chromosomes added to common wheat. Plant J. 14: 489-495.

Schwarzacher, T., K. Anamthawat-Jónsson, G.E. Harrison, A.K.M.R. Islam, J.Z. Jia, I.P. King, A.R. Leitch, T.R. Miller, S.M. Reader, W.J. Rogers, M. Shi and J.S. Heslop-Harrison, 1992. Genomic *in situ* hybridization to identify alien chromosomes and chromosome segments in wheat. Theor. Appl. Genet. 84: 778-786.

Serizawa, N., S. Nasuda, F. Shi, T.R. Endo, S. Prodanovic, I. Schubert and G. Künzel, 2001. Deletion-based physical mapping of barley chromosome 7H. Theor. Appl. Genet. 103: 827-834.

Shi, F., and T.R. Endo, 1999. Genetic induction of structural changes in barley chromosomes added to common wheat by a gametocidal chromosome derived from *Aegilops cylindrica*. Genes Genet. Syst. 74: 49-54.

Shi, F., and T.R. Endo, 2000. Genetic induction of chromosomal rearrangements in barley chromosome 7H added to common wheat. Chromosoma 109: 358-363.

Singh, R.J., and T. Tsuchiya, 1982a. An improved Giemsa N-banding technique for the identification of barley chromosomes. J. Hered. 73: 227-229.

Singh, R.J., and T. Tsuchiya, 1982b. Identification and designation of telocentric chromosomes in barley by means of Giemsa N-banding technique. Theor. Appl. Genet. 64: 13-24.

Smith, L., 1941. An inversion, a reciprocal translocation, trisomics, and tetraploids in barley. J. Agr. Res. 63: 741-750.

Smith, L., 1951. Cytology and genetics of barley. Bot. Rev. 17: 1-51, 133-202 and 285-355.

Song, Y.C., L.H. Liu, Y. Ding, X.B. Tian, Q. Yao, L. Meng, C.R. He and M.S. Xu, 1994. Comparisons of G-banding patterns in six species of the *Poaceae*. Hereditas 121: 31-38.

Sorokin, A., F. Marthe, A. Houben, U. Pich, A. Graner and G. Künzel, 1994. Polymerase chain reaction mediated localization of RFLP clones to microisolated translocation chromosomes of barley. Genome 37: 550-555.

Stoinova, J., 1994. Characteristics of meiosis in some winter barley varieties. Cytologia 59: 423-426.

Suoniemi, A., K. Anamthawat-Jónsson, T. Arna and A.H. Schulman, 1996. Retrotransposon *BARE-1* is a major, dispersed component of the barley (*Hordeum vulgare* L.) genome. Plant Mol. Biol. 31: 1321-1329.

Svitashev, S., T. Bryngelsson, A. Vershinin, C. Pedersen, T. Säll and R. von Bothmer, 1994. Phylogenetic analysis of the genus *Hordeum* using repetitive DNA sequences. Theor. Appl. Genet. 89: 801-810.

Taketa, S., H. Ando, K. Takeda, G.E. Harrison and J.S. Heslop-Harrison, 2000. The distribution, organization and evolution of two abundant and widespread repetitive DNA sequences in the genus *Hordeum*. Theor. Appl. Genet. 100: 169-176.

Taketa, S., M. Choda, R. Ohashi, M. Ichii and K. Takeda, 2002. Molecular and physical mapping of a barley gene on chromosome arm 1HL that causes sterility in hybrids with wheat. Genome 45: 617-625.

Taketa, S., G.E. Harrison and J.S. Heslop-Harrison, 1999. Comparative physical mapping of the 5S and 18S-25S rDNA in nine wild *Hordeum* species and cytotypes. Theor. Appl. Genet. 98: 1-9.

Taketa, S., J. Kato and K. Takeda, 1995. High crossability of wild barley (*Hordeum spontaneum* C. Koch) with bread wheat and the differential elimination of barley chromosomes in the hybrids. Theor. Appl. Genet. 91: 1203-1209.

Taketa, S., and K. Takeda, 1997. Expression of dominant marker genes of barley in wheat-barley hybrids. Genes Genet. Syst. 72: 101-106.

Taketa, S., and K. Takeda, 2001. Production and characterization of a complete set of wheat-wild barley (*Hordeum vulgare* ssp. *spontaneum*) chromosome addition lines. Breed. Sci. 51: 199-206.

Tjio, J.H., and A. Hagberg, 1951. Cytological studies on some x-ray mutants of barley. Anal. Estaciōn Expl. Aula Dei 2: 149-167.

Tsuchiya, T., 1960. Cytogenetic studies of trisomics in barley. Jap. J. Bot. 17: 177-213.

Tsuchiya, T., 1991. Chromosome mapping by means of aneuploid analysis in barley. In: P.K. Gupta and T. Tsuchiya (eds), Chromosome Engineering in Plants: Genetics, Breeding, Evolution. Part A. Elsevier, Amsterdam, pp. 361-384.

Tsujimoto, H., Y. Mukai, K. Akagawa, K. Nagaki, J. Fujigaki, M. Yamamoto and T. Sasakuma, 1997. Identification of individual barley chromosomes based on repetitive sequences: Conservative distribution of Afa-family repetitive sequences on the chromosomes of barley and wheat. Genes Genet. Syst. 72: 303-309.

Tuleen, N., 1971. Translocation-gene linkages from F_2 seedlings in barley. In: Barley Genetics II. Proc. 2nd Int. Barley Genet. Symp., Pullman, WA, USA, pp. 208-212.

Tuleen, N., 1973. Karyotype analysis of multiple translocation stocks of barley. Can. J. Genet. Cytol. 15: 267-273.

Uozu, S., T. Nakazaki and T. Tanisaka, 1997. Characterization of the late replication banding patterns in barley chromosomes. Genes Genet. Syst. 72: 125-130.

Vershinin, A., E.G. Alkhimova and J.S. Heslop-Harrison, 1996. Molecular diversification of tandemly organized DNA sequences and heterochromatic chromosome regions in some Triticeae species. Chromosome Res. 4: 515-525.

Vershinin, A.V., E.A. Salina, V.V. Solovyov and L.L. Timofeyeva, 1990. Genomic organization, evolution, and structural peculiarities of highly repetitive DNA of *Hordeum vulgare*. Genome 33: 441-449.

Vershinin, A., S. Svitashev, P.O. Gummesson, B. Salomon, R. von Bothmer and T. Bryngelsson, 1994. Characterization of a family of tandemly repeated DNA sequences in Triticeae. Theor. Appl. Genet. 89: 217-225.

Vosa, C.G., 1976. Chromosome banding patterns in cultivated and wild barleys (*Hordeum* spp.). Heredity 37: 395-403.

Xu, J., and K.J. Kasha, 1992. Identification of a barley chromosomal interchange using N-banding and *in situ* hybridization techniques. Genome 35: 392-397.

Xu, J., and J.W. Snape, 1988. The cytology of hybrids between *Hordeum vulgare* and *H. bulbosum* revisited. Genome 30: 486-494.

Diversity in Barley *(Hordeum Vulgare)*
Roland von Bothmer et al (Editors).

Chapter 7

Molecular diversity of the barley genome

Andreas Graner[a], Åsmund Bjørnstad[b], Takeo Konishi[c] and Frank Ordon[d]

[a]Institute of Plant Genetics and Crop Plant Research (IPK), D-06466 Gatersleben, Germany
[b]Department of Horticulture and Crop Sciences, Agricultural University, N-1432 Ås, Norway
[c]294 Okada, Mabi-cho, Kibi-gun, Okayama 710-1311, Japan
[d]Institute of Crop Science and Plant Breeding I, Justus-Liebig-University, D-35390 Giessen, Germany

Introduction

Cultivated and wild barley grow in a wide range of natural and man-made habitats. Therefore, the limits of barley cultivation on our globe reach to higher altitudes and latitudes and deeper into semi-arid areas than those of most other crop species.

Evidently, the underlying scheme of this morphological, physiological and functional adaptability is a large genetic diversity, which facilitated colonisation of primary and secondary habitats throughout the Fertile Crescent and in a range of most diverse environments. With the advent of molecular marker technologies and the first marker-maps, tools became available to study the diversity present on the seven barley chromosomes of cultivars, landraces and wild barley, *Hordeum vulgare* ssp. *spontaneum*, at a better resolution than was previously possible at the trait level.

The use of molecular marker techniques will further complete our understanding of the genetic basis of environmental adaptation, the process of domestication and the subsequent spread of cultivated barley. Their systematic deployment will also provide information on genetic erosion and the formation of distinct genepools. In addition, detailed knowledge of the molecular diversity of barley germplasm will become an important piece of information to be used in barley breeding programmes all over the world. The present paper will summarise past and recent achievements for three different marker systems: isoenzymes, storage proteins and DNA markers and provide an outlook on future developments in this area of research.

Isozymes

Isozyme loci and alleles

At least 60 isozyme loci have been detected in barley, with an average of 3.3 alleles per locus (Brown *et al.*, 1978a, 1989; Brown, 1983; Nielsen and Johansen, 1986). The most polymorphic is *Est2* containing 15 alleles in *Hordeum vulgare* ssp. *spontaneum* from Israel (Brown *et al.*, 1978a), of which only five alleles were frequently found in cultivated barley (Hvid and Nielsen, 1977).

Graner, A., Å. Bjørnstad, T. Konishi and F. Ordon, 2003. Molecular diversity of the barley genome. In: R. von Bothmer, Th. van Hintum, H. Knüpffer and K. Sato (eds), Diversity in Barley (*Hordeum vulgare*), pp. 121-141. Elsevier Science B.V., Amsterdam, The Netherlands.

Table 7.1. Chromosomal location of the isozyme and hordein loci.

Chromosome	Loci	Mean number of alleles/locus	Major loci
1H	8	2.1	*Gpi1, Mdh1, Pgd2, Ndh3, Hor1, Hor2, Hor3, Hor4, Hor5*
2H	4	2.3	*Est7, Gpd1, Idh2, Ndh4*
3H	10	4.7	*Est1, Est2, Est4, Mdh2*
4H	7	3.7	*Acp2, Adh1, Cat1, Pgm1*
5H	3	2.3	*Gdh1, Sdh1, Est9*
6H	8	4.3	*Aat2, Aco1, Acp3, α-Amy1*
7H	7	3.8	*α-Amy2, Est3, Est5, Prx4*
Unknown	13	1.8	*Acp1, Cat1, Pgd1, Prx1*
Total	60	3.3	

More than 75% of the loci have been chromosomally located (Table 7.1). Three commonly studied esterase loci, *Est1*, *Est2* and *Est4* are tightly linked as a complex locus on chromosome 3H (Konishi and Matsuura, 1987a), and are arranged in the order *Est2–Est1–Est4* with a recombination value of 0.59% between *Est2* and *Est4* (Kahler and Allard, 1970).

Allelic variation has been examined using natural and experimental populations and accessions collected from different locations. Brown and Munday (1982) estimated genetic diversity in barley landraces sampled at 12 locations in Iran. Of 25 loci examined, seven (*Acp3, Ndh1, Idh2, Est4, Est1, Est2* and *Aco1*) were markedly variable, four above average, eight weakly polymorphic, and six invariant. Marked differences among populations were found at some loci (e.g., *Cat*, the *Est2-1-4* complex, *Est6*, *Ndh1* and *Aco1*). Liu *et al.* (1999) estimated the genetic diversity indices (H) at six loci in the East Asian Barley Core Collection, where the *Aco1* locus was most polymorphic. Using the same accessions, the genetic diversity indices at different loci were reestimated from the data of the World Barley Collection of Okayama University (Konishi, 1995). The highest index was detected at the *Est1* locus (0.522) followed by *Est2* and *Est4* (Table 7.2).

Table 7.2. Genetic diversity index (H)[‡] in East Asian landraces of the Barley Core Collection at ten isozyme loci.

Country	*Est1* (4)[#]	*Est2* (4)	*Est4* (3)	*Aat3* (2)	*Pgd1* (2)	*Pgd2* (2)	*Gpi1** (2)	*Aco1** (3)	*Aco2** (2)	*Ndh2** (1)	No. of landraces
Japan	.474	.595	.320	.000	.000	.113	.000	.461	.039	.000	50
Korea	.495	.585	.113	.000	.000	.000	.000	.497	.000	.000	50
China	.460	.647	.640	.000	.000	.000	.000	.442	.047	.000	85
Bhutan	.498	.498	.604	.000	---	---	.000	.320	.000	.000	15
Nepal	.615	.295	.417	.423	.000	.000	.000	.495	.000	.000	50
India	.591	.211	.463	.211	.039	.000	.040	.420	.000	.000	50
Average	.522	.472	.426	.106	.008	.023	.007	.439	.014	.000	

[#]) Number of alleles detected in the landraces; *) After Liu *et al.* (1999); [‡]) According to Nei (1973).

Geographical distribution of esterase genotypes

Numerous studies on geographical distribution of alleles at individual loci have been conducted. However, there are only a few surveys using the world barley collection. Kahler and Allard (1981) examined worldwide patterns of genetic variation among *Est1*, *Est2*, *Est4* and *Est5* using the USDA World Barley Collection. They demonstrated that patterns of geographical distribution of alleles were non-random over both small and large geographical areas, including differences on a continental scale. As many alleles were differentiated at each locus, one allele (usually the most frequent) was designated Allele *1* and all other alleles were combined into a "synthetic" class designated Allele *2*. Among the expected 16 combinations of alleles *1* and *2* at the four loci, eight were frequent in North America, Europe, Central Asia (mainly Iran, Afghanistan and India) and East Asia (China, Korea and Japan). Four gametic types (*1221*, *1222*, *2112*, and *2111*) were conspicuously and consistently in excess in all of the four regions, the first two in East Asia, and the latter two in Central Asia. A cluster analysis revealed that the four regions could be classified into two groups, Oriental and Occidental.

Konishi and co-workers made further surveys of esterase genotypes at the *Est1*, *Est2* and *Est4* loci, using mainly landraces collected from different regions of the world and preserved at the Barley Germplasm Centre of Okayama University (Konishi, 1988, 1995). As shown in Table 7.3, esterase genotypes were classified into ten major types (tentatively called A to J) and others. Their distribution was congruent with the geographical location of the Vavilovian centres of diversity: the highest diversity in Southwest Asia where cultivated barley had its origin, decreasing elsewhere. It may be seen that in Nepal, India and Pakistan, four types (D to G) could be found frequently. East Asia (all six-rowed barley) appeared clearly different (types A to D, type I probably introduced from Occidental regions; Takahashi, 1955 and later publications). Furthermore, four other types (F to I) were often found in Turkey and Europe and only two types (F and G) were predominantly detected in Ethiopia. More recently, Cross (1994) detected a geographical trend only at the *Est4* locus in spring barleys, but not at the *Est2* and *Est5* loci. He considered that the inconsistent distribution reflected effects of extensive migration of barley varieties.

Compared with cultivated barley, *H. vulgare* ssp. *spontaneum* shows significantly larger variation in esterase genotypes and 13 more genotypes were detected in a sample comprising 180 accessions (Table 7.3). This may result from mutation as well as rare allelic recombinations by repeated outcrosses (Brown *et al.*, 1978b). Thus, the mutated *Sp* allele is specific to wild barley, while in Southwest Asian and Ethiopian accessions of cultivated barley a mutation to the null (*ne*) allele has occurred at *Est2* and *Est4*. Kahler *et al.* (1984) estimated that spontaneous mutation rates for isozyme loci in barley were too low to have much effect on the short-term evolutionary

Table 7.3. Geographical distribution of esterase genotypes in the Old World.

Genotype	A	B	C	D	E	F	G	H	I	J	*	No. of acc.
Est1	***Af***	***Ca***	***Ca***	***Ca***	***Pr***	***Al***	***Pr***	***Ca***	***Ca***	***Ca***		
Est2	*ne*	*ne*	***Un***	***Un***	***Fr***	***Fr***	***Fr***	***Fr***	***Fr***	***Dr***		
Est4	***Su***	***Nz***	***Su***	***Nz***	***At***	***At***	***Su***	***At***	***Su***	***Nz***		
Japan (6-row)	⦿	○	⦿	○	⚹	⚹	⚹	–	◆	–	⚹	563
Korea (6-row)	⦿	⚹	⦿	⚹	⚹	⚹	⚹	⚹	◆	–	⚹	462
China (6-row)	○	⦿	○	◆	◆	⚹	⚹	⚹	○	–	⚹	183
Nepal (6-row)	–	–	⚹	◆	○	⦿	◆	⚹	⚹	–	⚹	364
India-Pakistan	⚹	–	–	⦿	⦿	⦿	◆	⚹	⚹	⚹	⚹	333
SW Asia	◆	⚹	◆	⚹	○	⦿	○	⚹	○	–	○	205
Turkey	–	–	–	◆	⚹	○	⦿	○	○	⚹	⚹	135
Europe	–	⚹	⚹	⚹	⚹	◆	⦿	○	⦿	⚹	◆	439
Ethiopia	–	–	–	–	⚹	⦿	⦿	⚹	–	–	⚹	231
ssp. *spontaneum*	⚹	⚹	⚹	⚹	–	⚹	–	⦿	⚹	⚹	⦿	180

* Other genotypes; ⦿: more than 20%; ○: more than 10%; ◆: more than 5%; ⚹: less than 4.9%; –: none.

dynamics of barley populations, but spontaneous mutations have been important in creating variability in the genepool of wild barley.

Isozyme diversity in Ethiopian barley

On the basis of morphological diversity Vavilov (1926) distinguished two principal centres of the diversity of cultivated barley: one in the mountainous districts of Ethiopia, especially rich in forms of hulled barley, and the other in Southeast Asia including China, Japan and the regions adjoining Tibet, where varieties of naked barley, short-awned, awnless, or with hooded ears, are concentrated. However, Schiemann (1951) asserted that Ethiopia is only a secondary centre of origin of barley. Therefore, it is of great interest to compare the diversity of this region by means of neutral markers such as isozymes.

Bekele (1983) surveyed allelic variations at the *Est1*, *Est2*, *Est4, Est5* and *Acp1* loci in 158 landraces in Ethiopia. The diversity detected was very high: five alleles at *Est1*, four at *Est2*, six at *Est4,* six at *Est5* and six at *Acp1.* The null allele at *Est2* was abundant in two regions. Averages of genetic diversity (H) at the four loci ranged from 0.10 to 0.59 among regions. Around 45% of the total genetic diversity was due to genetic variation within populations and genetic distance did not follow a geographical pattern. Demissie and Bjørnstad (1997) surveyed 15 isozyme and two hordein loci in 51 populations collected from ten regions. They observed lower diversity than in Bekele's investigations, in agreement with the picture of Ethiopian barleys presented in Table 7.3. Correlations between morphological, isozyme and RFLP diversity were low in this study (Demissie *et al.*, 1998). Ecogeographical associations were absent, as in Bekele's investigations.

Isozyme diversity in Himalayan barley

According to Vavilov (1926), another gene diversity centre of barley is Southeast Asia including Tibet, Bhutan and Nepal. Especially, Tibetan barley in particular has received much attention, because Åberg (1938) discovered a new six-rowed barley with brittle rachis, named *H. vulgare* ssp. *agriocrithon*, in a wheat seed sample collected by a Swedish team on a botanical exploration in Eastern Tibet. Åberg (1940) also classified the Tibetan barley collections into various types of naked barley including four new varieties and 11 new strains and concluded that a large diversity of cultivated barley had been preserved in Tibet.

Konishi and Matsuura (1991) surveyed isozyme variations at seven loci among 650 Himalayan landraces collected from Tibet, Nepal, North India and North Pakistan. Large diversities were detected at the *Est1*, *Est2*, *Est4* and *Aat3* loci, little at *Pgd1* and *Pgd2* and none at *Aat2*. The allelic combinations were non-randomly distributed. The covered barleys predominant in southern regions of the Himalayas were characterised principally by the esterase *Al-Fr-At* (F) genotype, combined with the rare *Mo* allele at the *Aat3*, while the naked barleys found mainly in northern regions (including Tibet and Ladakh) possessed the genotypes *Ca-Un-Nz* (D) or *Pr-Fr-At* (E), together with the common *Eg* allele at the *Aat3*.

The Tibetan accessions have been repeatedly studied. Dai and Zhang (1989) investigated 463 accessions of Tibetan cultivated barley. Extensive variation was detected at *Est1*, *Est2*, *Est4* and *Est5*, whereas *Acp1* and *Aat3* were monomorphic. Approximately 95% of the total variation was recognised within regions, only 4% among regions. The multilocus genotypes *Pr-Fr-At* (type E in Table 7.3) and *Pr-Fr-Su* (G) were most frequent. Konishi (1995) also examined esterase genotypes of 16 Tibetan barley accessions classified by Åberg (1940). He found that genotypes of 12 accessions were *Pr-Fr-At* (E), two were *Ca-Un-Nz* (D) and the other two were *Pr-Fr-Su* (G), which indicates a lower diversity in esterases than expected from morphological diversity. In a further study esterase genotypes in 17 accessions of *H. vulgare* ssp. *agriocrithon*, collected from Tibet by different expedition teams were compared with cultivated accessions from Tibet, North India, and North Pakistan (Konishi, unpublished results). It is noticeable that the genotypes of both groups were completely identical (types D, E and G in Table 7.3), without any genotypes of *H. vulgare* ssp. *spontaneum*, even though ssp. *agriocrithon* of Tibet was assumed to originate from spontaneous hybridisation between Tibetan six-rowed cultivated barley and *H. vulgare* ssp. *spontaneum*. Zhang *et al.* (1994) conducted a comparative analysis of genetic polymorphism at the same esterase loci in wild and cultivated barley from Tibet. Here, genetic diversity decreased in the order *H. vulgare* ssp. *spontaneum*, ssp. *agriocrithon* to cultivated barley.

Isozyme variation and ecogeographical associations in H. vulgare *ssp.* spontaneum

The evolutionary significance of genetic diversity in wild relatives of barley has been emphasised by several authors. Nevo, Brown and co-workers have extensively investigated isozyme variation in *H. vulgare* ssp. *spontaneum* collected from Israel. Using 28 populations of ssp. *spontaneum* collected from seven regions in Israel, Brown *et al.* (1978a) surveyed the isozyme variation at 28 isozyme loci of which 25 were polymorphic, and six loci (*Est2*, *Acp2*, *Est4*, *Acp3*, *Ndh1* and *Est5*) were very diverse. The average diversity was apportioned into 17% between regions, 32% between populations within regions, and 51% within populations. Nevo *et al.* (1979) found that the genetic diversity and polymorphism were associated with annual rainfall at the 28 locations in Israel (R^2=0.53*** and 0.74***, respectively). The associations were maximal for the rainfall range of 200-600 mm, while they were down to zero at two locations with rainfall exceeding 1000 mm. Chalmers *et al.* (1992) also reported that distribution of grain isozyme alleles detected in ssp. *spontaneum* from Israel was non-random and correlated with a range of ecogeographical factors. Kahler and Allard (1981) compared the USDA World Barley Collection and ssp. *spontaneum* from Israel and Turkey in isozyme variations at four esterase loci. Two alleles were specific to domesticated barley, whereas three alleles were specific to ssp. *spontaneum*, again confirming the general picture of less diversity in the domesticated genepool, but also opening the possibility of new alleles having arisen since domestication.

Isozyme variation and ecogeographical associations in composite cross barley populations

Ecogeographical associations have also been observed in composite cross populations, synthesised from many cultivars representing all of the major barley growing regions of the world (Harlan and Martini, 1929). Allard and his colleagues (summarised in Allard, 1988, 1992) extensively investigated changes in allelic frequencies of esterase isozymes (as well as other traits, like scald resistance). Detailed analyses showed highly significant changes in allelic frequencies at numerous marker loci, which were significantly correlated with environment (particularly variation in amount of rainfall). In the Composite Cross II (CCII) population grown at Davis, California, only one gametic type *2112*, increased steadily in a frequency to 95% in F_{53}. On the other hand, grown at Bozeman, Montana, for 19 generations from F_{27} to F_{46}, *2112* declined rapidly, whereas gametic types *1212*, *1112* and *2121* increased in frequency. In this connection, all commercial cultivars of the Central Valley of California carry gametic type *2112*, while *2121* is the most common gametic type of North Plain and Prairie Province cultivars examined.

Another case of unintentional shift has been detected in Japanese two-rowed barley cultivars, where one esterase genotype predominant until the 1950s (*Ca-Dr-Nz* derived from 'Golden Melon') was almost completely replaced by *Pr-Fr-Su* from 'Prior' (Konishi and Matsuura, 1987b). In this case, selection for an adaptive QTL allele may be hypothesised.

Hordeins

The loci

Hordeins – the storage proteins of barley – comprise a group of proteins similar in biological function, but differentiated to various degrees as to structure and chemical properties. They may be differentiated by electrophoresis (mostly by SDS-PAGE, but also IEF or by starch gels) or by HPLC. They have been classified based on differences in electrophoretic mobility, their amino acid or nucleotide sequences (for extensive review, see Kreis and Shewry, 1992). Five loci, termed *Hor1* to *Hor5*, have been identified. (The locus terminology may be confused with that used for designating their electrophoretically separated gene products, designated as Hordein C, B, D, G and F/γ, respectively. A corresponding locus designation – *HorF, HorG*, etc. – has been used mainly in Russian papers). All the loci are located on the short arm of chromosome 1H, except *Hor3*, which is located on the long arm. Moreover, *Hor1* and *Hor4* are closely linked, as are *Hor2* and *Hor5*. The average linkage distance between these two clusters is *ca.* 12cM. Both *Hor1* and *Hor2* are complex loci (multigene families), derived by duplication and divergence from a common ancestral gene (Shewry *et al.*, 1981; Faulks *et al.*, 1981). Moreover, the multiple allelic mildew resistance locus *Ml-a* is located between *Hor1* and *Hor2,* while the resistance loci *Ml-k* and *Yr4* are closely linked to *Hor1*, towards the centromere (Shewry *et al.*, 1980). These features of hordeins – complex storage protein loci closely linked to resistance genes – have striking homoeologies in oat and maize (Yu *et al.*, 1996).

As to the locus differentiation, Shewry and Burgess (1991) showed that although the *Hor4* locus is located proximally to *Hor1*, it cross-hybridises with *Hor2*, which it resembles most closely in terms of protein products. The authors suggested that it has arisen from a translocation of part of *Hor2* into the *Hor1* region in 'Elgina' or one of its ancestors. The hordein copy numbers vary strongly between loci, with 20-30 gene copies at each of *Hor1* and *Hor2*, while *Hor4* and *Hor5* have low copy numbers. In contrast, *Hor3* has one single copy and is structurally more similar to the HMW glutenin subunits than to the other loci. The copy number is strongly correlated with the mRNA and protein quantities present in the mature seed, and gene expression is transcriptionally controlled (Kreis and Shewry, 1992).

Hordein alleles

The loci *Hor1* and *Hor2* have very high numbers of electrophoretically distinguishable alleles. In contrast, only two alleles have been recorded at each of *Hor3*, *Hor4* and *Hor5*, respectively (Netsvetaev and Sozinov, 1982, 1984; Shewry *et al.*, 1983). At the *Hor4,* one rare allele has been recorded from Russian barleys related to 'Elgina'.

The high numbers of alleles at loci with the highest copy numbers, as well as the probable homoeologous nature of the various loci, point to a common diversity-generating mechanism, most probably unequal crossing-over events. Shewry *et al.* (1990) demonstrated an intralocus recombination event between subfamilies of genes at the *Hor2* locus in 281 F_2 seeds, both at the protein and DNA levels. The number of gene copies exchanged was not determined.

Hordein diversity

Due to their diversity as well as their quantitatively greater expression and hence ease of detection, *Hor1* and *Hor2* have almost exclusively been 'the' hordein loci used in diversity studies. A notable exception is Cross (1994), who also studied *Hor3* in his study of diversity in 1,118 accessions from the world spring barley collection. A widely used protocol is the half-seed procedure described by Doll and Andersen (1981). Most studies have recorded (probable) alleles

Hor1 and *Hor2*, including heterozygotes. Asfaw (1989), however, scored individual gel bands in a binary manner. The number of alleles recorded is probably impossible to determine exactly from the various studies. However, some conclusions may be drawn.

No total 'count' of alleles at Hor1 and *Hor2 has ever been done.* In the most comprehensive study of diversity to date, based on 1,118 accessions from the world spring barley collection, Cross (1994) used the recommended allelic designations and reference cultivars described by Johansen and Shewry (1986), which include 17 and 25 alleles at *Hor1* and *Hor2,* respectively. Such a cautious approach is clearly commendable in diversity studies, but may have contributed to the discrepancy between this and other studies. For example, based on 242 Ethiopian accessions, Cross detected only two alleles at *Hor1*, whereas Bjørnstad *et al.* (1997) found 10 alleles and Asfaw (1989) was able to distinguish between 68 different two-locus combinations based on a similar sample size. Although the sampling within the country may have been much broader in the latter studies, such discrepancies are hard to explain. Given the diversity recorded from the limited data available from ssp. *spontaneum*, the *Hor1* and *Hor2* still rank among the most polymorphic loci ever recorded in plants, possibly only surpassed by certain SSRs (Saghai-Maroof *et al.*, 1994).

Diversity in hordeins is greater than in isozymes. In the study by Doll and Brown (1979), the diversity in ssp. *spontaneum* was of an order of magnitude higher than the average isozyme diversity. This seems not to be so pronounced in ssp. *vulgare.* In his study of diversity in 1,118 accessions from the world spring barley collection, Cross (1994) found diversity estimates at *Est2* (0.59) close to that of *Hor1* (0.63), but lower than *Hor2* (0.85). In a comparison of Ethiopian and Iranian landraces, Bjørnstad *et al.* (1997) found that *Acp3* had a diversity comparable to *Hor1* and *Hor2*, although hordein diversity was clearly higher in Iran. In a set of 66 cultivars from the 'Danish list of cultivars, 1983/84' the corresponding figures were 13 and 15 alleles at *Hor1* and *Hor2*, respectively, while no other locus had more than four alleles (Nielsen and Johansen, 1986). In contrast to this, Cross (1994) found that the *Hor3* diversity was limited, with one allele fixed or nearly fixed in most regions, except India, Nepal and Ethiopia.

Diversity in ssp. spontaneum *is greater than in ssp.* vulgare. Using starch gels, Doll and Brown (1979) detected 20 alleles (or, more precisely, phenotypes) at *Hor1* and 17 at *Hor2* in three populations of ssp. *spontaneum*, each with 32 analysed individuals. The populations differed strongly in diversity and the diversity estimate was 0.84 in the most polymorphic population, from Tabigha in Israel. In contrast, from 60 individuals from Composite Cross XXI, nine and 11 alleles were detected at *Hor1* and *Hor2*, respectively, still with a diversity estimate of 0.80. However, when Nevo *et al.* (1983) increased the sample size from Tabigha to 123, the number of detected alleles increased from 11 to 15 at *Hor1*, and from nine to 16 at *Hor2*. They also found strong correlations between hordein phenotypes and ecogeographical variables at the different sites, possibly due to natural selection for the closely linked resistance loci.

Diversity in landraces is not always greater than in modern cultivars of ssp. vulgare. In a sample of 184 European cultivars, Shewry *et al.* (1979) detected fewer alleles at both loci in landraces than in modern cultivars. In the most 'intact' landrace, 'Nuremberg', they only found three – otherwise common – alleles. The authors attributed this finding to be due to the linkage with mildew resistance and an effect of widespread introgression of resistance from exotic sources.

DNA Markers

Marker systems to study DNA diversity in barley – a brief overview

With the emergence of DNA marker technologies, tools became available to analyse genetic diversity at a refined level. From a technical point of view, DNA markers can be categorised into hybridisation-based RFLP (Restriction Fragment Length Polymorphism) markers and Polymerase Chain Reaction (PCR)-based marker systems. Comprehensive reviews on basic marker technologies have been provided by Karp and Edwards (1997), Karp *et al.* (1998), and Gupta and Varshney (2000). Besides these generic methods, derived marker techniques have been developed for barley based on the high copy number of retrotransposons especially BARE-1, i.e., Sequence Specific Amplification Polymorphisms (S-SAP, Waugh *et al.*, 1997; Ramsay *et al.*, 1999) as well as IRAPs (Inter-Retrotransposon Amplified Polymorphisms) and REMAPs (Retrotransposon-Microsatellite Amplified Polymorphisms; Kalendar *et al.*, 1999).

Historically, the analysis of RFLPs was the first technique to study genetic diversity at the DNA-level in barley (Saghai-Maroof *et al.*, 1984). However, because of their inherent advantages, PCR-based markers increasingly replaced RFLPs. In this respect, RAPDs have to be considered as a simple and easy-to-handle method, bearing in mind the constraints that a high level of standardisation is needed for reproducible and comparable results. In contrast to the majority of RFLPs, RAPDs mostly follow a dominant mode of inheritance and therefore are less informative. Since barley is regarded as a rather strict, homozygous inbreeder, this limitation seems to be of minor importance. More recently, the RAPD technique has been replaced by AFLPs (Amplified Fragment Length Polymorphisms), which to some extent combine the RFLP and the PCR technique and lead to the simultaneous detection of 100 to 200 fragments. Due to this multi-locus detection and their high potential for automation, AFLPs represent an efficient tool for diversity studies.

In contrast to the PCR-based marker systems mentioned above, sequence information is required for the detection of SSR (Simple Sequence Repeats) or microsatellite polymorphisms. These are locus-specific, multi-allelic, follow a codominant mode of inheritance and are a robust marker system, which can be easily exchanged between laboratories. In addition to initial efforts (Saghai-Maroof *et al.*, 1994; Becker and Heun, 1995; Liu *et al.*, 1996), several hundred barley SSR markers have been developed in the context of a systematic approach and were subsequently integrated into a molecular marker map (Ramsay *et al.*, 2000). The salient features of the individual DNA marker systems are summarised in Table 7.4.

Taken together, the results of a large number of diversity studies including cultivars, landraces and wild barley (*H. vulgare* ssp. *spontaneum*) provided evidence of a large variability present in the primary genepool of barley, which may be surprising given the strongly inbreeding propagation of this species. Evidently, the underlying scheme of this genetic diversity is the

Table 7.4. Selected features of the most common genetic marker systems used in diversity studies of barley.

	Isozymes	Hordeins	RFLPs	RAPDs	AFLPs	SSRs
Number of loci	up to 60	2-3	>1000	Very high	Very high	>500
Polymorphism	Limited	Very high	Limited	Limited	High	Very high
Gene action	Codominant	Codominant	Codominant	Dominant	Dominant	Codominant
Genome coverage	Limited	Very limited	High	Very high	Very high	High
No. of loci per assay	1 to few	2-3	1-few	1-10	50-100	1-few
Transferability between laboratories	High	High	High	Low	Medium	High
High throughput detection	No	No	No	No	Yes	Yes

morphological, physiological and functional adaptation of wild barley, which facilitated colonisation of primary and secondary habitats throughout the Fertile Crescent in a range of most diverse environments. However, it has to be taken into account that genetic similarity estimates obtained by these techniques depend on the marker system used and the size and nature of the sample being investigated. If the portion of polymorphic bands is taken as a measure to compare marker systems, SSR markers show the highest polymorphism (100%), followed by RFLPs (83.2%), RAPDs (66.3%) and AFLPs (46.6%) (Russell *et al.*, 1997). In several studies, a low correlation between different marker types was reported (Sanchez de la Hoz *et al.*, 1996; Noli *et al.*, 1997; Russell *et al.*, 1997; Papa *et al.*, 1998). This may be attributed to the fact that different portions of the genome were sampled by different marker types or that differences in the observed marker diversity result from different selection pressures exerted on coding and non-coding regions of the genome (i.e., cDNAs and genomic sequences). Another parameter strongly influencing the absolute figures is the selection of markers within the marker system under consideration. Evidently, markers that have been preselected for being polymorphic yield higher diversity estimates than those selected at random (Bjørnstad *et al.*, 1997). There is no general rule on how to minimise the bias resulting from the use of a given marker system, but the analysis of a sufficient number of informative markers as well as an even marker coverage of the genome represent prerequisites for obtaining meaningful results.

*Wild barley (*H. vulgare *ssp.* spontaneum*)*

In accordance with its wide adaptation, wild barley is characterised by a large genetic variability as it is frequently found only in outbreeding species. Typical estimates for genetic diversity range from 45% to 86%, depending on the sample size and composition as well as on the marker system used (Saghai-Maroof *et al.*, 1994; Davila *et al.*, 1998). It seems noteworthy that the highest levels of genetic diversity can be expected for populations from the most stressful environments such as arid-cold or arid-hot (Nevo *et al.*, 1998).

Analysis of allelic frequencies at the *Rrn1* and *Rrn2* ribosomal RNA loci revealed that wild barley shows a larger number of intergenic spacer length variants than can be detected in cultivated forms. Moreover, those alleles that are the most prevalent ones in wild barley do not necessarily show the highest frequency in cultivars reflecting the selection response to domestication and breeding efforts (Saghai-Maroof *et al.*, 1990). While it can be argued that diversity values at the two ribosomal RNA loci might not be representative for assessing the genetic diversity on a genome-wide level, an RFLP study based on 16 markers derived from all barley chromosomes except 3H revealed that the wild barley sample showed a higher number of RFLP than could be detected in cultivars (Saghai-Maroof *et al.*, 1995). Similar to RFLP markers, SSR markers underscore the large variability present in ssp. *spontaneum* (Saghai-Maroof *et al.*, 1994). The combined analysis of three SSR markers, two of which are located within intron 2 and intron 3 and the third close to its 5′end of the *Mlo* gene conferring resistance to powdery mildew, revealed 24 different haplotypes in a sample of wild barley and landrace accessions. Of these, only three were present in a set of cultivated barley accessions (L. Cardle, pers. comm.). The common pattern emerging from these studies is that the variability decreases in the order wild barley > landraces > cultivars. Within the limitations set by sampling, most of the alleles being infrequent in wild barley are not found in cultivated barley. On the other hand, those alleles that are frequent in wild barley are still present in cultivated germplasm (Saghai-Maroof *et al.*, 1994).

Knowledge of the genetic structure of a genepool is a prerequisite for the optimisation of both conservation and utilisation programmes. In this regard, partitioning of the genetic diversity present within and between wild barley populations provides valuable information. A comparison of wild barley accessions from Israel, Turkey and Iran using isoenzymes showed 54% of diversity within and 39% among populations. Another 8% of the genetic variability occurred between countries (Nevo, 1998). In a RAPD study of similar material, the within-region genetic distances closely resembled those between regions (Nevo *et al.*, 1998). Dawson *et al.* (1993) observed an average of 57% within-population diversity in 10 ssp. *spontaneum* populations from various parts of Israel. An even higher within-population diversity of 75% was observed by Baum *et al.* (1997). Hence, it seems that the major part of the genetic diversity seen in wild barley is due to the genetic variation present at the collection site. The low value of 35% within-population variation observed by Saghai-Maroof *et al.* (1990) is probably due to the bias resulting from their estimate being based on only two loci (*Rrn1, Rrn2*).

Although the direct partitioning of genetic diversity into within- and between-population components does not reveal geographic effects, i.e., populations from increasingly distant collection sites are not necessarily characterised by an increasing between-population genetic distance, phenograms provide circumstantial evidence that genetic diversity of wild barley is broadly correlated with its geographic distribution (Song and Henry, 1995). In an AFLP analysis of wild barley accessions from several parts of the Fertile Crescent accessions were grouped according to their origin with south-western genotypes being more related to those of the northern area than to those of the south-eastern region (Pakniyat *et al.*, 1997). Similar results were obtained by Petersen *et al.* (1994), where accessions of both cultivated and wild barley were grouped according to their geographic origin. However, these effects are increasingly blurred in cultivated barley by seed exchange as they occurred with increased frequency. This is exemplified in a study by Ordon *et al.* (1997; Fig. 7.1), where ssp. *spontaneum* accessions from Turkey and Israel formed distinct clusters according to their geographic origin, while the geographic pattern was less clear for the cultivated barleys analysed. In particular, a set of Japanese malting barley cultivars, whose pedigrees include European progenitors, grouped along with European lines indicating the influence of seed exchange and cross-breeding on the structure of local genepools.

The geographic effects described above are restricted to samples where allelic frequencies are mainly a function of natural selection and do not reflect human activities during the processes of domestication and breeding. Additional reasons are that passport data may suggest a false origin or the geographic location of a donor institution is confused with the (frequently unknown) collection site, leading to wild barley accessions assigned to, e.g., USA, Latvia or China (Struss and Plieske, 1998). Hence, the more advanced the germplasm under investigation, the less it is to assign it to a clear origin, owing to the complex pedigrees of modern barley germplasm, frequently including progenitors from most diverse geographic areas (e.g., Strelchenko *et al.*, 1999).

Landraces and cultivars

Similar to wild barley, landraces are characterised by a large within-population diversity. Analysis of genetic diversity of populations from Sardinia revealed that only 11% of the diversity detected by RAPD markers occurred between populations (Papa *et al.*, 1998). These results were confirmed by the analysis of a set of landraces originating from various parts of Ethiopia (Demissie *et al.*, 1998). The large within-population diversity disagrees with the highly inbreeding nature of barley, which would lead to the rapid fixation of alternate alleles between populations. It can only be explained by a high level of seed exchange between farmers leading to heterogeneity of the individual populations or by the intentional selection of cultivar mixtures, which, under extensive farming conditions, are superior to their pure stands. This genetic structure of landrace

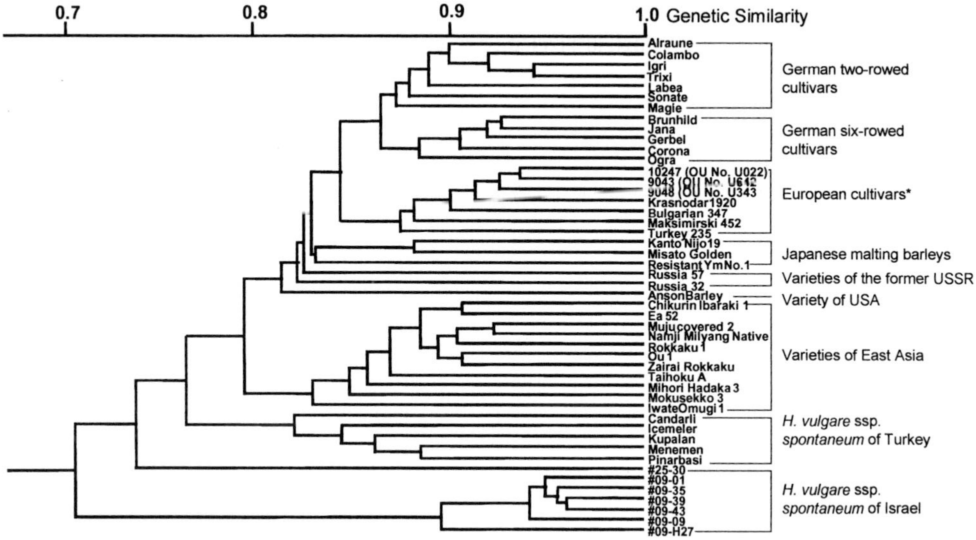

Figure 7.1. Formation of genepools in wild and cultivated barley originating from different geographic regions (according to Ordon *et al.*, 1997). The cluster analysis is based on 20 RAPD primers corresponding to 544 bands. The majority of the accessions analysed form groups according to their geographic origin. However, because of European progenitors in their pedigree, Japanese malting barleys group separately from the remaining East Asian germplasm. *: For additional information, see Takahashi *et al.*, 1983.

populations needs to be kept in mind regarding the collection and the conservation of the corresponding germplasm. Taking into account the genetic structure of wild barley and landrace populations, the genetic diversity seems to be better represented in a smaller number of larger populations from a few ecogeograpically distinct sites instead of holding single accessions from many sites. However, more data need to be collected to develop ideas of the required size of the corresponding populations.

Regarding the genetic variability present within cultivated barley, Dahleen (1997) observed a diversity index of 0.419, within a set of cultivars from North America, which compares well to a value of 0.479 obtained by Saghai-Maroof *et al.* (1994). A lower genetic distance (0.3) was estimated in a set of 37 European spring and winter cultivars by Casas *et al.* (1998), whereas Graner *et al.* (1990) observed 35% diversity in a set of 48 European spring and winter cultivars. The latter estimate is conservative, since it is based on a random set of unmapped RFLP clones, which were not preselected for revealing polymorphism. Using Random Amplified Microsatellite Polymorphic DNA (RAMP) markers Davila *et al.* (1998) observed an average diversity of 35.7% in a set of 70 barley cultivars and landraces. The lowest estimate (8.1%) was obtained by Schut *et al.* (1997) using a set of 681 AFLP fragments to analyse a set of barley cultivars mainly from Europe. This high level of genetic similarity is due to the previously mentioned tendency of AFLPs to reveal less polymorphism.

Undoubtedly, it is difficult to get a good estimate for the diversity present in cultivated barley, since continuous breeding efforts have caused the formation of regional and temporal diversity patterns, which reflect several factors such as cropping system (winter *vs*. spring), end use (feed *vs*. malting) and the strategy of individual breeders to rely on distinct progenitors in developing their germplasm (Fischbeck, 1992). The variation in the diversity figures reported above may thus be attributed to the analysis of non-representative cross-sections of germplasm which originate from distinct geographic regions and which are based on a diverse set of progenitors.

cp-DNA diversity

Due to the high level of within-species diversity seen at the genomic level, the vast majority of diversity studies has been limited to genomic DNA or to products encoded by it. However, intricate interactions between nuclear and chloroplast genomes have been demonstrated for a number of physiological and biochemical processes. For example, ribulose 1,5-bisphosphate carboxylase-oxygenase (Rubisco) representing a key enzyme in CO_2 assimilation taking place in the chloroplast, is made up of subunits, which are encoded both in the chloroplast and in the nucleus. The lack of recombination, the low level of nucleotide substitutions and the maternal inheritance make cp-DNA a useful tool to study evolution at an interspecific level. Moreover, the corresponding DNA polymorphisms reflect maternal lineages and the genetic structure of barley populations reflects gene flow due to seed – not pollen – dispersal. A survey of the occurrence of polymorphic recognition sites of 16 restriction enzymes has revealed three polymorphic sites in the cp-genome of wild barley representing 30 populations from Israel and Iran. The three haplotypes observed define three different lineages, while apart from one off-type no variation was detected in a set of 50 accessions from *H. vulgare.* Variability was present both between and within populations and seems to follow a geographic pattern (Neale *et al.*, 1988).

Similar to genomic DNA, SSRs were shown to be present on chloroplast DNA of many plant species, including barley, where cp-SSR markers are primarily due to variation in the length of single nucleotide repeats. The level of cp-SSR polymorphism detected in wild barley reached twice that detected by RFLP probes (Provan *et al.*, 1999). Nevertheless, a set comprising 101 European cultivars was monomorphic at all seven cp-SSR loci tested in that study. Even the 125 landraces that were examined could be separated into two haplotypes only, which in turn were a subset of the haplotypes found in ssp. *spontaneum* (Provan *et al.*, 1999). These data provide further evidence of the loss of genetic diversity during the process of barley domestication and breeding. There are significant cytonuclear disequilibria in wild barley that suggest a coadaptation of the two genomes to different ecogeographical environments (Saghai-Maroof *et al.,* 1992). The availability of molecular markers will now allow the systematic combination in cultivated barley of a given cp-haplotype with various backgrounds of nuclear DNA, to study potential effects that may emerge from novel combinations.

Adaptive diversity

As shown above, marker diversity frequently reflects geographic patterns. Apart from founder effects and genetic drift, the formation of distinct genepools in wild barley is probably a result of local adaptation producing a change in allele frequencies or even the elimination of alleles in a population. In that context, allelic shifts that were observed at the *Rrn1* and *Rrn2* loci during the multiplication of a composite cross in the Mediterranean provided experimental proof that these loci undergo adaptive variation (Saghai-Maroof *et al.*, 1984).

Cluster analysis of 10 *H. vulgare* ssp. *spontaneum* populations spread over Israel revealed that 58% of the distribution of the frequencies of three RAPD haplotypes was accounted for by four ecogeographical parameters (Dawson *et al.*, 1993). Using an extended dataset comprising populations from Israel, Turkey and Iran, marker associations were found with rainfall, altitude, annual and mean January temperature and longitude. Within the subsample from Israel, additional correlations were established for Thornthwaite's index of humidity, number of rainy days and number of tropical days (Nevo *et al.*, 1998). A comparison of several ssp. *spontaneum* populations sampled from six microsites located at the slopes of the Evolution Canyon (Israel), revealed clear differences between individual populations in terms of RAPD polymorphism. These differences were congruent with microclimatic conditions such as radiation, temperature, and humidity. The ensuing selective forces led to a shift in allelic frequencies that in extreme cases resulted in the fixation of opposite RAPD alleles in two different microenvironments (Owuor *et al.*, 1997). Similar to the results obtained at the Evolution Canyon, RAPD analysis of populations from a second microsite (Tabigha) located near the Sea of Galilee (Northern Israel) disclosed the association of individual marker alleles and multilocus patterns with two different soil types and transects lending strength to the hypothesis of an edaphically differentiated genetic structure of the corresponding populations (Owuor *et al.*, 1999). These studies demonstrated that genotypes from the same microsite can exhibit very similar marker patterns, while sharp genetic differences were observed between populations separated by relatively short distances only. In addition to edaphic and microclimatic associations, AFLP analysis of 39 wild barley accessions from several micro- and macrogeographic regions from the Fertile Crescent identified a series of markers that were associated with salt tolerance. In this study, regression analysis revealed that 60% of the variation observed for shoot Na-content could be accounted for by only three AFLP markers (Pakniyat *et al.*, 1997).

The results listed above provide preliminary evidence of the identification and genetic localisation of adaptive traits by means of association genetics. This strategy may provide an interesting alternative to conventional linkage mapping requiring the establishment and subsequent analysis of segregating populations. However, more data are needed to validate the approach. In particular, genetic evidence is required that the alleles that are unique to specific environments are linked to particular traits that were selected in the environment under consideration. The outcome of this approach will not only depend on the even distribution of markers across the genome but also on the marker density required to efficiently detect linkage disequilibrium in populations of wild barley.

Impact of plant breeding on the structure

Barley breeding rests on the targeted generation of genetic variability and the subsequent identification of favourable alleles that are introgressed into elite germplasm with the ultimate goal of combining all positive alleles in one genotype. Undoubtedly, the knowledge of the genetic diversity present in a given set of cultivars and breeding lines, the occurrence and distribution of specific alleles and the possibility of grouping any new accession in an existing set of characterised germplasm opens new doors to genetic improvement. Therefore, a series of studies using almost any marker system has been undertaken to examine genetic relatedness in collections of barley cultivars, even before the first molecular marker maps became available (Graner *et al.*, 1990). In a number of studies phenetic analysis of adapted germplasm led to the identification of distinct groups consisting of spring and winter types or subgroups consisting of two-rowed or six-rowed types or discernible germplasm groups that were formed by cultivars released during a distinct period of time (Melchinger *et al.*, 1994; Ellis *et al.*, 1997; Hayes *et al.*, 1997; Casas *et al.*, 1998). Since these phenological and morphological traits are simply inherited, they cannot explain the formation of distinct groups. As an example, the observed differentiation of European barleys in terms of ear type and growth habit results from the fact that selection within these groups has been targeted at different traits (Noli *et al.*, 1997). While two-rowed spring barley is preferentially used in malt production, two- and six-rowed winter types are mainly used for feed production.

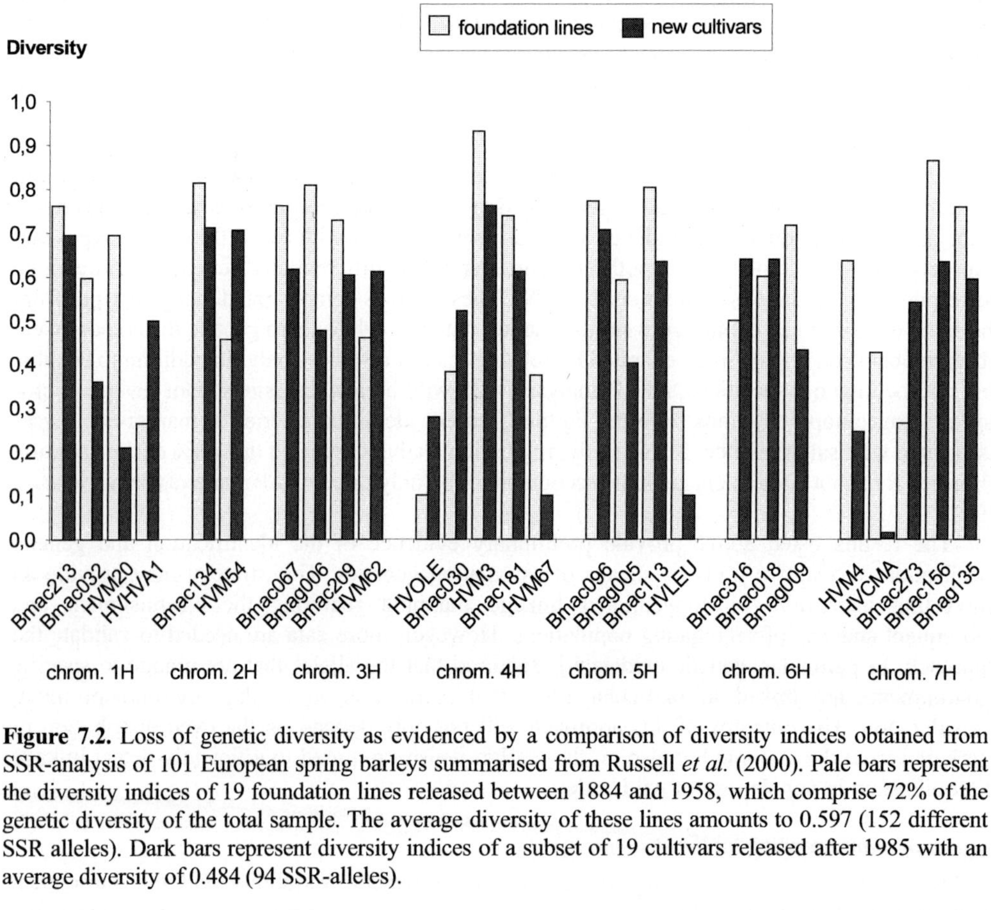

Figure 7.2. Loss of genetic diversity as evidenced by a comparison of diversity indices obtained from SSR-analysis of 101 European spring barleys summarised from Russell *et al.* (2000). Pale bars represent the diversity indices of 19 foundation lines released between 1884 and 1958, which comprise 72% of the genetic diversity of the total sample. The average diversity of these lines amounts to 0.597 (152 different SSR alleles). Dark bars represent diversity indices of a subset of 19 cultivars released after 1985 with an average diversity of 0.484 (94 SSR-alleles).

There is concern throughout the world about the genetic narrowness of elite barley germplasm. Recently Russell *et al.* (2000) have analysed a comprehensive set of landraces and cultivars released between 1884 and 1998. A comparison of the genetic diversity present in the set of cultivars released before 1958 and those released after 1985 provided evidence of genetic erosion (Figure 7.2). Reasons given by Fischbeck (1992) include (1) the limited number of landraces from various geographic regions that were used to select superior genotypes during the initial phase of barley breeding; (2) the small number of outstanding cultivars that were repeatedly used as progenitors in cross-breeding programmes during the last seven decades, and (3) the limited use of exotic germplasm that was used in recycling breeding programmes, mainly to introgress major disease resistance genes. Similar to European barley germplasm, the genetic basis of principal U.S. cultivars grown since the beginning of the 20th century could be traced back to a base population consisting of less than 20 exotic accessions (Eslick and Hockett, 1974, cited in Hayes *et al.*, 1997). Frequently, the analysis of pedigrees revealed that cultivars forming distinct clusters share common progenitors. This in turn leads to an increase in allelic frequencies derived from these ancestors and a simultaneous reduction of genetic diversity within the cluster under consideration. On the other hand, genetic diversity of Finnish six-rowed barley cultivars released before and after 1937 is similar. In this case, the looming loss of diversity was counteracted by gene flow from outside the Nordic genepool (Manninen and Nissilä, 1997).

Although the different subgroups obtained in multivariate analyses can be roughly associated with the occurrence of principal progenitors, the information on the genetic diversity obtained from pedigrees and the coefficient of coancestry (*f*, Malécot, 1948) derived from them correlates only poorly with the genetic similarity (*GS*) values based on DNA marker data. A series of studies in which *f* was compared to *GS* values obtained from RAPDs (Manninen and Nissilä, 1997), RFLP (Graner *et al.*, 1994; Davila *et al.*, 1998) or AFLP data (Ellis *et al.*, 1997; Schut *et al.*, 1997) revealed only poor correlations (0.1-0.25). However, the correlation improved from 0.4 to 0.65, after more distantly related cultivars had been included in the study. Moderate correlations (r = 0.51) between *f* and *GS* were observed by Tinker *et al.* (1993) using RAPD and Casas *et al.* (1998) using RFLP markers (r = 0.54-0.67).

There are several reasons for the poor congruence of these two measures of genetic relatedness. Besides sampling effects, resulting from too small a number of markers and a non-random distribution across the genome, mistakes in band scoring can affect the accuracy of *GS*. While type A errors (two identical bands are scored as being different) and type B errors (two different bands are scored as being identical) may occur equally for unrelated pairs of cultivars, the relative importance of type A errors is expected to increase with increasing coancestry (Graner *et al.*, 1994). Moreover, type A errors are inherent in all marker systems, since two DNA fragments of the same mobility are considered as being identical by descent although most point mutations may be present. Only the analysis of diversity at the sequence level will circumvent this limitation. To cover a sufficient number of genetic loci in this way, a systematic analysis of single nucleotide polymorphisms (SNPs) will be required. On the other hand, most assumptions underlying the calculation of *f* are unrealistic for breeding material as effects of selection, mutation or genetic drift are not taken into account. Moreover, pedigree information itself is prone to errors as cultivars may be known by more than one name (synonyms) or one name has been used for more than one cultivar or genotype (homonyms, e.g., landraces). In addition, pedigrees may be wrongly recorded and not informative, especially if encoded lines are included (Ellis *et al.*, 1997).

Conclusions and outlook

Since the early days of agriculture, farmers and breeders have been acting successfully to adapt barley to a multitude of environments, which frequently exceed the ecogeographical extremes of the natural habitats of its centre of origin. As the data on genetic diversity reveal, the various selective forces imposed during the processes of domestication, the development of landraces and of today's cultivars have brought about a considerable loss of allelic diversity. While the majority of the alleles that have been lost during this process may have been of no importance for barley improvement, other alleles that contribute positively to quantitative traits were lost since they remained undetected while being masked by deleterious alleles. Hence, it will be the challenge of the future to define the portion of useful allelic variability being deposited on the shelves of our seed-banks by using molecular marker-based strategies (Tanksley and McCouch, 1997).

Although a considerable amount of marker data has been generated by now using any of the three marker systems described in this chapter, individual datasets pertain only to a restricted number of accessions and, regarding DNA markers, are usually based on a unique marker set. Hence, it is at present not possible to pool data in order to further complete the picture of the structure and diversity of the genepool.

Despite some advantages such as the high information content of hordein markers or the low costs per data point with many isoenzyme systems, DNA marker systems are superior in terms of the available marker number together with their even genome coverage and the amenability to high throughput studies, which will make them the markers of choice for the analysis of the large number of barley accessions kept in *ex situ* collections (Table 7.4). In order to exploit their potential fully, it will be important to initiate systematic large-scale studies, using modern high throughput technologies in conjunction with the generation of relational databases that provide quick access to the relevant information.

The genetic diversity as it is measured on the DNA level seems to provide an accurate picture of the relatedness of barley cultivars that can be exploited for parent selection, the identification of subgenomic regions for the targeted increase in variability and to assess the impact of cross-breeding on the formation of distinct genepools in cultivated barley. In addition to analysing genetic relatedness on a whole genome level, the application of molecular markers can focus on the analysis of distinct subgenomic regions, which might be preferentially manipulated in a breeding programme.

There are manifold applications for fingerprinting data, ranging from the management of individual germplasm collections to the identification of appropriate germplasm for a breeding programme. However, the outcome of the efforts will mainly depend on the ability of individual laboratories to generate compatible datasets. This is both a chance and a challenge that requires international coordination addressing the issues of a standardised marker system, standardised laboratory protocols and quality checks as well as a standardised data management.

Because of their inherent advantages, a comprehensive set of highly informative SSR markers has been developed for barley that could serve as a common resource for systematic fingerprinting (Ramsay *et al.*, 2000). On the other hand, the most accurate estimate of genetic diversity can be obtained only by direct sequence comparisons. While Petersen and Seberg (1996) were not able to identify DNA sequence polymorphism in the internal transcribed spacer regions of ribosomal DNA, systematic approaches to identify point mutations, commonly denoted as single nucleotide polymorphisms or SNPs, on a genome-wide level are underway in several laboratories. In conjunction with the application of DNA-chip technologies, or mini sequencing procedures, the analysis of genetic diversity may thus reach an unprecedented accuracy. In addition to the analysis of genetic diversity *sensu lato,* a rapidly increasing number of genes are currently identified by the generation of barley ESTs (expressed sequence tags). The concomitant construction of genetic and physical maps will allow one to examine the genetic diversity of selected chromosomal regions and the corresponding genes thus permitting the identification of novel alleles and their efficient manipulation in the context of a genetic study or a breeding programme.

References

Åberg, E., 1938. *Hordeum agriocrithon*, a wild six-rowed barley. Ann. Agr. Coll. Sweden 6: 159-216.

Åberg, E., 1940. The taxonomy and phylogeny of *Hordeum* L. sect. *Cerealia* Ands., with special reference to Thibetan barleys. Symb. Bot. Upsalienses 4(2): 1-156.

Allard, R.W., 1988. Genetic changes associated with the evolution of adaptedness in cultivated plants and their wild progenitors. J. Heredity 79: 225-238.

Allard, R.W., 1992. Predictive methods for germplasm identification. In: H.T. Stalker and J.P. Murphy (eds), Plant Breeding in the 1990s. CAB International, Wallingford, UK, pp. 119-146.

Asfaw, Z., 1989. Variation in hordein polypeptide patterns within Ethiopian barley, *Hordeum vulgare* L. (Poaceae). Hereditas 110: 185-191.

Baum, B.R., E. Nevo, D.A. Johnson and A. Beiles, 1997. Genetic diversity in wild barley (*Hordeum spontaneum* C. Koch) in the Near East: a molecular analysis using Random Amplified Polymorphic DNA (RAPD) markers. Genet. Res. Crop Evol. 44: 147-157.

Becker, J., and M. Heun. 1995. Barley microsatellites: Allele variation and mapping. Plant Mol. Biol. 27: 835-845.

Bekele, E., 1983. Some measures of gene diversity analysis on land race populations of Ethiopian barley. Hereditas 98: 127-143.

Bjørnstad, Å., A. Demissie, A. Kilian and A. Kleinhofs, 1997. The distinctness and diversity of Ethiopian barleys. Theor. Appl. Genet. 94: 514-521.

Brown, A.H.D., 1983. Barley. In: S.D. Tanksley and T.J. Orton (eds), Isozymes in Plant Genetics and Breeding, Part B. Elsevier Science Publishers B.V., Amsterdam, pp. 57-77.

Brown, A.H.D., G.L. Lawrence, M. Jenkin, J. Douglass and E. Gregory, 1989. Linkage drag in backcross breeding in barley. J. Heredity 80: 234-239.

Brown, A.H.D., and J. Munday, 1982. Population-genetic structure and optimal sampling of landraces of barley from Iran. Genetica 58: 85-96.

Brown, A.H.D., E. Nevo, D. Zohary and O. Dagan, 1978a. Genetic variation in natural populations of wild barley (*Hordeum spontaneum*). Genetica 49: 97-108.

Brown, A.H.D., D. Zohary and E. Nevo, 1978b. Outcrossing rate and heterozygosity in natural populations of *Hordeum spontaneum* Koch in Israel. Heredity 41: 49-62.

Casas, A.M., E. Igartua, M.P. Valles and J.L. Molina-Cano, 1998. Genetic diversity of barley cultivars grown in Spain, estimated by RFLP, similarity and coancestry coefficients. Plant Breed. 117: 429-435.

Chalmers, K.J., R. Waugh, J. Watters, B.P. Forster, E. Nevo, R.J. Abbott and W. Powell, 1992. Grain isozyme and ribosomal DNA variability in *Hordeum spontaneum* populations from Israel. Theor. Appl. Genet. 84: 313-322.

Cross, R.J., 1994. Geographical trends with a diverse spring barley collection as identified by agro/morphological and electrophoretic data. Theor. Appl. Genet. 88: 597-603.

Dahleen, L.S., 1997. Mapped clone sequences detecting differences among 28 North American barley cultivars. Crop Sci. 37: 952-957.

Dai, X., and Q. Zhang, 1989. Genetic diversity of six isozyme loci in barley from Tibet. Theor. Appl. Genet. 78: 281-286.

Davila, J.A., M.P. Sanchez de la Hoz, Y. Loarce and E. Ferrer, 1998. The use of random amplified microsatellite polymorphic DNA and coefficients of parentage to determine genetic relationships in barley. Genome 41: 477-486.

Dawson, I.K., K.J. Chalmers, R. Waugh and W. Powell, 1993. Detection and analysis of genetic variation in *Hordeum spontaneum* populations from Israel using RAPD markers. Mol. Ecol. 2: 151-159.

Demissie, A., and Å. Bjørnstad, 1997. Geographical, altitude and agro-ecological differentiation of isozyme and hordein genotypes of landrace barleys from Ethiopia: implications to germplasm conservation. Genet. Res. Crop Evol. 44: 43-55.

Demissie, A., Å. Bjørnstad and A. Kleinhofs, 1998. Restriction fragment length polymorphisms in landrace barleys from Ethiopia in relation to geographic, altitude, and agro-ecological factors. Crop Sci. 38: 237-243.

Doll, H., and B. Andersen, 1981. Preparation of barley storage protein, hordein, for analytical sodium dodecyl sulphate-polyacrylamide gel electrophoresis. Analytical Biochemistry 115: 61-66.

Doll, H., and A.H.D. Brown, 1979. Hordein variation in wild (*Hordeum spontaneum*) and cultivated (*H. vulgare*) barley. Can. J. Genet. Cytol. 21: 391-404.

Ellis, R.P., W. McNicol, E. Baird, A. Booth and P. Lawrence, 1997. The use of AFLPs to examine genetic relatedness in barley. Mol. Breed. 3: 359-369.

Eslick, R.F., and E.A. Hockett, 1974. Genetic engineering as a key to water use efficiency. Agric. Meteor. 14: 13-23.

Faulks, A.J., P.R. Shewry and B.J. Miflin, 1981. The polymorphism and structural homology of storage polypeptides (hordein) coded by the *Hor-2* locus in barley (*Hordeum vulgare* L.). Biochem. Genet. 19: 841-858.

Fischbeck, G., 1992. Barley cultivar development in Europe – success in the past and possible changes in the future. In: Barley Genetics VI. Proc. 6th Int. Barley Genet. Symp., Helsinborg, Sweden, pp. 885-901.

Graner, A., W.F. Ludwig and A.E. Melchinger, 1994. Relationships among European barley germplasm: II. Comparison of RFLP and pedigree data. Crop Sci. 34: 1199-1205.

Graner, A., H. Siedler, A. Jahoor, R.G. Herrmann and G. Wenzel, 1990. Assessment of the degree and type of restriction fragment length polymorphism in barley (*Hordeum vulgare*). Theor. Appl. Genet. 80: 826-832.

Gupta, P.K., and R.K. Varshney, 2000. The development and use of microsatellite markers for genetic analysis and plant breeding with emphasis on bread wheat. Euphytica 113: 163-185.

Harlan, H.V., and M.L. Martini, 1929. A composite hybrid mixture. J. Amer. Soc. Agron. 21: 407-490.

Hayes, P.M., J. Cerono, H. Witsenboer, M. Kuiper, M. Zabeau, K. Sato, D.K. Kleinhofs, A. Kilian, M. Saghai-Maroof and D. Hoffman, 1997. Characterizing and exploiting genetic diversity and quantitative traits in barley (*Hordeum vulgare*) using AFLP markers. J. Quantitative Trait Loci 3: http://probe.nalusda.gov:8000/otherdocs/jqtl

Hvid, S., and G. Nielsen, 1977. Esterase isoenzyme variants in barley. Hereditas 87: 155-162.

Johansen, H.B., and P.R. Shewry, 1986 Recommended designations for hordein alleles. Barley Genet. Newsl. 16: 9-11.

Kahler, A.L., and R.W. Allard, 1970. Genetics of isozyme variants in barley. I. Esterase. Crop Sci. 10: 444-448.

Kahler, A.L., and R.W. Allard, 1981. World-wide patterns of genetic variation among four esterase loci in barley (*Hordeum vulgare* L.). Theor. Appl. Genet. 59: 101-111.

Kahler, A.L., R.W. Allard and R.D. Miller, 1984. Mutation rates for enzyme and morphological loci in barley (*Hordeum vulgare* L.). Genetics 106: 729-734.

Kalendar, R., T. Grob, M. Regina, A. Suoniemi and A. Schulman, 1999. IRAP and REMAP: two new retrotransposon-based DNA fingerprinting techniques. Theor. Appl. Genet. 98: 704-711.

Karp, A., and K.J. Edwards, 1997. Molecular techniques in the analysis of the extent and distribution of genetic diversity. In: W.G. Ayad, T. Hodgkin, A. Jaradat and V.R. Rao (eds), Molecular Genetic Techniques for Plant Genetic Resources. Report of an IPGRI Workshop, Rome, Italy, pp. 11-22.

Karp, A., P.G. Isaac and D.S. Ingram, 1998. Molecular Tools for Screening Biodiversity. Chapman and Hall, London.

Konishi, T., 1988. Genetic differentiation and geographical distribution in barley. In: S. Suzuki (ed), Crop Genetic Resources of East Asia. Proc. Int. Workshop Crop Genet. Res. East Asia. IBPGR, Rome, pp. 237-243.

Konishi, T., 1995. Geographical diversity of isozyme genotypes in barley. Kyushu Univ. Press, Fukuoka, Japan.

Konishi, T., and S. Matsuura, 1987a. Linkage analysis of *Est4* locus for esterase isozyme-4 in barley. Barley Genet. Newsl. 17: 68-70.

Konishi, T., and S. Matsuura, 1987b. Variation of esterase isozyme genotypes in a pedigree of Japanese two-rowed barley. Japan. J. Breed. 37: 412-420.

Konishi, T., and S. Matsuura, 1991. Geographical distribution of isozyme alleles of Himalayan barley (*Hordeum vulgare*). Genome 34: 704-709.

Kreis, M., and P.R. Shewry, 1992. The control of protein synthesis in developing barley seeds. In: P.R. Shewry (ed), Barley Genetics, Biochemistry, Molecular Biology and Biotechnology. CAB International, pp. 319-334.

Liu, F., R. von Bothmer and B. Salomon, 1999. Genetic diversity among East Asian accessions of the barley core collection as revealed by six isozyme loci. Theor. Appl. Genet. 98: 1226-1233.

Liu, Z.W., R.M. Biyashev and M.A. Saghai-Maroof, 1996. Development of simple sequence repeat DNA markers and their integration into a barley linkage map. Theor. Appl. Genet. 93: 869-876.

Malécot, G., 1948. Les mathématiques de l'hérédité. Masson et Cie., Paris.

Manninen, O., and E. Nissilä, 1997. Genetic diversity among Finnish six-rowed barley cultivars based on pedigree information and DNA markers. Hereditas 126: 87-93.

Melchinger, A.E., A. Graner, M. Sing and M.M. Messmer, 1994. Relationships among European barley germplasm: I. Genetic diversity among spring and winter cultivars revealed by RFLPs. Crop Sci. 34: 1191-1199.

Neale, D.B., M.A. Saghai-Maroof, R.W. Allard, Q. Zhang and R.A. Jørgensen, 1988. Chloroplast DNA diversity in populations of wild and cultivated barley. Genetics 120: 1105-1110.

Nei, M., 1973. Analysis of gene diversity in subdivided populations. Proc. Natl. Acad. Sci. USA 70: 3321-3323.

Netsvetaev, V.P., and A.A. Sozinov, 1982. Linkage studies of genes *Gle1* and *HrdF* in barley chromosome 5. Barley Genet. Newsl. 12: 13-18.

Netsvetaev, V.P., and A.A. Sozinov, 1984. Location of a hordein G locus, *HrdG*, on chromosome 5 of barley. Barley Genet. Newsl. 14: 4-6.

Nevo, E., 1998. Genetic diversity in wild cereals: regional and local studies and their bearing on conservation *ex situ* and *in situ*. Genet. Res. Crop Evol. 45: 355-370.

Nevo, E., B. Baum, A. Beiles and D.A. Johnson, 1998. Ecological correlates of RAPD DNA diversity of wild barley, *Hordeum spontaneum*, in the Fertile Crescent. Genet. Res. Crop Evol. 45: 151-159.

Nevo, E., A. Beiles, N. Storch, H. Doll and B. Andersen, 1983. Microgeographic edaphic differentiation in hordein polymorphisms of wild barley. Theor. Appl. Genet. 64: 123-132.

Nevo, E., D. Zohary, A.H.D. Brown and M. Haber, 1979. Genetic diversity and environmental associations of wild barley, *Hordeum spontaneum*, in Israel. Evolution 33: 815-833.

Nielsen, G., and H.B. Johansen, 1986. Proposal for the identification of barley varieties based on the genotypes for 2 hordein and 39 isoenzyme loci of 47 reference varieties. Euphytica 35: 717-728.

Noli, E., S. Salvi and R. Tuberosa, 1997. Comparative analysis of genetic relationships in barley based on RFLP and RAPD markers. Genome 40: 607-616.

Ordon, F., A. Schiemann and W. Friedt, 1997. Assessment of the genetic relatedness of barley accessions (*Hordeum vulgare* L. s.l.) resistant to soil-borne mosaic-inducing viruses (BaMMV, BaYMV, BaYMV-2) using RAPDs. Theor. Appl. Genet. 94: 325-330.

Owuor, E.D., T. Fahima, A. Beharav, A. Korol and E. Nevo, 1999. RAPD divergence caused by microsite edaphic selection in wild barley. Genetica 105: 177-192.

Owuor, E.D., T. Fahima, A. Beiles and A. Korol, 1997. Population genetic response to microsite ecological stress in wild barley, *Hordeum spontaneum*. Mol. Ecol. 6: 1177-1187

Pakniyat, H., W. Powell, E. Baird, L.L. Handley, D. Robinson, C.M. Scrimgeour, E. Nevo, C.A. Hackett, P.D.S. Caligari and B.P. Forster, 1997. AFLP variation in wild barley (*Hordeum spontaneum* C. Koch) with reference to salt tolerance and associated ecogeography. Genome 40: 332-341.

Papa, R., G. Attene, G. Barcaccia, A. Ohgata and T. Konishi, 1998. Genetic diversity in landrace populations of *Hordeum vulgare* L. from Sardinia, Italy, as revealed by RAPDs, isozymes and morphophenological traits. Plant Breed. 117: 523-530.

Petersen, G., and O. Seberg, 1996. ITS regions highly conserved in cultivated barleys. Euphytica 90: 233-234.

Petersen, L., H. Østergaard and H. Giese, 1994. Genetic diversity among wild and cultivated barley as revealed by RFLP. Theor. Appl. Genet. 89: 676-681.

Provan, J., R. Russell, A. Booth and W. Powell, 1999. Polymorphic chloroplast simple sequence repeat primers for systematic and population studies in the genus *Hordeum*. Mol. Ecol. 8: 505-511.

Ramsay, L., M. Macaulay, R. Cardle, M. Morgante, S. Gegli-Ivanissevich, E. Maestri, W. Powell and R. Waugh, 1999. Intimate association of microsatellite repeats with retrotransposons and other dispersed repetitive elements in barley. Plant J. 17: 415-425.

Ramsay, L., M. Macaulay, S. Gegli-Ivanissevich, K. MacLean, L. Cardle, J. Fuller, K.J. Edwards, S. Tuvesson, M. Morgante, A. Massari, E. Maestri, N. Marmiroli, T. Sjakste, M. Ganal, W. Powell and R. Waugh, 2000. A simple sequence repeat-based linkage map of barley. Genetics 156: 1997-2005.

Russell, J.R., R.P. Ellis, W.T.B. Thomas, R. Waugh, J. Provan, A. Booth, J. Fuller, P. Lawrence, G. Young and W. Powell, 2000. A retrospective analysis of spring barley germplasm development from 'foundation genotypes' to currently successful cultivars. Mol. Breed. 6: 553-568.

Russell, J.R., J.D. Fuller, M. Macaulay, B.G. Hatz, A. Jahoor, W. Powell and R. Waugh, 1997. Direct comparison of levels of genetic variation among barley accessions detected by RFLPs, AFLPs, SSRs and RAPDs. Theor. Appl. Genet. 95: 714-722.

Saghai-Maroof, M.A., R.W. Allard and Q. Zhang, 1990. Genetic diversity and ecogeographical differentiation among ribosomal DNA alleles in wild and cultivated barley. Proc. Natl. Acad. Sci. USA 87: 8486-8490.

Saghai-Maroof, M.A., R.M. Biyashev, G.P. Yang, Q. Zhang and R.W. Allard, 1994. Extraordinarily polymorphic microsatellite DNA in barley: Species diversity, chromosomal locations and population dynamics. Proc. Natl. Acad. Sci. USA 91: 5466-5470.

Saghai-Maroof, M.A., K.M. Soliman, R.A. Jørgensen and R.W. Allard, 1984. Ribosomal DNA spacer-length in barley: Mendelian inheritance, chromosomal location, and population dynamics. Proc. Natl. Acad. Sci. USA 81: 8014-8018.

Saghai-Maroof, M.A., Q. Zhang and R. Biyashev, 1995. Comparison of restriction fragment length polymorphisms in wild and cultivated barley. Genome 38: 298-306.

Saghai-Maroof, M. A., Q. Zhang, D.B. Neale and R.W. Allard, 1992. Associations between nuclear loci and chloroplast DNA genotypes in wild barley. Genetics 131: 225-231.

Sanchez de la Hoz, M.P., J.A. Davila, Y. Loarce and E. Ferrer, 1996. Simple sequence repeat primers used in polymerase chain reaction amplifications to study genetic diversity in barley. Genome 39: 112-117.

Schiemann, E., 1951. New results on the history of cultivated cereals. Heredity 5: 305-320.

Schut, J.W., X. Qi and P. Stam, 1997. Association between relationship measures based on AFLP markers, pedigree data and morphological traits in barley. Theor. Appl. Genet. 95: 1161-1168.

Shewry, P.R., and S. Burgess, 1991. RFLP analysis of the *Hor4* (*HrdG*) locus encoding a B hordein-like polypeptide. Barley Genet. Newsl. 20: 52-54.

Shewry, P., A.J. Faulks, R.A. Pickering, I.T. Jones, R.A. Finch and B.J. Miflin, 1980. The genetic analysis of barley storage proteins. Heredity 44: 383-389.

Shewry, P., R.A. Finch, S. Parmar, J. Franklin and B.J. Miflin, 1983. Chromosomal location of *Hor3*, a new locus governing storage proteins. Heredity 50: 179-189.

Shewry, P.R., E.J.L. Lew and D.D. Kasarda, 1981. Structural homology of storage proteins coded by the *Hor-1* locus of barley (*Hordeum vulgare* L.). Planta 153: 246-253.

Shewry, P.R., S. Parmar, J. Franklin and S.R. Burgess, 1990. Analysis of a rare recombination event within the multigenic *Hor2* locus of barley (*Hordeum vulgare* L.). Genet. Res., Cambr. 55: 171-176.

Shewry, P.R, H.M. Pratt, A.J. Faulks, S. Parmar and B.J. Miflin, 1979. The storage protein (hordein) of barley (*Hordeum vulgare* L.) in relation to varietal identification and disease resistance. J. Natl. Inst. Agric. Bot. 15: 34-50.

Song, W., and R.J. Henry, 1995. Molecular analysis of the DNA polymorphism of wild barley (*Hordeum spontaneum*) germplasm using the polymerase chain reaction. Genet. Res. Crop Evol. 42: 273-281.

Strelchenko, P., O. Kovalyova and K. Okuno, 1999. Genetic differentiation and geographical distribution of barley germplasm based on RAPD markers. Genet. Res. Crop Evol. 46: 193-205.

Struss, D., and J. Plieske, 1998. The use of microsatellite markers for detection of genetic diversity in barley populations. Theor. Appl. Genet. 97: 308-315.

Takahashi, R., 1955. The origin and evolution of cultivated barley. In: M. Demerec (ed), Advances in Genetics 7: 227-266. Academic Press, New York.

Takahashi R., S. Yasuda, J. Hayashi, T. Fukuyama, I. Moriya and T. Konishi, 1983. Catalogue of barley germplasm preserved in Okayama University. Institute of Agricultural and Biological Sciences, Okayama University, Kurashiki, Japan, 217 p.

Tanksley, S.D., and S.R. McCouch, 1997. Seed banks and molecular maps: unlocking genetic potential from the wild. Science 277: 1063-1066.

Tinker, N.A., M.G. Fortin and D.E. Mather, 1993. Random amplified polymorphic DNA and pedigree relationships in spring barley. Theor. Appl. Genet. 85: 976-984.

Vavilov, N.I., 1926. Studies on the origin of cultivated plants. Bull. Appl. Bot. Genet. Pl. Breed. U.S.S.R. 16(2): 1-248. (In Russian with English summary).

Waugh, R., K. McLean, A.J. Flavell, S.R. Pearce, A. Kumar, W.T.B. Thomas and W. Powell, 1997. Genetic distribution of Bare-1-like retrotransposable elements in the barley genome revealed by sequence-specific amplification polymorphisms (S-SAP). Mol. Gen. Genet. 253: 687-694.

Yu, G.X., A.L. Bush and R.P. Wise, 1996. Comparative mapping of homoeologous group 1 regions and genes for resistance to obligate biotrophs in *Avena*, *Hordeum* and *Zea mays*. Genome 39: 155-164.

Zhang, Q., G.P. Yang, X. Dai and J.Z. Sun, 1994. A comparative analysis of genetic polymorphism in wild and cultivated barley from Tibet using isozyme and ribosomal DNA markers. Genome 37: 631-638.

Diversity in Barley *(Hordeum Vulgare)*
Roland von Bothmer et al (Editors).

Chapter 8

Diversity in resistance to biotic stresses

Jens Weibull[a], Ursula Walther[b], Kazuhiro Sato[c], Antje Habekuß[b], Doris Kopahnke[b] and Gerhard Proeseler[b]

[a]Swedish Biodiversity Centre, PO Box 54, SE-230 53 Alnarp, Sweden
[b]Bundesanstalt für Züchtungsforschung an Kulturpflanzen, Institut für Epidemiologie und Resistenz, Theodor-Römer-Weg 1-4, D-06449 Aschersleben, Germany
[c]Barley Germplasm Centre, Research Institute for Bioresources, Okayama University, Kurashiki, 710-0046, Japan

Introduction

Barley, the fourth largest cereal crop globally (FAO, 1999) suffers naturally from damage caused by a large number of organisms. Mathre (1982), in his compendium on barley diseases, lists altogether four bacterial, 36 fungal and 37 viral diseases, including six nematode species, as regular or casual agents inflicting crop damage. Similarly, a substantial number of insect species is also known to attack barley (Starks and Webster, 1985). For the purpose of this volume, emphasis will be placed on the diseases and pests inflicting the most severe crop losses on a worldwide scale.

Several attempts have been carried out to estimate the level of crop losses in barley, on both national and global scales. Oerke and Dehne (1997), reviewing altogether 15,700 literature references and 3,700 field trials, concluded that the actual global crop losses in barley caused by diseases and pests amounted to 10.5 and 8.5%, respectively. Thus, while the efficacy of crop protection was estimated at 37% (herbicides included), overall one-fifth of the world production of barley is lost every year due to these two groups of organisms.

Terminology

The defence of a plant to biotic stress is a very important key for germplasm evaluation. Regions where the host and the pathogen are indigenous and have co-evolved are birthplaces for the development of different defence mechanisms and, consequently, of a high diversity. Nelson (1984) described the resistance as an active, dynamic response of a host to a parasite and classified the resistance to diseases into two major types.

The first type is resistance to infection (synonyms being *hypersensitivity, race-specific, vertical* or *major gene resistance)*. In barley this resistance is associated with obligate fungi such as rusts and mildew, it is race-specific and works as a gene-to-gene interaction between host

---

Weibull, J., U. Walther, K. Sato, A. Habekuß, D. Kopahnke and G. Proeseler, 2003. Diversity in resistance to biotic stresses. In: R. von Bothmer, Th. van Hintum, H. Knüpffer and K. Sato (eds), Diversity in Barley (*Hordeum vulgare*), pp. 143-178. Elsevier Science B.V., Amsterdam, The Netherlands.

genes for resistance and pathogen genes for avirulence. The expression of vertical resistance is often influenced by genetic background and by environmental conditions.

The second type allows the pathogen to colonise the host and to reproduce itself. This type is characterised by a long latency period, a slow disease development and a limited area of leaf subjected to attack and, hence, a reduction in the amount and rate of disease severity. Synonyms are *quantitative* or *partial resistance*, (*slow rusting*), *field resistance*, *durable resistance*, *non-race-specific resistance*, and *horizontal* or *minor gene resistance*. This resistance is common for most saprophytic fungi, but also for obligate fungi. These two kinds of resistance seem to form a continuous scale ranging from the highly resistant reaction of 'hypersensitivity' to more or less susceptibility. Other types of resistance, such as passive resistance, will not be dealt with in this text.

This chapter aims to review the situation concerning diversity in the genus *Hordeum* with respect to resistance or tolerance to important diseases or pests. In particular, emphasis will be placed on descriptions of the taxonomic, the geographic and, where known, the genomic level of diversity. The present state of the art regarding plant breeding to control the agent will also be discussed as well as the current situation of use or future prospects for breeding resistant varieties. Following Franckowiak *et al.* (1997b), three-letter gene symbols are used throughout the text.

Diversity in resistance to fungal diseases

Valuable reviews have been published by Jørgensen (1988), who listed the locus names and the location of genes on the chromosomes for the important barley diseases, and Munck *et al.* (1999), who provided a survey of the present state and possibilities of resistance research and breeding.

Powdery mildew

In most barley growing regions mildew (*Blumeria graminis* (syn.: *Erysiphe graminis*) (D.C. Speer) f. sp. *hordei*) is very common and yield losses can be dramatic. The high economic importance of this disease is based on the ability of the fungus to adapt to new resistance genes and to multiply quickly. Therefore, resistance breeding began in the thirties and the search for new resistance genes has become more and more extensive during the last 30 years. Intensive studies were carried out on barley landraces from Ethiopia (Negassa, 1985a, b), Jordan and Syria (van Leur *et al.*, 1989) and other countries of the Near East (Weltzien, 1988), Europe (Honecker, 1938), India (Freisleben, 1940), Japan (Hiura, 1960). Accessions worldwide (Moseman, 1955; Nover and Mansfeld, 1955, 1956; Hoffmann and Nover, 1959; Rigina, 1966; Wiberg, 1974a, 1974b; Moseman and Smith, 1976) and natural populations of wild barley (*Hordeum vulgare* ssp. *spontaneum*) in Israel (Wahl *et al.*, 1978; Moseman *et al.*, 1981, 1983; Jahoor and Fischbeck, 1987a, 1987b; Segal *et al.*, 1987) have also been studied.

The *Mla* locus (chromosome 1HS) has the highest number of alleles among the powdery mildew and other disease resistance loci. This locus carries 32 (semi)dominant *Mla* alleles (*Mla1-Mla34)* that govern resistance against pathotypes of *E. graminis* f. sp. *hordei.* The highest variability hitherto has been found in Israeli wild barley populations. Israel is part of the Fertile Crescent defined by Harlan and Zohary (1966) and of the Near East megacentre defined by Zeven and Zhukovsky (1975). These authors described these centres as origins of cultivated barley following Vavilov's (1926, 1951) theory about the centres of origin of cultivated plants. The *H. vulgare* ssp. *spontaneum* populations are at present the most important and richest genepool for this kind of resistance genes. Another potential resource is *Hordeum bulbosum* from the secondary genepool (Pickering and Morgan, 1983; Bothmer, 1992). More than 60 major resistance genes to mildew have been described (Table 8.1) and some have been used to develop near isogenic lines (Kølster *et al.*, 1986).

Winter barley is generally poor in resistance genes contrary to spring barley in Europe. Breeding for mildew resistance in European spring barley varieties has hitherto been based on a few resistant parents: 'Weihenstephan CP 127422' - *Mlg*, 'Algerian' - *Mla*, 'Ricardo' - *Mla3*, 'Arabic' - *Mla12*, 'Monte Cristo' - *Mla9* and 'Rupee' - *Mla13*, *H. distichon* var. *laevigatum* - *MlLa*, 'L100' (HOR 2556), 'L92' (HOR 1504), 'Grannenlose Zweizeilige' (HOR 2937), with the spontaneous allele *mlo9* from Ethiopia, and the 'Diamant Mutant' carrying the induced '*mlo9*'. At least three lines formed the source of *Mla7*: 'Lyallpur 645' (CI 3395), 'Multan' (CI 3401) and 'Long Glumes' (CI 6168). In contrast, there are only three major resistance genes, which were first of all introduced in winter barley: *Mlh* ('Ragusa B') in Yugoslavia, *Mla6* ('*H. spontaneum nigrum*') in Russia and *Mla8* ('Hanna') in Germany. Some older resistance genes were transferred from spring to winter barley using spring × winter barley crosses (Fischbeck *et al.*, 1976; Moseman and Craddock, 1976; Torp *et al.,* 1978; Schwarzbach and Fischbeck, 1981; Brown and Jørgensen, 1991; Jensen *et al.*, 1992; Jørgensen, 1992). Apart from these genes, there are a few additional but generally uncharacterised genes in cultivars.

The main history of incorporation of powdery mildew resistance genes in cultivated barley and the exploration of their genetic diversity in Europe is well covered by Wolfe and Schwarzbach (1978). Since the beginning of the nineties and the breaking of many of the resistance genes mentioned above, there is now a strong dominance for using *mlo* resistance.

The diversity of genes for partial resistance to powdery mildew on national and worldwide levels is unclear, due to the difficulties in identifying the kind and number of participating genes, since the character is mostly quantitatively inherited and each gene has minor effects. The knowledge of the mode of operation of these genes influencing traits such as infection frequency, latency period, colony size and sporulation capacity is rather limited. Generally, many assessments of partial resistance to pathogens in barley have been made with powdery mildew (Asher and Thomas, 1983, 1984, 1987; Wright and Heale, 1984; Anderson and Torp, 1986; Carver, 1986; Heun, 1986; Geiger and Heun, 1989; Newton, 1990; Kmecl *et al.*, 1995).

Scald

Scald, caused by the imperfect fungus *Rhynchosporium secalis* (Oudem.) J.J. Davis f. sp. *hordei,* is a major foliar disease of barley grown in areas of the world where the climate is cooler and more humid because the leaf remains wet for a longer period. Reports about the pathogenic variability of *R. secalis* are numerous (Brown, 1985; Mc Donald *et al.*, 1999), and in some countries an increase in genetic variability of the fungus has been described (Tekauz, 1991). Numerous papers report pathogenic variability of scald on all five continents (Hansen and Magnus, 1973; Williams and Owen, 1973; Ali *et al.*, 1976; Jackson and Webster, 1976; Metcalfe *et al.*, 1977; Ceolini, 1980; Brown, 1985; Cromey, 1987; Jørgensen and Smedegaard-Petersen, 1995), which directly influences the estimates of the diversity of scald resistance genes in the regions of the world. It is clear that the virulence spectrum of scald populations within regions of barley cultivation is able to change in short periods of time (Tekauz, 1991) and correlations indicate that the variability of scald populations reflect the cultivar composition.

Several evaluations of barley for resistance to scald have been carried out (Fukuyama *et al.*, 1998; Yitbarek *et al.*, 1998), and many resistance genes have also been described (Table 8.2). However, reviews about scald and corresponding resistance genes (Bockelman *et al.*, 1977; Beer, 1991) have been confused, on account of contradictory results about reaction patterns between

Table 8.1. Main resistance genes to powdery mildew.

Chromosome	Gene	Accession/cultivar	Origin	Author(s)
1H	*Mla1*	Algerian (CI 1179)	Algeria	Briggs and Stanford (1938)
1H	*Mla2*	Black Russian (CI 2202)	Georgia	Schaller and Briggs (1955)
1H	*Mla3*	Ricardo (CI 6306)	Uruguay	Moseman and Schaller (1960)
1H	*Mla5*	Gopal (CI 1091)	India	Baker (1964)
1H	*Mla6*	*H. spontaneum nigrum* H. 204 (CI 8825)	Russia	Hoffmann and Kuckuck (1938), Moseman *et al.* (1965)
1H	*Mla7*	Lyallpur 3645	Pakistan	Scholz and Nover (1967)
1H	*Mla8*	Heils Hanna (CI 682)	Germany	Hiura (1960), Moseman and Jørgensen (1971)
1H	*Mla9*	Monte Cristo (CI 1017)	India	Hiura (1960), Moseman and Jørgensen (1971)
1H	*Mla10*	Iso 12 (Durani CI 6316)	Afghanistan	Moseman and Jørgensen (1971)
1H	*Mla11*	A 222 (CI 11555)	Japan	Moseman and Jørgensen (1971)
1H	*Mla12*	Emir (CI 11790)	Netherlands	Wiberg (1974a), Giese *et al.* (1981)
1H	*Mla13*	Rupee (CI 4355)	India	Moseman and Jørgensen (1971), Giese *et al.* (1981)
1H	*Mla14*	Iso 20R = Franger/4(F15) Man (CI 16151)	Germany	Giese *et al.* (1981)
1H	*Mla16*	D × 1B-54B	Israel	Jahoor and Fischbeck (1987a)
1H	*Mla17*	RS170-47 × Kieb.B	Israel	Jahoor and Fischbeck (1987a)
1H	*Mla18*	RS20-1 × Kieb.B	Israel	Jahoor and Fischbeck (1987a)
1H	*Mla19*	D × 1B-86B	Israel	Jahoor and Fischbeck (1987a)
1H	*Mla20*	RS145-39 × Kieb.B	Israel	Jahoor and Fischbeck (1987a)
1H	*Mla21*	D × 1B-152B	Israel	Jahoor and Fischbeck (1987a)
1H	*Mla22*	HOR 1657	Pakistan	Favret (1960), Jørgensen (1991)
1H	*Mla23*	HOR 1402	Turkey	Nover (1972), Jørgensen (1991)
1H	*Mla24*	Engledow India (CI 7555)	India	Favret (1960), Jahoor *et al.* (1990)
1H	*Mla25*	RS170-10 × Piccolo A	Israel	Jahoor and Fischbeck (1993)
1H	*Mla26*	D × 1B-20	Israel	Jahoor and Fischbeck (1993)
1H	*Mla27*	RS1-8 × Piccolo E	Israel	Jahoor and Fischbeck (1993)
1H	*Mla28*	D × 1B-151	Israel	Jahoor and Fischbeck (1993)
1H	*Mla29*	110-4 × Sonja	Israel	Kintzios *et al.* (1995)
1H	*Mla30*	Nigrate (CI 2444)	China	Hiura (1960); Jørgensen (1993)
1H	*Mla31*	Turkey 290	Turkey	Jørgensen (1993)
1H	*Mla32*	142-29 × Dura	Israel	Kintzios *et al.* (1995)
1H	*Mla33*	RS90-13 × Kieb.	Israel	Dheeranupattana (1995)
1H	*Mla34*	RS70-29 × Piccolo	Israel	Dheeranupattana (1995)
1H	*Mlat*	Atlas (CI 4118)	USA	Moseman and Schaller (1960)
4H	*mlBO*	mutant in Bomi (N182)	Denmark	Heun (1984)
4H	*MlCP*	Weihenstephan CP127422	Germany	Freisleben and Metzger (1942)
?	*mld*	Duplex (CI 2433)	Russia	Nilan (1964)
7H	*Mlf*	RS137-28 × Elgina	Israel	Schönfeld *et al.* (1996)
7H	*Mlf2*	141-10	Israel	Schönfeld (1997)
7H	*Mlf3*	101-49	Israel	Schönfeld (1997)
7H	*Mlf4*	81-B-8	Israel	Schönfeld (1997)
4H	*Mlg*	Weihenstephan CP127422	Germany	Honecker (1931)
1H	*Ml(Ga)*	Galleon (CI 3576)	Egypt	Hossain and Sparrow (1991)

Chromosome	Gene	Accession/cultivar	Origin	Author(s)
6H	*Mlh*	Hanna (CI 906)	Germany	Hayashi and Heta (1985)
1H	*Mli*	RS42-8 × Or.A	Israel	Jahoor and Fischbeck (1987a)
5H	*Mlj*	HSY-78 × Ar.	Israel	Schönfeld *et al.* (1996)
5H	*mlj2*	RS164-6 × Ar.	Israel	Schönfeld (1997)
1H	*Mlk*	Kwan (CI 1016)	India	Briggs and Stanford (1938)
?	*Mln*	Nepal (CI 595)	Nepal	Moseman (1971)
4H	*mlo1*	Mutante 66 (CI 15217) (Haisa CI 9855)	Denmark (Germany)	Freisleben and Lein (1942)
4H	*mlo2*	H3502 (CI 15223) (Probstdorfer Vollkorn CI 15222)	Denmark (Germany)	Hänsel and Zakovsky (1956)
4H	*mlo3*	MC20 (CI 15225) (Malteria Heda CI 15224)	Denmark (Argentina)	Favret (1965)
4H	*mlo4*	SR1 (Foma)	-	Wiberg (1973)
4H	*mlo5*	R5678 CI 15219 (Carlsberg II CI 15218)	Denmark	Jørgensen (1975)
4H	*mlo6*	R6018 CI 15220 (Carlsberg II)	Denmark	Jørgensen (1975)
4H	*mlo7*	R7085 CI 15662 (Carlsberg II)	Denmark	Jørgensen (1975)
4H	*mlo8*	R7372 CI 15221 (Carlsberg II)	Denmark	Jørgensen (1975)
4H	*mlo9*	SZ5139 (CI 15227) (Diamant CI 15226)	Denmark	Schwarzbach (1967)
4H	*mlo10*	SR7 (Foma)	-	Wiberg (1973)
4H	*mlo11*	HOR 2937	Ethiopia	Nover (1972)
4H	*Mlo12*	Mutante No. 4122 (Elgina)	Germany	Hentrich (1977)
4H	*mlo13*	Mutante No. 2018 (Plena)	Germany	Hentrich (1977)
4H	*mlo14*	Mutante No.2029 (Plena)	Germany	Hentrich (1977)
4H	*mlo15*	Mutante No. 4123 (Elgina)	Germany	Hentrich (1977)
4H	*mlo16*	Mutante No. 2267 (Alsa)	Germany	Hentrich (1977)
4H	*mlo17*	Mutante No. 2034 (Plena)	Germany	Hentrich (1977)
4H	*mlo18*	Mutante ML-3A (Azuma Golden)	Japan	Yamaguchi and Yamashita (1979)
4H	*mlo19*	Mutante ML-4F (Fuji Nijo)	Japan	Yamaguchi and Yamashita (1979)
4H	*mlo20*	Mutante ML-9F (Fuji Nijo)	Japan	Yamaguchi and Yamashita (1979)
4H	*mlo21*	Mutante ML-13F (Fuji Nijo)	Japan	Yamaguchi and Yamashita (1979)
4H	*mlo22*	Mutante B1012 (Bomi)	Denmark	Röbbelen and Heun (1991)
4H	*Mlo23*	Mutante B1101 (Bomi)	Denmark	Röbbelen and Heun (1991)
4H	*mlo24*	Mutante B1865 (Bomi)	Denmark	Röbbelen and Heun (1991)
4H	*mlo25*	Mutante N105 (Bomi)	Denmark	Röbbelen and Heun (1991)
1H(?)	*Mlp1*	Psaknon (CI 6305)	Australia	Stanford and Briggs (1940)
1H(?)	*Mlp2*	RS145-1 × Ar.B	Germany	Jahoor *et al.* (1989)
1H(?)	*Mlp3*	RS170-35 × Ar.B	Germany	Jahoor *et al.* (1989)
?	*Mlr*	Rabat	Morocco	Patterson and Shands (1957)
1H	*Mlra*	Weihenstephan 41/145	Dalmatia	Wiberg (1974a), Doll and Jensen (1986)
1H	*MlRu2*	Iso42R (Rupee CI 4355)	India	Wiberg (1974a)
(1H)	*mls*	Spiti (CI 4343-1)	Tibet	Favret (1960)
7H	*mlt*	RS42-6 × Oriol	Israel	Schönfeld *et al.* (1996)
7H	*mlt2*	101-23	Israel	Schönfeld (1997)
7H	*mlt3*	92-49	Israel	Schönfeld (1997)
5H	*Ml(Tr)*	TR306	-	Falak *et al.* (1999)
?	*Mlw*	West China	Argentina	Smith 1951
?	*Mly*	Arlington Awnless	Arabia	Smith 1951

Table 8.2. Main resistance genes to scald.

Chromo-some	Gene	Accession/cultivar	Origin	Author
3H	*Rrs1*	Brier CI 7157	USA	Bryner (1957)
7H	*Rrs2*	Atlas CI 4118	Germany	Dyck and Schaller (1961)
3H	*Rrs3*	Turk CI 5611-2	Turkmenistan	Dyck and Schaller (1961)
3H	*Rrs4*	La Mesita CI 7565, Trebi CI 936, Osiris CI 1622	USA Germany	Dyck and Schaller (1961)
3H?	*Rrs5*	Turk CI 5611	Turkmenistan	Dyck and Schaller (1961)
?	*rrs6*	Jet CI 967, Steudelli CI 2226	Ethiopia	Baker and Larter (1963)
?	*rrs7*	Jet CI 967, Steudelli CI 2226	Ethiopia	Baker and Larter (1963)
?	*rrs8*	Nigrinudum CI 2222	Ethiopia	Wells and Skoropad (1963)
?	*Rrs9*	Abyssinian CI 668, Kitchin CI 1296	Ethiopia USA	Baker and Larter (1963)
?	*Rrs10*			Habgood and Hayes (1971)
?	*Rrs11*			Habgood and Hayes (1971)
7H	*Rrs12*			Abbott *et al.* (1995)
6H	*Rrs13*			Abbott *et al.* (1995)

lines possessing the same resistance. This is most probably due to the use of local pathotypes, a known problem for many foliar diseases in barley. Also, a set of standard differentials to distinguish pathotypes is necessary for the systematic analysis of resistance.

Rust diseases

Barley is currently being attacked by three rust species: leaf rust (*Puccinia hordei* Otth), stripe rust (*P. striiformis* West. f. sp. *hordei*) and stem rust (*P. graminis* Pers.: Pers. f. sp. *tritici* Eriks. et Henn.).

Leaf rust (*Puccinia hordei* Otth) is considered as a serious threat to barley production in many areas. Like mildew, leaf rust will remain a significant pathogen since new pathotypes occur frequently and the fungus quickly adapts to new resistance genes (Golan *et al.*, 1978; Parlevliet *et al.*, 1981, both cited by Clifford, 1985; Steffenson *et al.*, 1993; Walther, 1996).

The release of mildew-resistant cultivars with unknown rust resistance gave the later appearing leaf rust a chance to develop without restriction and induced an intensive breeding for resistance. The origins of resistance sources were similar to those of mildew-resistant landraces and wild relatives from the Mediterranean region where both the host and the pathogen are indigenous and have co-evolved (Anikster and Wahl, 1979). Israel, in particular, is part of the centre of origin and genetic variation of the wild native *Hordeum* species *H. vulgare* ssp. *spontaneum*, *H. bulbosum* and *H. murinum* (Wahl *et al.*, 1988; Kandawa-Schulz, 1996). Interesting material has also been found in Azerbaijan and Turkmenistan (Bakhteev collection) and Iran (Kuckuck collection) (Nover and Lehmann, 1974; Walther and Lehmann, 1980, both cited by Clifford, 1985). Clifford (1985) summarised the activities in evaluation and in breeding for major genes and partial resistance until 1985, including pioneering work in the 1920s. Evaluations have continued into recent years (Reinhold and Sharp, 1986; Yahyaou *et al.*, 1988; Khokhlova *et al.*, 1989; Jin *et al.*, 1995; Alemayehu and Parlevliet, 1997; Lukyanova and Terentyeva, 1997). Franckowiak *et al.* (1997a) have provided an overview of the recommended allele symbols. Until today, sixteen genes have been described (Table 8.3).

Table 8.3. Resistance genes for leaf rust of barley.

Chromosome	Gene	Accession/cultivar	Origin	Author(s)
2H	*Rph1*	Oderbrucker[1], Sudan[2], Speciale[3]	[1]G,[2]Sudan, [3]USA	Roane and Starling (1967)*
5H	*Rph2*	Reka 1[4], Weider, Ricardo[5], Peroga[1], Peruvian[3], Betna, Quinn (+*Rph5*)[3], Bolivia (+*Rph6*)[3], Ackermanns MGZ[1]	[1]G, [3]USA, [4]Australia, [5]Uruguay	Frecha (1970)*; Brückner (1971)*; Clifford and Roderik (1978)*
7H	*Rph3*	Estate[6], Rika × F_1[1], Aim[3], HOR 679-3[7], HOR 1873[7]	[1]G, [3]USA, [6]Egypt, [7]Peloponnes	Roane and Starling (1967)†; Johnson (1968)*; Brückner (1971)*; Clifford (1976)*; Tan (1978)*; Walther (1984)*
1H	*Rph4*	Gold[8], Lechtaler[9], Franger[10]	[8]DK, [9]Austria, [10]Israel	Roane and Starling (1967, 1970)†; Jensen (1979)
5H	*Rph5*	Quinn (+*Rph2*)[3], Magnif[11]	[3]USA, [11]Argentina	Roane and Starling (1967)*; Frecha (1970)*; Tan (1978)*; Jin *et al.* (1995)
	Rph6	Bolivia (+*Rph2*)	USA	Roane and Starling (1970)*
3H	*Rph7*	La Estanzuela[5], Gondar[12], H 2212[12], HOR 4279, HOR 5943[12], HOR 5944[12]	[5]Uruguay, [12]Ethiopia	Johnson (1968)*; Parlevliet (1976)*; Clifford (1976)*; Tan (1978)*; Walther and Lehmann (1980)*
	Rph8	Egypt 4	USA	Tan (1977, 1978)†
5H	*Rph9*	HOR 2596, HOR 1859, HOR 1643, HOR 2927, HOR 2926, HOR 1633	Ethiopia	Tan 1977†
3H	*Rph10*	Clipper BC 8 (*H. spontaneum*)	Israel	Feuerstein *et al.* (1990)
6H	*Rph11*	Clipper BC 67 (*H. spontaneum*)	Israel	Feuerstein *et al.* (1990)
5H	*Rph12*	Trumpf	Germany	Jin *et al.* (1993a); Steffenson (1999)
	Rph13			Jin *et al.* (1996)
	Rph14			Jin *et al.* (1996)
2H	*rph16*	*H. spontaneum* 680	Israel	Ivandic *et al.* (1998)
2H	*RphHb*	*H. bulbosum*		Pickering *et al.* (1997)
1,2,3H	QTLs	HOR 1063	Turkey	Kicherer *et al.* (1996); Qui *et al.* (1998)

G-Germany, DK-Denmark; *cited by Prochnow (1997); †cited by Clifford (1985)

Since the number of effective resistance genes in barley landraces has decreased due to adaptation of the fungus, the partial resistance genes in *H. vulgare* ssp. *spontaneum* have become increasingly important (Manisterski *et al.*, 1986; Jin and Steffenson, 1994; Prochnow, 1997). Investigations of Walther *et al.* (2000) have also shown that *H. bulbosum* is an effective source of leaf rust resistance genes.

Sources of major resistance genes are restricted in modern cultivars. Walther *et al.* (1999) identified *Rph1* and *Rph2* in German winter barley cultivars and *Rph12*, *Rph3* (incl. their combination) and *Rph7* (in 'Hanka') in German spring barley cultivars. Over the years, the use of partial resistance has become more popular in spring barley than in winter barley cultivars. In

Table 8.4. Genes for resistance to barley stripe rust.

Chromosome	Gene	Accession/cultivar	Origin	Author(s)
	yr1 (*BYR1*)	Bigo[1], BBA 2890, Abyssinian 14[2]	[1]NL, [2]Ethiopia	Nover and Scholz (1969)*; Upadhyay and Prakash (1977)*; Meadway *et al.* (1991)
	yr2 (*BYR2*)	Abed Binder	Denmark	Nover and Scholz (1969)*; Meadway *et al.* (1991)
	yr3 (*BYR3*)	I 5	Hungary	Nover and Scholz (1969)*; Meadway *et al.* (1991)
	Yr4	Deba Abed, Cambrinus[1], Europa[3]	Denmark, [1]NL, [3]G	Johnson (1968)*; Luthra (1988)
5H, 7H	3 QTL			Hayes *et al.* (1996)
3H	1 QTL			Toojinda *et al.* (1998)

G-Germany, NL-The Netherlands; *cited by Stubbs (1985)

the future, the combination of partial resistance and effective major gene resistance, for instance from *H. vulgare* ssp. *spontaneum*, will be necessary to achieve a stable resistance in practical breeding programmes.

Stripe rust (*Puccinia striiformis* West. f. sp. *hordei*) is known in most of the barley growing regions. In some regions epidemics are rare, because favourable environmental conditions are generally lacking. In Europe the last severe epidemic occurred in 1961. The disease was introduced from Europe to South America (Dubin and Stubbs, 1986) but was not observed in Mexico before 1987 (Sandoval-Islas *et al.*, 1998). Since 1991 the disease has spread rapidly over the entire USA (Chen and Line, 1999). Stubbs (1985) provided a general review of the work until 1985 and the resistance genes determined so far are listed in Table 8.4.

Chen and Line (1999), determining resistance genes against North American isolates, found dominant and recessive reaction patterns, respectively, on the first and second leaf of young 'Cambrinus' and 'Mazurka' plants. Earlier, Xu and Snape (1989) detected a new resistance in *H. bulbosum*.

Extensive evaluations and/or breeding have been carried out in some countries in search of new resistance genes (Nover and Lehmann, 1966, 1970, 1975; Upadhyay and Prakash, 1977; all cited by Stubbs, 1985; Leur *et al.*, 1989; Okunowski, 1990; Luthra *et al.*, 1992; Hill *et al.*, 1995). Many contemporary European cultivars possess the resistance of 'Trumpf' (donor S 3170) while other genes for stripe rust resistance (in particular minor genes from 'Emir' and 'S 3192') have also been used in breeding programmes, thereby giving resistance to other barley diseases as well.

Stem rust (*Puccinia graminis* Pers.: Pers. f. sp. *tritici* Eriks. et Henn.) occurs on cultivated barleys world-wide and is mainly caused by *P. graminis* f. sp. *tritici* in Europe, but also by f. sp. *secalis* in North America and Australia, and by f. sp. *hordei*, a hybrid of these two, in Australia (Luig, 1985). Harder and Legge (1996) have reviewed the present state of the art and the genes described to date; these are listed in Table 8.5.

The most widespread cultivars carrying the gene *Rpg1* have until now not been severely attacked (Steffenson, 1992). Extensive breeding for stem rust resistance has been reported in Canada (Anonymous, 1972) and India (Krishnamurthy *et al.*, 1972).

Table 8.5. Resistance genes to stem rust.

Chromo-some	Gene	Accession/cultivar	Origin	Author(s)
1H	*Rpg1*	Chevron (CI 1111), Peatland (CI 5267)	USA	Jin *et al.* (1993b, 1994); Jedel *et al.* (1989)
	Rpg2	Hietpas 5 (CI 7124)	USA	Jin *et al.* (1993b, 1994); Jedel *et al.* (1989)
	Rpg3	PI 392313		Jin *et al.* (1993b, 1994); Jedel *et al.* (1989)
	rpg4	Q 21861		Jin *et al.* (1994)
	RpgU			Fox and Harder (1996)
	1 semi-dominant gene	Resistant to *P. graminis* f. sp. *secalis*		Sun *et al.* (1996)
	2 recessive			Jin *et al.* (1994)

Diseases caused by Drechslera *spp.*

Net blotch of barley (*Pyrenophora teres* (Died.) Drechsl. f. *teres*) is caused by the ascomycete *Pyrenophora teres* f. *teres* (perfect state). The imperfect state is *Drechslera teres* (Sacc.) f. *teres* Shoemaker (syn.: *Helminthosporium teres* Sacc.). A second form of *D. teres* is *D. teres* (Sacc.) Shoemaker f. *maculata* Smedegaard-Petersen (perfect stage: *P. teres* Drechsl. f. *maculata* Smedegaard-Petersen). The latter form causes ‘spot’ symptoms (Smedegaard-Petersen, 1971). Net blotch is a common disease of barley everywhere the crop is grown and is particularly prevalent in cool and wet growing regions.

Probably the most effective means of controlling net blotch is through the use of resistant cultivars. The focus on a limited number of high-yielding cultivars favours the outbreak of epidemics (Khan and Tekauz, 1982). Pathogen specialisation in *P. teres* and the occurrence of new pathotypes have been reported from most areas where net blotch is a problem (Smedegaard-Petersen, 1971; Tekauz and Millis, 1974; Tekauz and Buchannon, 1977; Tekauz, 1978; Khan, 1982; Afanasenko *et al.*, 1995). The first studies to identify resistant forms were carried out by Heschele (1928). Khan and Boyd (1969b) gave the first clear definition of physiological races of the net type by means of two differential cultivars.

Sato and Takeda (1991, 1993) reported on differences between Canadian and Japanese isolates of *P. teres* f. *teres* in temperature requirements for conidia formation and in the virulence spectrum by the responses to barley genotypes on diverse resistance sources.

Intensive studies among wild species and landraces have been carried out on barley accessions by many authors (Schaller and Wiebe, 1952; Buchannon and McDonald, 1965; Gaike, 1970; Metcalfe *et al.*, 1978; Smirnova and Trofimovskaya, 1985; Proeseler *et al.*, 1989; Lukyanova, 1990; Faiad *et al.*, 1996). Sato and Takeda (1994) studied the variation of host resistance by inoculating each of four *P. teres* f. *teres* (net form) and two *P. teres* f. *maculata* (spot form) isolates from Japan and Canada onto 2,233 accessions of the barley world collection. On average, sources of resistance were abundant in Ethiopia, North Africa and Korea, while resistance sources for the spot form were particularly abundant in the Far East and North Africa. New sources with resistance to several (up to eight) races of *P. teres* were found among Peruvian landrace accessions (Afanasenko *et al.*, 2000). Earlier, Sato and Takeda (1992) found that the two-rowed malting barleys were significantly more susceptible than six-rowed barleys. In recent years, wild barley (*H. vulgare* ssp. *spontaneum*) has been increasingly used as a resistance source against net and spot blotch (Lehmann, 1988; Gustafsson and Bothmer, 1994; Jana and Bailey, 1995; Sakti and Bailey, 1995; Sato and Takeda, 1997).

Table 8.6. Resistance genes for net blotch.

Chromosome	Gene	Accession/cultivar	Origin	Author
2H	*Rpt3d*	CI 7584	USA	Bockelman *et al.* (1977)
3H	*Rpt1a*	Tifang, CI 14373	China	Bockelman *et al.* (1977)
3H	*Rpt1b*	CI 9819	Ethiopia	Bockelman *et al.* (1977)
3H	*Pt,a*	Igri	Ethiopia	Graner *et al.* (1996)
5H	*Rpt2c*	CI 9819	Ethiopia	Bockelman *et al.* (1977)
7H	*Rpt4*	Galleon	Australia	Williams *et al.* (1999)
2H,3H,4H,5H,6H,7H	QTL	Steptoe/Morex	USA	Steffenson *et al.* (1996)
3H,4H,5H,6H,7H	QTL	Harrington/TR306	USA	Spaner *et al.* (1998)

The many genetic studies on net blotch resistance include those by Schaller (1955), Frecha (1958), Mode and Schaller (1958), Gray (1966), Khan and Boyd (1969a), Afanasenko and Kushnirenko (1989), Hartmann (1993) and Choo *et al.* (1994). They concluded that dominant or incompletely dominant gene(s) conditioned the resistance of the different cultivars, but they were unable to find a relation between the genes described, because a standardised international differential set of barleys to characterise the pathotypes was lacking. That resistance to net blotch is probably quantitatively inherited has been reported by Keeling and Banttari (1975), Douglas and Gordon (1985), Arabi *et al.* (1990), Steffenson and Webster (1992), Steffenson *et al.* (1991, 1996), Cherif and Harrabi (1993), and Richter *et al.* (1998). The genes hitherto determined are listed in Table 8.6.

Recently, a large number of QTLs has been found in the adult plant resistance to this disease (Steffenson *et al.*, 1996, Spaner *et al.*, 1998), indicating that many genes with minor effects can affect net blotch resistance. Because of the high variability observed in *P. teres* and the difficulty in finding lines with resistance to all isolates, Douiyssi *et al.* (1998) suggested that breeding for resistance should emphasise pyramiding of resistance genes.

Barley stripe disease (*Pyrenophora graminea* Ito & Kuribayashi) is caused by the ascomycete *Pyrenophora graminea* (sexual state), the asexual state being *Drechslera graminea* (Rabenhorst ex Schlechtendal) Shoemaker (syn.: *Helminthosporium gramineum* Rabh.).

Although barley stripe is effectively controlled by seed treatments, the breeding of resistant cultivars also provides control. Physiological races of *P. graminea* occur, and cultivars resistant in one region are not necessarily resistant elsewhere (Mathre, 1982; Tekauz, 1983; Skou and Haahr, 1985).

Sources resistant to barley stripe have been detected by extensive evaluation work (Gaike, 1973; Baigulova and Pitonya, 1979; Nettevich and Vlasenko, 1985; Skou and Haahr, 1985; Leur, 1989; Su *et al.*, 1989; Lukyanova, 1990; Bisht and Mithal, 1991), and have later been incorporated in breeding programmes (Rodriguez, 1973; Garkavyi *et al.*, 1974; Ceccarelli *et al.*, 1976; Kirdoglo, 1990; Lukyanova, 1990; Skou *et al.*, 1992, 1994). It seems that two-rowed cultivars and lines are more resistant than six-rowed ones, and that spring barleys are more resistant than winter types (Bobes *et al.*, 1974; Giuliari *et al.*, 1984; Delogu *et al.*, 1989).

In an early study on the genetics behind resistance, Isenbeck (1930) concluded that resistance or susceptibility to *P. graminea* was a heritable trait and that several dominant genes conditioned resistance. Arny (1945) established that resistance was dominant, and three genetic factors were probably involved in some crosses and incompletely dominant and conditioned by many factors in other crosses. Søgaard and Wettstein-Knowles (1987) described three resistance genes in barley (*Rhg1*, *Rhg2* and *Rhg3*) but neither the localisation on the chromosome nor the number of alleles have yet been identified. Boulif and Wilcoxson (1988) also found that resistance to *P. graminea* might be monogenic or oligogenic.

Delogu *et al.* (1995) characterised the resistance of barley to *P. graminea* as being complex, involving both a 'horizontal' multigenic component and a 'vertical' mono- or oligogenic one. Pecchioni *et al.* (1996) found one major QTL on chromosome 7H, linked to the naked caryopsis (*nud*), and another one on the P arm of chromosome 2H. In addition they reported two minor QTLs. Finally, Thomson *et al.* (1997) found a dominant resistance gene on the long arm of chromosome 2H and designated it as *Rdg1.a.*

Common root rot and spot blotch on barley (*Cochliobolus sativus* (Ito & Kurib.) Drechsler ex Dastur. is caused by the ascomycete *Cochliobolus sativus* (perfect stage). The imperfect state is *Bipolaris sorokiniana* (Sacc.) Shoemaker (syn.: *Helminthosporium sativum* (Pamm) King et Bakke; *H. sorokinianum* Sacc. ex Sorok.).

Numerous accessions have been evaluated for resistance (Banttari *et al.*, 1975; Velibekova, 1981; Rochev and Levitin, 1986; Lehmann and Bothmer, 1988; Lukyanova, 1990; Gilchrist *et al.*, 1995; Semeane, 1995; Faiad *et al.*, 1996).

In genetic studies of this disease, investigators have concluded that resistance to spot blotch is due to a relatively small number of genes (Hayes and Stakman, 1921; Hayes *et al.*, 1923; Griffee, 1925; Arny, 1951; Luthra and Rao, 1973; Wilcoxson *et al.*, 1990). Hitherto five genes have been described and are listed in Table 8.7.

Other fungal diseases

In addition to the most important fungal diseases of barley worldwide, a number of other fungal pathogens are also significant.

Ear diseases

***Ustilaginales* – The smuts.** Three species of cereal smut affect barley: *Ustilago nuda* (Jens.) Rostr. (*U. segetum* var. *nuda*), *U. nigra* Tapke (*U. segetum* var. *avenae*) and *U. hordei* (Pers.) Lagerh. (*U. segetum*). Yield losses due to smut are currently controlled by seed treatments and by the use of resistant cultivars.

Numerous evaluations have been carried out (Shchelko, 1969; Nover *et al.*, 1976; Damania and Porceddu, 1981; Oniskova, 1987; Dunaevskij *et al.*, 1989, Surin, 1989; Lukyanova, 1990; Dubey and Mishra, 1992) identifying sources of resistance from Ethiopia, Yemen, Asia (especially Tibet), Canada and USA. Research was very extensive in the former USSR and Krivchenko (1984) has produced a review of the work on smut diseases. Investigations on the inheritance of smut resistance (Buivids, 1977; Emara and Freake, 1981; Bakhareva, 1987; Khokhlova, 1987, 1990; Mathur and Siradhana, 1990; Kozera and Roszko, 1995) and on the use of resistance genes in breeding (Czembor and Ralski, 1972; Anonymous, 1974; Rodina and

Table 8.7. Resistance genes for spot blotch.

Chromosome	Gene	Accession/cultivar	Author
2H	*(hl1*) Rcs1*		Steffenson *et al.* (1996)
5H	*(hl2) Rsc2*		*Ibid.*
7H	*(hl3) Rsc3*		*Ibid.*
	(hl4) Rsc4		*Ibid.*
1H	*Rcs5*	Steptoe	*Ibid.*

*former designation by Søgaard and Wettstein-Knowles (1987)

Soshnikova, 1986; Wicke and Weltzien, 1986; Surin and Lyakhova, 1993) have also been carried out.

Mitosporic *Ascomycota* – The fusarioses. Fusarioses on barley, most often caused by *Fusarium graminearum* Schwabe (*Gibberella zeae* (Schw.) Petch) and *F. culmorum* (W.G. Sm.) Sacc., are important as head (scab) and root diseases. Besides the negative influences on yield, the mycotoxin production of *Fusarium* spp. is of great importance. In light of the epidemics of 1991-1996, McMullen *et al.* (1997) published a review about scab on wheat and barley as a re-emerging problem in the USA. Many authors have documented inter-cultivar differences (Grigoryev *et al.*, 1988; Leur, 1989; Gu, 1989; Corazza *et al.*, 1990; Khatskevitch and Benken, 1990; Lukyanova, 1990; Takeda, 1992; Filippova *et al.*, 1993; Nelson and Burgess, 1994; Perkowski *et al.*, 1995, 1997) and valuable sources of resistance seem to be particularly frequent in the East Asian region (Takeda and Heta, 1989). The resistance, more common in two-rowed than in six-rowed barley, appears to be non-race-specific and governed by additive factors (Takeda and Heta, 1989). Interspecific hybrids with *Roegneria (Elymus) tsukushiensis* var. *transiens* (Wan *et al.*, 1997) and *H. chilense* (Rubiales *et al.*, 1996) have also produced promising results and may act as an alternative in the near future.

Root diseases

Eyespot (*Cercosporella herpotrichoides* Fron.), take all (*Gaeumannomyces graminis* (Sacc.) Arx & Oliv., *Ophiobolus graminis* Sacc.) and sharp eyespot (*Rizoctonia cerealis* van der Hoeven) are the most common barley root diseases. Another frequent disease in some cold-temperate regions is damping off/winter kill due to *Typhula incarnata* Lasch.

As for other diseases in barley, resistance can be found both among cultivars (Bojarczuk and Bojarczuk, 1973; Chung *et al.*, 1978; Cavelier and Maroquin, 1980; Cunningham, 1980; Benada and Vanova, 1981; Clulov and Wale, 1984; Hollins and Scott, 1986) as well as in more distant genepools (Jensen and Jørgensen, 1976a, 1976b; Mielke, 1980). In particular, *H. chilense* and *H. jubatum* have demonstrated interesting properties.

Leaf diseases

In addition to the most important leaf diseases explained in previous sections, some others have been reported in barley. *Septoria nodorum* Berk. (*Leptosphaeria nodorum* Müll.) and *Selenophoma donacis* (Pass.) Sprague & A.G. Johnson are observed as threats in some regions and have therefore been subjected to screening and evaluation tests (Magnus, 1974; Brokenshire and Cooke, 1975, 1978; Cooke and Martin, 1977; Pluck *et al.*, 1981; Sharma and Brown, 1983; Cunfer *et al.*, 1984).

Diversity in resistance and tolerance to virus diseases

Viruses cause very important diseases of barley all over the world. In particular, the aphid-borne *barley yellow dwarf viruses* (BYDV-PAV, -MAV) and the *cereal yellow dwarf virus* (CYDV-RPV) are distributed in all regions where cereal and grass crops are grown. In contrast to BYDV/CYDV the fungus-borne barley mosaic virus complex occurs only in East Asia and Europe.

Cultivated barley is also a natural host for several other viruses not described in the following chapters, for example *barley stripe mosaic virus*, *barley yellow streak mosaic virus* and *wheat dwarf virus*.

The mosaic virus complex

One of the most serious diseases of winter barley is due to the mosaic virus complex. Depending on geographic region and specificity of the contaminated field, the disease is caused by *barley mild mosaic virus* (BaMMV) and *barley yellow mosaic virus* (BaYMV), both members of the genus *Bymovirus*. The vector of these viruses is the soil-borne fungus, *Polymyxa graminis* Led. In East Asia, mainly in Japan, BaYMV predominates (Saeki *et al.*, 1999), whereas in Europe winter barley is usually attacked by a mixed infection (Huth and Adams, 1990).

BaMMV and BaYMV clearly differ in serological properties, nucleotide sequence of the capsid proteins and their reaction to barley genotypes. Hence, the viruses are differentiated into strains that are distinguishable by the spectrum of pathogenicity to barley genotypes. In Japan, six strains of BaYMV have been classified (Kashiwazaki *et al.*, 1989). In Europe, two strains were isolated, BaYMV-1 and BaYMV-2 (Huth, 1989). Furthermore, two strains of BaMMV have been described from Japan (Nomura *et al.*, 1996).

Cultivation of resistant barley cultivars is the most effective measure of virus control. The evaluation of genetic resources, the analysis of resistance by conventional and molecular biological methods and the introgression of resistance genes into modern cultivars are therefore very important.

In Europe the reaction of barley genotypes can be classified into the following main groups (although exceptions occur):

- susceptibility to the whole virus complex
- resistance to BaMMV, only
- resistance to BaMMV and BaYMV-1, and
- resistance to the whole virus complex.

Resistance to the mosaic virus complex can usually be described as qualitative, because plants of resistant genotypes react without symptoms in a contaminated field and no virus can be detected in the plants by either biological, serological or electron microscopic methods. Other authors describe this type of resistance as immunity. However, some genotypes may also show elements of quantitative resistance (variable number of plants with symptoms, reduced intensity of symptoms or extended incubation period). Furthermore, some accessions susceptible to BaMMV following mechanical inoculation cannot be infected in fields contaminated with virus, which suggests that resistance is conferred by an impaired translocation of the virus within the plant.

Two centres of diversity include East Asia (the whole virus complex) and the Balkan area (BaMMV and BaYMV-1). As resistant genotypes are also located in other geographic regions, a rich source of genetic variability exists today. Yasuda and Rikiishi (1997) evaluated a total of 4,342 barley accessions from the world collection on a field in Japan infected with strain I (Kashiwazaki *et al.*, 1989) of BaYMV. The percentage of asymptomatic cultivars (degree 0) was highest among Ethiopian cultivars followed by those from Japan. Cultivars showing severe disease symptoms (degrees 3 and 4) were frequently found among Chinese, Nepalese, Southeast Asian, North African, North American and European barleys. Over the years genetic analyses have revealed a number of resistance genes (Table 8.8).

The gene *rym4* represents the major source of resistance present in European barley germplasm. Another gene, *rym5*, also located on chromosome 3H in close linkage to *rym4,* is of special interest to many breeders because it conditions resistance to BaYMV-2. The pyramiding of multiple resistance genes, for example *rym3* and *rym5*, into a single cultivar appears to be the best way to delay breakdown of the resistance (Ordon *et al.*, 1999). The reason why 'Mokusekko

Table 8.8. Resistance genes to the barley mosaic virus complex.

Chromosome	Resistance gene	Resistant to	Susceptible to	Accession designation	Origin	Author
4H	*rym1*	M, Y1/Y2		Mokusekko 3	Japan	Takahashi *et al.* (1973)*
7H	*Rym2*	M, Y1/Y2		Mihori Hadaka 3	Japan	Takahashi *et al.* (1973)*
5HS	*rym3*	all strains of Y	M	Mutant Ea 52	Japan	Ukai & Yamashita (1980)*
3HL	*rym4*	M + Y1	Y2	Ragusa	Yugoslavia	Graner & Bauer (1993)†
3HL	*rym5*	M, Y1/Y2	Y III	Resistant Ym No. 1	Japan	Konishi *et al.* (1989)*; Graner *et al.* (1999)
3HL	*rym6*	M, Y1/Y2		Miho Golden	Japan	Kaiser & Friedt (1989)*; Iida *et al.* (1999)
5HS	*rmm7*	M	Y1/Y2	HHOR 3365	Soviet Union	Graner *et al.* (1995)†
4HL	*rym8*	M, Y1	Y2	10247	Yugoslavia	Bauer *et al.* (1997)†
4HL	*rmm9*	M	Y1/Y2	Bulgarian 347		Bauer *et al.* (1997)†
3HL	*rym10*	M, Y1	Y2	Hiberna	Germany	Graner (pers. comm.), quoted in Schiemann (1999)
4HL	*rym11*	M, Y1/Y2		Russia 57	Soviet Union	Bauer *et al.* (1997)†

M = BaMMV, Y1 = BaYMV-1, Y2 = BaYMV-2, Y III = strain of BaYMV in Japan;
* cited by Saeki *et al.*, 1999; † cited by Graner *et al.*, 1999

3' is completely resistant in Japan to all strains of BaYMV, including strain III, is that this cultivar contains two different resistance genes, *rym1* and *rym5* (Saeki *et al.*, 1999). In Europe, *rym5* alone confers resistance to all strains of the BaMMV/BaYMV complex (Graner *et al.*, 1999).

Compared to cultivated barley, the number of resting spores of *Polymyxa graminis* in the roots is lower in ssp. *spontaneum* and extremely rare in *H. bulbosum* (Ordon *et al.*, 1997). Different genotypes of ssp. *spontaneum* from Israel and Turkey reacted either with resistance to BaMMV or the whole virus complex. Ruge *et al.* (2000) reported that resistance to the barley mosaic virus complex and other pathogens has been transferred from *H. bulbosum* into common barley.

Barley yellow dwarf viruses and cereal yellow dwarf virus

The virus disease generally known as barley yellow dwarf is distributed worldwide and can cause serious yield losses in all cereals and grasses (Lister and Ranieri, 1995), for instance in barley up to 25% (Pike, 1990). Based on the transmission efficiency by different aphid species, five strains of barley yellow dwarf viruses have been distinguished (Rochow, 1969; Rochow and Muller, 1971). With increasing knowledge of the serological relationships, ultrastructural symptomatology (Gill and Chong, 1979) and the sequences of the nucleic acids (Vincent *et al.*, 1991) the viruses have been divided into two subgroups. Following the latest virus taxonomy (Mayo and D'Arcy, 1999) these two groups are now described as different viruses of the family Luteoviridae, namely the barley yellow dwarf viruses (BYDV) and the cereal yellow dwarf virus

(CYDV).

Chemical control of aphids and specific agronomic practices are normally used to control the disease. However, because of the wide host range of the viruses the most reliable and environment-friendly method is the cultivation of resistant or tolerant cultivars. Already in the fifties, when the barley yellow dwarf was detected as a virus disease (Oswald and Houston, 1951), researchers started screening programmes to select resistant or tolerant sources. The first recessive resistance gene detected, *yd1*, was found in the spring barley 'Rojo' (Suneson, 1955) but was not included in practical breeding because of its low level of resistance/tolerance. Field resistance was also detected in several Ethiopian barleys (Schaller *et al.*, 1964). An incomplete dominant gene named *Yd2,* and later renamed *Ryd2*, was localised on the long arm of chromosome 3 near the centromere (Schaller *et al.*, 1963; Collins *et al.*, 1996). As the effectiveness of this gene depends, however, on the genetic background, the tested virus and the environmental conditions (Schaller, 1984), it is not effective in late ripening barleys (Jones and Catherall, 1970). *Ryd2* has been successfully used in breeding resistant spring barleys and has also been introduced into winter barley cultivars, such as 'Vixen', 'Wysor' and 'Venus' (see references in Burnett *et al.*, 1995). Several marker systems have been developed for the detection of the *Ryd2*, e.g., Holloway and Heath (1992) detected a polypeptide that was specifically expressed in plants with this gene. An important achievement for a rapid and reliable detection of the gene was recently reached with the development of an allele-specific PCR marker (Ford *et al.*, 1998).

The genetic basis of the resistance of other barleys described as 'moderately resistant' or 'BYDV-tolerant' is still unknown, although a polygenic inheritance is supposed. Research is currently being carried out to find QTL markers (Ordon, pers. comm.).

Diversity in resistance to pests and nematodes

Background

In their monumental review concerning the estimated crop losses in some of the world's most important crops Oerke *et al.* (1994) concluded that control measures against pests in barley are often uneconomic. This is not to say that insects or nematodes can indeed cause serious damage to barley crops around the world, locally or even regionally, but that losses in general are relatively limited (8.8% in the given estimates). Furthermore, many of the attacking insect pests in fact cause greater damage as vectors for plant viruses, notably barley yellow dwarf virus (BYDV). Consequently, many regions report that various aphid species are the most important pests affecting barley production. In this chapter, we will, therefore, place particular emphasis on the aphids, while mainly summarising the information available for other insect groups and nematodes.

Table 8.9. List of aphid pests of barley.

Species	Vernacular name	Main distribution area
Diuraphis noxia (Mordv.)	Russian wheat aphid	N America, Asia, S Africa
Metopolophium dirhodum (Walk.)	Rose-grain aphid	Worldwide
Rhopalosiphum maidis (Fitch)	Corn leaf aphid	Worldwide
Rhopalosiphum padi (L.)	Bird cherry-oat aphid	Worldwide
Schizaphis graminum (Rond.)	Greenbug	N America, Asia
Sipha flava (Forbes)	Yellow sugarcane aphid	The Americas
Sitobion avenae (Fabr.)	(English) grain aphid	Worldwide

Table 8.10. Summary of evaluations of barley genepools for resistance against aphid species.

Aphid species	Genepool studied		
	Primary (*H. vulgare* ssp. *vulgare* & ssp. *spontaneum*)	Secondary (*H. bulbosum*)	Tertiary (all other *Hordeum* spp.)
Diuraphis noxia	✓	✓	✓
Metopolophium dirhodum	✓		
Rhopalosiphum maidis	✓		
Rhopalosiphum padi	✓		✓
Schizaphis graminum	✓		✓[1]
Sitobion avenae	✓		

[1]*H. chilense*; see references by Castro *et al.* 1994

Aphids

Barley is regularly being attacked by several aphid species (Homoptera: Aphididae; Table 8.9), causing both direct damage to the crop and acting as vectors for a range of plant viruses, BYDV in particular. Investigations carried out in Great Britain (Vickerman and Wratten, 1979 – cited by Oerke *et al.*, 1994), however, have shown that aphid damage alone in spring barley can be as high as 14.5%. Because of their economic importance and rapid population development in the fields, due to their parthenogenetic reproductive behaviour, effective control is indispensable. One such control measure has been to utilise [genotypic] host-plant resistance, either expressed as non-preference (antixenosis), antibiosis or even tolerance (i.e., the plant's ability to withstand an infestation).

The search for resistant barley genotypes has been carried out in all three genepools (*sensu* Bothmer *et al.*, 1995, cf. also Chapter 2), although slightly dependent on the aphid species (Table 8.10). The primary genepool has been investigated extensively for resistance to *Diuraphis noxia* (Calhoun *et al.*, 1991; Webster *et al.*, 1991), *Rhopalosiphum maidis* (Narang and Rana, 1999) and *R. padi* (Rautapää, 1970; A'Brook and Dewar, 1977; Marrewijk and Dieleman, 1980; Weibull and Hanson, 1986). While this work has revealed useful genotypes on the cultivar level to be utilised immediately in breeding programmes to provide resistance to the first two species, good resistance to *R. padi* has mainly been restricted to *H. vulgare* ssp. *spontaneum* (Weibull, 1994). This vast source of genetic variation was also explored by Kindler *et al.* (1993), but, surprisingly, in a collection of 876 ssp. *spontaneum* accessions, no genotype was found to be resistant to *D. noxia*.

Inheritance of the resistance factor(s) to *Schizaphis graminum* and *D. noxia* varies from monogenically (Smith *et al.*, 1962; Robinson *et al.*, 1992) to possibly digenically (Nieto-Lopez and Blake, 1994), indicating a rather limited genetic base available for resistance breeding. On the other hand, resistance to *R. maidis* and *R. padi* appears to be governed by several genes expressing minor effects (Weibull, 1994; Moharramipour *et al.*, 1996, 1997).

The secondary genepool, i.e., *H. bulbosum*, has so far only been evaluated for resistance to *D. noxia* (Clement and Lester, 1990; Kindler and Springer, 1991; Kindler *et al.*, 1993) and has as such been found to possess highly valuable resistance traits. This interesting species, with novel resistance to various diseases, hybridises easily with cultivated barley (Subrahmanyam and Bothmer, 1987) and ought to be looked at in more detail regarding resistance to other aphid species.

Table 8.11. Summary of resistance levels of the third genepool taxa reported in the literature.

Section	Species	Aphid species		
		D. noxia	*R. padi*	*S. graminum*
Anisolepis	*chilense*	++		+++
	cordobense			
	euclaston			
	flexuosum		+++	
	intercedens			
	muticum			
	pusillum			
	stenostachys	+		
Critesion	*arizonicum*			
	comosum		+++	
	halophilum			
	jubatum	++	++	
	lechleri	-	++	
	procerum	++		
	pubiflorum			
Hordeum	*murinum s. lat.*	+		
Stenostachys	*bogdanii*	+++	+++	
	brachyantherum s. lat.	(+)+	+++	
	brevisubulatum s. lat.	++	++	
	capense			
	depressum			
	erectifolium			
	fuegianum			
	guatemalense			
	marinum s. lat.	+		
	parodii			
	patagonicum			
	roshevitzii		++	
	secalinum	++	++	
	tetraploidum			

Range given: +++ = highly resistant; - = susceptible

The tertiary genepool, finally, has been subjected to extensive studies by several authors (Weibull, 1987; Clement and Lester, 1990; Kindler and Springer, 1991; Kindler *et al.*, 1993) although much information is still lacking (Table 8.11). Particular focus has been placed on *H. chilense* and the intraspecific variation in host suitability to *S. graminum* (Castro *et al.*, 1994, 1995). Also resistant to *D. noxia*, this species has been used in developing amphiploids with *Triticum aestivum* and *T. turgidum* (Martín and Cubero, 1981). Recent work has concluded that resistance to both aphid species is in fact expressed in the amphiploids (Castro *et al.*, 1998) and that genes for effects influencing *S. graminum* life history parameters are located on different chromosomes (Castro *et al.*, 1996).

Very few studies have attempted to shed light on the possible interactions between the presence of resistance to aphids and genome structure. Weibull (1987), relating *R. padi* fecundity to ploidy levels of a set of wild barley species and interspecific hybrids, concluded that diploids (2x) were less suitable hosts than either tetra- or hexaploids. The observation that cultivated barley is diploid, and yet a good host, was explained by the long domestication history in *H.*

Table 8.12. Insect pests of barley - other than aphids - of possible economic importance.

Order	Species	Common name	Resistance available
Orthoptera	*Melanoplus* spp.	Grasshoppers	Yes
Hemiptera	*Laodelphax striatellus*	Small brown leafhopper	Yes
	Blissus leucopterus	Chinch bug	Yes
	Eurygaster integriceps	Sunn bug	No
Thysanoptera	*Limothrips denticornis*	Barley thrips	Not known
Lepidoptera	*Pseudaletia/Spodoptera*	Armyworms	Not known
	Agrotis/Heliothis	Cutworms	Not known
Diptera	*Chlorops pumilionis*	Gout fly	Not known
	Oscinella frit	Frit fly	Yes
	Opomyza florum	Yellow cereal fly	Yes
	Agromyza s. lat.	Mining flies	Yes
	Mayetiola destructor	Hessian fly	Yes
	Tipula sp.	Crane flies	Not known
Coleoptera	*Oulema melanopus*	Cereal leaf beetle	Yes
	Phyllotreta vittula	Barley flea beetle	Not known
Hymenoptera	*Cephus s. lat.*	Stem sawflies	Yes
	Tetramesa [Harmolita] hordei	Barley jointworm	Yes

vulgare and subsequent loss of unfavourable alleles. Kindler *et al.* (1993) reviewed the genome relationships within perennial Triticeae and their possible contribution to resistance to *D. noxia*, but were unable establish any correlation between the present genomic classification (H, I, X and Y; Bothmer *et al.*, 1995) and degree of resistance.

Few, if any, barley cultivars grown today have been bred specifically for resistance to aphids in spite of all the literature reporting valuable and highly significant resistance. This may possibly be explained by the laborious nature of aphid screening and breeding work. However, because certain aphid species continue to be serious pests of barley - notably *D. noxia* - the description of resistance genes or gene complexes, and their molecular markers, will greatly facilitate future breeding of resistant cultivars.

Other insect pests

Although many other insect species attack barley (Table 8.12), this brief review will only deal with those for which there is sufficient information available relating to barley diversity. From scanning the literature two major conclusions can be drawn: first of all, very little work has actually been done over the last 20 years to elucidate the nature of resistance for pests other than aphids and, secondly, knowledge of the genetics behind each known case of resistance is often imperfect. Monogenic resistance appears, however, to be involved in most of the cases studied so far: *Oulema melanopus* (Hahn, 1968), *Mayetiola destructor* (Olembo *et al.*, 1966) and *Tetramesa (Harmolita) hordei* (Sterling and Maclaren, 1975) although genetic analysis is partly lacking. Resistance to *M. destructor* is, moreover, associated with an incomplete factor Hf_2 working at higher temperatures which makes screening work cumbersome.

The identified resistance genes have so far been found exclusively in the primary genepool, e.g., in landraces such as described for *M. destructor* and *T. hordei*, often originating from areas where the pest is present. Typically, many of these ('Nile', 'Abusir', 'Mianwali', CIho 6469 and CIho 6671; USDA-ARS, 1999) were collected in the 1920-30s and represent exciting

Table 8.13. Summary of resistance genes against the cereal cyst nematode.

Resistance gene	Source	Origin	Marker(s)	Reference(s)
Ha1	Drost	Denmark*		
Ha2	KVL 191 (*H. pallidum*)	Unknown	pau RFLP AWBMA21 & MWG694	Andersen & Andersen (1973) Kretschmer *et al.* (1997)
Ha3	Allelic to *Ha2*			
Ha4	Courier New		RFLP XYL	Barr *et al.* (1998)

*The pedigree of 'Drost' includes both Swedish and Czech landraces

heterogeneous material often used for initial screening work. It is important to stress that resistance to new or up-and-coming pests is frequently found in closely related germplasm of the primary genepool, and that it is initially not necessary to extend the search for resistance genes to more distantly related material.

At present, few insect species other than aphids pose real threats to the world's barley production, which is not to say that no new problems may arise. In the case of future barley pests of significance, we can foresee an accelerating use of a range of molecular markers that will enhance and simplify screening and breeding work.

Nematodes

Barley, like most other cereal crops, suffers from damage by parasitic nematodes. Important species, for which information bearing on barley diversity is available, include the cereal cyst nematode (*Heterodera avenae* Woll.) and members of the root-knot nematodes (*Meloidogyne* spp.).

The cereal cyst nematode has been in focus for resistance studies in Denmark since the 1950s (Andersen, 1961). Somewhat later, Andersen and Andersen (1968) concluded from crossing experiments that two independently inherited major resistance genes, *Ha1* and *Ha2*, were available in 'Drost' and 'KVL 191', the latter of which is a primitive six-rowed form ('*Hordeum pallidum*'; Cotten and Hayes, 1969). Through linkage studies (Andersen and Andersen, 1973) both genes were found to be located on chromosome 2H. Because *H. avenae* continues to be a problem in barley-growing areas, notably Australia, a successful search for novel resistance genes has been pursued. A new gene, *Ha4*, was recently mapped to the long arm of chromosome 5H of 'Galleon' (Barr *et al.*, 1998). The present state of the art concerning *H. avenae*, including known markers, is summarised below (Table 8.13).

While the primary genepool has been important for genes resistant to *H. avenae*, more distantly related species within the tertiary genepool have been investigated for resistance to *Meloidogyne* spp. In this respect, *H. chilense* appears to stand out as one of the most promising sources of valuable genes (Person-Dedryver *et al.*, 1990). In a single study on genome effects within Triticeae, Jensen and Griffin (1994) concluded that the genomes J (*Thinopyrum*) and N (*Psathyrostachys*) were more resistant to *M. chitwoodi* than genomes P (*Agropyron*), S (*Pseudoroegneria*) and H (*Hordeum*). Their results confirmed the value of *H. chilense*. Further studies will show how these promising sources can be employed in future breeding.

Conclusions and outlook

The last century has seen a remarkable development in the ability to combat important diseases and pests. We have gained insight into intricate host-parasite relationships to a degree previously unthinkable, and we have also been able to greatly extend the spheres from which resistance genes can be brought. Nevertheless, as the world's barley crops continue to be haunted by organisms, the breeding work must proceed.

The vast number of examples reported here shows us that the desired diversity is often available in the primary genepool and may readily be utilised in breeding programmes. On the other hand, to manage a few difficult problems we have had to turn to the diversity found in more distant genepools, thereby accepting the obstacles involved in bringing the desired genes into adapted cultivars. Despite a few successful examples, it is obvious that we need to attain higher success in these attempts and work on developing the methods for the transfer of exotic genes. Solving the problems associated with interspecific hybridisation in *Hordeum* is therefore a main challenge for future breeders and geneticists.

Resistance breeding is often a cumbersome task demanding sharp selection tools to be efficient. With the arrival of new marker techniques, however, selection methodology has taken a great leap forward and in many instances revolutionised the breeding work. Illustrative examples in this respect are the mosaic viruses and the cereal cyst nematode. It is our firm belief that this development will continue and include many more organisms to facilitate the task of breeders carrying out their important mission.

Acknowledgements

The invaluable assistance of the following persons in writing sub-sections of this chapter is gratefully acknowledged: M. Schönfeld – powdery mildew, S. Yasuda – barley yellow mosaic virus. I. Terentyeva, St. Petersburg, provided information regarding research in Russia and the former Soviet Union.

References

A'Brook, J.A., and A.M. Dewar, 1977. Evaluation of different screening techniques for testing resistance in barley to cereal aphids. IOBC/WPRS Bull. 1977/3: 31-35.

Abbott, D.C., E.S. Lagudah and A.H.D. Brown, 1995. Identification of RFLPs flanking a scald resistance gene on barley chromosome 6. J. Heredity 86: 152-154.

Afanasenko, O.S., H. Hartleb, N.N. Guseva, V. Minarikova and M. Janosheva, 1995. A set of differentials to characterise populations of *Pyrenophora teres* Drechs. for international use. J. Phytopath. 143: 501-507.

Afanasenko, O.S., and I.Yu. Kushnirenko, 1989. Inheritance of resistance to the net blotch pathogen in some barley varieties. Genetika Moskva 25: 1994-2000.

Afanasenko, O.S., I.A. Terentyeva and I.N. Makarova, 2000. Landraces from Peru – new sources of resistance to net blotch of barley. In: Barley Genetics VIII. Proc. 8th Int. Barley Genet. Symp., Adelaide, Australia, pp. 71-72.

Alemayehu, F., and J.E. Parlevliet, 1997. Variation between and within Ethiopian barley landraces. Euphytica 94, 183-189.

Ali, S.M., A.H. Mayfield and B.G. Clare, 1976. Pathogenicity of 203 isolates of *Rhynchosporium secalis* on 21 barley cultivars. Physiol. Plant Pathol. 9: 135-143.

Andersen, K., and S. Andersen, 1968. Inheritance of resistance to *Heterodera avenae* in barley. Nematologica 14: 128-130.

Andersen, S., 1961. Resistens mod havreål, *Heterodera avenae*. Ph. D. thesis, Dansk Videnskabs Forlag, Copenhagen. 179 p. (In Danish).

Andersen, S., and K. Andersen, 1973. Linkage between marker genes on barley chromosome 2 and a gene for resistance to *Heterodera avenae*. Hereditas 73: 271-276.

Anderson, J.B., and J. Torp, 1986. Quantitative analysis of the early powdery mildew infection stages on resistant barley genotypes. J. Phytopathol. 115: 173-185.

Anikster, Y., and I. Wahl, 1979. Coevolution of rust fungi on Gramineae and Liliaceae and their hosts. Ann. Rev. Phytopathol. 117: 367-430.

Anonymous, 1972. Research Branch Report 1971, Canada, 377 p.

Anonymous, 1974. Dept. of Agriculture Research Branch Report 1973, I. Crop Plants, Canada, 376 p.

Arabi, M.I., A. Sarrafi, G. Barrault and L. Albertini, 1990. Inheritance of partial resistance to net blotch in barley. Plant Breed. 105: 150-155.

Arny, D.C., 1951. Inheritance of resistance in barley to headblight incited by *Helminthosporium sativum*. Ph.D. thesis, University of Minnesota, St. Paul, 83 p.

Arny, R.E., 1945. Inheritance of resistance to barley stripe. Phytopathology 35: 781-804.

Asher, M.J.C., and C.E. Thomas, 1984. Components of partial resistance to *Erysiphe graminis* in spring barley. Plant Pathol. 33: 123-30.

Asher, M.J.C., and C.E. Thomas, 1987. The inheritance of mechanisms of partial resistance to *Erysiphe graminis* in spring barley. Plant Pathol. 36: 66-72.

Asher, M.J.C., and W.T.B. Thomas, 1983. The genetical control of incomplete forms of resistance to *Erysiphe graminis* in spring barley. Ann. Appl. Biol. 103: 149-156.

Baigulova, G.K., and A.A. Pitonya, 1979. *Helminthosporium* on barley. Uzbekiston Biologija Zurnali 4: 55-57. (In Russian).

Baker, E.P., 1964. Further evidence for an allelic series conditioning resistance to powdery mildew at the Algerian locus in barley. In: Barley Genetics I. Proc. 1st Int. Barley Genet. Symp., Pudoc Wageningen, The Netherlands, pp. 236-238.

Baker, R.J., and E.N. Larter, 1963. The inheritance of scald resistance in barley. Can. J. Genet. Cytol. 5: 445-449.

Bakhareva, Zh.A., 1987. Using the results of diallel analysis in breeding barley for resistance to loose smut. In: Principles and methods of breeding intensive cultivars of agricultural crops. Novosibirsk, pp. 85-89. (In Russian)

Banttari, E.E., W.H. Anderson and D.C. Rasmusson, 1975. *Helminthosporium* headblight resistance in six-row spring barleys. Plant Dis. Rep. 59: 274-277.

Barr, A.R., K.J. Chalmers, A. Karakousis, J.M. Kretschmer, S. Manning, R.C.M. Lance, J. Lewis, S.P. Jeffries and P. Langridge, 1998. RFLP mapping of a new cereal cyst nematode resistance locus in barley. Plant Breed. 117: 185-187.

Beer, W.W., 1991. Leaf blotch of barley (*Rhynchosporium secalis*). Zentralbl. Mikrobiol. 146: 339-358.

Benada, J., and M. Vanova, 1981. The resistance of winter wheat and winter barley to eyespot (*Cercosporella herpotrichoides*) and the effect of fungicides on yield. Ochrana Rostlin 17: 257-264.

Bisht, I.S., and S.K. Mithal, 1991. Evaluation of barley germplasm for resistance against stripe disease. Indian Phytopathol. 44: 516-517.

Bobes, I., N. Florea and E. Perseca, 1974. Response of varieties and lines of barley to attack by *Helminthosporium gramineum* Rabenh. Lucrari Stiintifice, Institutul Agronomic "Dr. Petru Groza", Agricultura 28: 251-256.

Bockelman, H.E., E.L. Sharp and R.F. Eslick, 1977. Trisomic analysis of several scald and net blotch resistance genes. Barley Genet. Newsl. 7: 11-15.

Bojarczuk, J., and M. Bojarczuk, 1973. Assessment of resistance of winter barley to eyespot (*Cercosporella herpotrichoides* Fron.). Hodowla Roślin, Aklimatyzacja i Nasiennictwo 17: 215-228.

Bothmer, R. von, 1992. The wild species of *Hordeum*. Relationships and potential use for improvement of cultivated barley. In: P.R. Shewry (ed), Barley. Genetics, Biochemistry, Molecular Biology and Biotechnology, CAB International, pp. 3-18.

Bothmer, R. von, N. Jacobsen, C. Baden, R.B. Jørgensen and I. Linde-Laursen, 1995. An ecogeographical study of the genus *Hordeum*. 2nd edition. Systematic and Ecogeographic Studies on Crop Genepools 7. IPGRI, Rome.

Boulif, M., and R.D. Wilcoxson, 1988. Inheritance of resistance to *Pyrenophora graminea* in barley. Plant Dis. 72: 233-238.

Briggs, F.N., and E.H. Stanford, 1938. Linkage of factors for resistance to mildew in barley. J. Genet. 37: 107-117.

Brokenshire, T., and B.M. Cooke, 1975. The reaction of spring barley cultivars to *Selenophoma donacis*. Trans. Brit. Mycol. Soc. 65: 443-449.

Brokenshire, T., and B.M. Cooke, 1978. The effect of *Selenophoma donacis* at different growth stages on spring barley cultivars. Ann. Appl. Biol. 89: 211-217.

Brown, J.K.M., and J.H. Jørgensen, 1991. A catalogue of mildew resistance genes in European barley varieties. In: J.H. Jørgensen (ed), Integrated control of cereal mildews. Virulence patterns and their change. Risø National Laboratory, Roskilde, Denmark, pp. 263-272.

Brown, J.S., 1985. Pathogenic variation among isolates of *Rhynchosporium secalis* from cultivated barley growing in Victoria, Australia. Euphytica 34: 120-133.

Bryner, C.S., 1957. Inheritance of scald resistance in barley. Ph.D. thesis, Pennsylvania State University, University Park, Pennsylvania.

Buchannon, K.W., and W.C. McDonald, 1965. Sources of resistance in barley to *Pyrenophora teres*. Can. J. Plant Sci. 45: 189-193.

Burnett, P.A., A. Comeau and C.O. Qualset, 1995. Host Plant tolerance or resistance for control of barley Yellow Dwarf. In: C.J. D'Arcy and P.A. Burnett (eds), Barley Yellow Dwarf – 40 Years of Progress. APS Press, St. Paul, pp. 321-343.

Buivids, K.P., 1977. Study of resistance of spring barley to loose smut. Ph. D. thesis abstract, Leningrad, 23 p. (In Russian).

Calhoun, D.S., P.A. Burnett, J. Robinson and H.E. Vivar, 1991. Field resistance to Russian wheat aphid in barley: I. Symptom expression. Crop Sci. 31: 1464-1467.

Carver, T.L.W., 1986. Histology of infection by *Erysiphe graminis* f. sp. *hordei* in spring barley lines with various levels of partial resistance. Plant Pathol. 35: 232-240.

Castro, A.M., A. Martín and L.M. Martín, 1996. Location of genes controlling resistance to greenbug (*Schizaphis graminum* Rond.) in *Hordeum chilense*. Plant Breed. 115: 335-338.

Castro, A.M., L.M. Martín and A.F.G. Dixon, 1995. Genetic variability in antibiotic resistance to the greenbug *Schizaphis graminum* in *Hordeum chilense*. Plant Breed. 114: 510-514.

Castro, A.M., L.M. Martín, A. Martín, H.O. Arriaga, N. Tobes and L.B. Almaraz, 1994. Screening for greenbug resistance in *Hordeum chilense* Roem. et Schult. Plant Breed. 112: 151-159.

Castro, A.M., A. Vasicek, S. Ramos, A. Martín, L.M. Martín and A.F.G. Dixon, 1998. Resistance against greenbug, *Schizaphis graminum* Rond., and Russian wheat aphid, *Diuraphis noxia* Mordvilko, in tritordeum amphiploids. Plant Breed. 117: 515-522.

Cavelier, M., and C. Maroquin, 1980. Varietal susceptibility of winter barley to *Typhula incarnata* Lasch and to barley yellow dwarf virus. Bull. Recherches Agron. Gembloux, Belgium 15:309-325.

Ceccarelli, S., E. Piano and S. Arcioni, 1976. Evaluation of selected F4 and F5 families in a pedigree programme in barley (*Hordeum vulgare* L.). In: Abstracts, XX Annual Meeting, Italian Society for Agricultural Genetics. Papers on miscellaneous crop plants 30: 91.

Ceolini, C., 1980. Race differentiation and search for sources of resistance to *Rhynchosporium secalis* in barley in Italy. Euphytica 29: 547-553.

Chen, X.M., and R.F. Line, 1999. Recessive genes for resistance to *Puccinia striiformis* f. sp. *hordei* in barley. Phytopathology 89: 226-232.

Cherif, M., and M. Harrabi, 1993. Transgressive segregation for resistance to *Pyrenophora teres* in barley. Plant Pathol. 42: 617-621.

Choo, T.M., A. Tekauz, K.M. Ho and R.A. Martín, 1994. Use of doubled haploids in studying net blotch resistance in barley. In: Proc. SABRAO 7th Intern. Congr. and WSAA Symp., Taipei, Taiwan, Vol. I. Spec. Publ. Taichung District Agric. Improvem. Stn. 35: 179-187.

Chung, B.J., J.M. Sung, K.S. Kim and K.S. Sung, 1978. The occurrence of *Ophiobulus graminis* Sacc. on wheat and barley in Korea. Kor. J. Plant Prot. 17: 49-51.

Clement, S.L., and D.G. Lester, 1990. Screening wild *Hordeum* species for resistance to Russian wheat aphid. Cer. Res. Comm. 18: 173-177.

Clifford, B.C., 1985. Barley leaf rust. In: A.P. Roelfs and W.R. Bushnell (eds), The Cereal Rusts. Vol. II. Diseases, distribution, epidemiology, and control, pp. 173-205.

Clulov, M., and S.J. Wale, 1984. Resistance of cultivars of wheat, barley and oats to eyespot. Ann. Appl. Biol. 104. Suppl., Test of Agrochemicals and Cultivars 5: 104-105.

Collins, N.C., N.G. Paltridge, C.M. Ford and R.H. Symons, 1996. The *Yd2* gene for barley yellow dwarf virus resistance maps close to the centromere on the long arm of barley chromosome 3. Theor. Appl. Genet. 92: 858-864.

Cooke, B.M., and S.I. Martin, 1977. The potential of *Septoria nodorum* on barley and *Septoria avenae* f. sp. *triticea* on wheat and barley to reduce grain yield and quality in spring and winter varieties. In: Progress Reports on Research and Development 1975/76, Univ. Wales, Cardiffe, UK. Highgate, London, pp. 38-40.

Corazza, L., V. Balmas, G. Chilosi and R. Nalli, 1990. Evaluation of resistance of wheat and barley varieties to attack by pathogenic species of *Fusarium*. Sement Elette 36: 25-28.

Cotten, J., and J.D. Hayes, 1969. Genetic resistance to the cereal cyst nematode (*Heterodera avenae*). Heredity 24: 593-600.

Cromey, M.G., 1987. Pathogenic variation in *Rhynchosporium secalis* on barley in New Zealand. New Zealand J. Agric. Res. 30: 95-99.

Cunfer, B.M., J.W. Johnson and A.R. Brown, 1984. Resistance of winter barley to *Leptosphaeria nodorum*. Phytopathology 74: 878.

Cunningham, P.C., 1980. Differential host preferences in *Pseudocercosporella herpotrichoides* with spring sown wheat and barley monoculture. Ann. Appl. Biol. 94: 33-39.

Czembor, H.J., and E. Ralski, 1972. Problems of breeding for disease resistance in the light of literature and the authors own work. Biul. Inst. Hodowli i Aklimatyzacji Roślin 3-4: 55-65.

Damania, A.B., and E. Porceddu, 1981. Screening YAR barley for disease resistance. Plant Genet. Res. Newsl. 48: 2-3.

Delogu, G., A. Infantino, A. Gatti, C. Delogu, A. Porta-Pulia and G. Vannacci, 1989. Variability of the sensitivity of barley genotypes to different isolates of *Drechslera graminea* Shoemaker. In: XII Eucarpia Congr., Göttingen, Germany. Book of poster abstracts, part I, pp. 3-16.

Delogu, G., A. Porta-Pulia, A.M. Stanca and G. Vannacci, 1995. Interaction between barley and *Pyrenophora graminea*. An overview of research in Italy. Rachis 14: 29-34.

Dheeranupattana, S., 1995. Untersuchungen zur Vererbung der Mehltauresistenz bei Gerste. Diss. TU München/Weihenstephan, 125 p.

Doll, H., and H.P. Jensen, 1986. Localization of powdery mildew resistance gene *Ml-ra* on barley chromosome 5. Hereditas 105: 61-65.

Douglas, G.B., and J.L. Gordon, 1985. Quantitative genetics of net blotch resistance in barley. New Zealand J. Agric. Research 28: 157-164.

Douiyssi, A., D.C. Rasmusson and A.P. Roelfs, 1998. Responses of barley cultivars and lines to isolates of *Pyrenophora teres*. Plant Dis. 82: 316-321.

Dubey, P., and B. Mishra, 1992. Evaluation of barley varieties against covered smut. J. Res. Birsa Agric. Univ. 4: 179-180.

Dubin, H.J., and R.W. Stubbs, 1986. Epidemic spread of barley stripe rust in South America. Plant Dis. 70: 141-144.

Dunaevskij, A.G., V.V. Sotnikov and E.V. Bibik, 1989. Research for sources of resistance to the main diseases in spring barley. Selektsiya i Semenovodstvo, Kiev 67: 14-19. (In Russian).

Dyck, P.L., and C.W. Schaller, 1961. Inheritance of resistance in barley to several physiologic races to the scald fungus. Can. J. Genet. Cytol. 3: 165-169.

Emara, Y.A., and G.W. Freake, 1981. Quantitative inheritance of host-parasite relationship in barley-*Ustilago* system. Abstract. Genetics 97, 1, Supplement s34.

Faiad, M.G.R., M.M.V.S. Wetzel, A.N. Salomao and R. Cunha, 1996. Evaluation of fungi in seed germplasm before long term storage. Seed Sci. Technol. 24: 505-511.

Falak, I., D.E. Falk, N.A. Tinker and D.E. Mather, 1999. Resistance to powdery mildew in a doubled haploid barley population and its association with marker loci. Euphytica 107: 185-192.

FAO, 1999. Annual statistics (data available at http://apps.fao.org/ in October 1999).

Favret, E.A., 1960. Spontaneous and induced mutations of barley for the reaction of mildew. Hereditas 46: 20-28.

Favret, E.A., 1965. Induced mutations in breeding for disease resistance. Radiation Bot. 5: 521-536.

Feuerstein, U., A.H.D. Brown and J.J. Burdon, 1990. Linkage of rust resistance genes from wild barley (*Hordeum spontaneum*) with isoenzyme markers. Plant Breed. 104: 318-324.

Filippova, G.G., L.A. Kashemirova and I.V. Chiburova, 1993. On seed testing in spring barley. Zashchita Rasteniy, Moskva 11: 29-30. (In Russian).

Fischbeck, G., E. Schwarzbach, Z. Sobel and I. Wahl, 1976. Mehltauresistenz aus israelischen Populationen der zweizeiligen Wildgerste (*Hordeum spontaneum*). Z. Pflanzenzüchtg. 76: 163-166.

Ford, C.M., N.G. Paltridge, J.P. Rathjen, R.L. Moritz, R.J. Simpson and R.H. Symons, 1998. Rapid and informative assays for *Yd2*, the barley yellow dwarf virus resistance gene, based on the nucleotide sequence of a closely linked gene. Mol. Breed. 4: 23-31.

Fox, S.L., and D.E. Harder, 1996. Resistance to stem rust on barley and inheritance of resistance to race QCC. Can. J. Plant Sci. 75: 781-788.

Franckowiak, J.D., Y. Jin and B.J. Steffenson, 1997a. Recommended allele symbols for leaf rust resistance genes in barley. Barley Genet. Newsl. 27: 36-41.

Franckowiak, J.D., U. Lundqvist and T. Konishi, 1997b. New and revised names for barley genes. Barley Genet. Newsl. 26: 4-8.

Frecha, J.H., 1958. Behavior of barleys in regard to the parasitic population of *H. teres* in Argentine. Inheritance of resistance of factors to *H. teres* of the barley vars. Rolo, Hoyo Epuyen and *H. spontaneum*. Rev. Invest. Agr. Buenos Aires 12: 41-45.

Freisleben, R., 1940. Die phylogenetische Bedeutung asiatischer Gersten. Züchter 12: 257-272.

Freisleben, R., and A. Lein, 1942. Über die Auffindung einer mehltauresistenten Mutante nach Röntgenbestrahlung einer anfälligen reinen Linie von Sommergerste. Naturwissenschaften 30: 608.

Freisleben, R.L., and J. Metzger, 1942. Genetische Studien zur Gerstenzüchtung. Z. Pflanzenzüchtg. 24: 507-522.

Fukuyama, T., S. Yamaji and H. Nakamura, 1998. Differentiation of virulence in *Rhynchosporium secalis* in the Hokuriku district and sources of resistance to the pathogen. Breed. Sci. 48: 23-28.

Gaike, M.V., 1970. Resistance of barley cultivars to net blotch. In: Genetic basis of disease resistance of field crops. Riga, pp. 70-74. (In Russian).

Gaike, M., 1973. A study of initial material for breeding barley at the Priekuli experimental breeding station. In: G. Egorova (ed), Breeding and Seed Production of Crop Plants. Zinatne, Riga, Latvia, pp. 29-35. (In Russian).

Garkavyi, P.F., T.A. Nechaeva and Ya.K. Nazarenko, 1974. Breeding winter barley for resistance to fungal diseases. Nauch.-tekhn. Byul. Vses. Selekts.-genet. In-ta No. 22: 57-61. (In Russian).

Geiger, H.H., and M. Heun, 1989. Genetics of quantitative resistance to fungal diseases. Ann. Rev. Phytopathol. 27: 317-341.

Giese, H., J.H. Jørgensen, H.P. Jensen and J. Jensen, 1981. Linkage relationship of ten powdery mildew resistance genes of barley chromosome 5. Hereditas 95: 43-50.

Gilchrist, S.L., F.H. Vivar, C.F. Gonzalez and C. Velazquez, 1995. Selecting sources of resistance to *Cochliobolus sativus* under subtropical conditions and preliminary loss assessment. Rachis 14: 35-39.

Gill, C.C., and J. Chong, 1979. Cytopathological evidence for the division of barley yellow dwarf virus isolates into two subgroups. Virology 95: 59-69.

Giuliari, S., G. Beni and S. Padovani, 1984. Barley variety trials in the Veneto region, 1983-84. Informatore Agrario 40: 47-49.

Golan, T., Y. Anikster, J.G. Moseman and I. Wahl, 1978. A new virulent strain of *Puccinia hordei*. Euphytica 27: 185-189.

Graner, A., B. Foroughi-Wehr and A. Tekauz, 1996. RFLP mapping of a gene in barley conferring resistance to net blotch (*Pyrenophora teres*). Euphytica 91: 229-234.

Graner, A., S. Streng, A. Kellermann, A. Schiemann, E. Bauer, B. Waugh, B. Pellio and F. Ordon, 1999. Molecular mapping and genetic fine structure of the *rym5* locus encoding resistance to different strains of the barley yellow mosaic virus complex. Theor. Appl. Genet. 98: 285-290.

Gray, G.G., 1966. Genetic system in the net blotch disease complex of barley. Ph.D. Thesis, North Dakota State University, Fargo, 65 p.

Griffee, F., 1925. Correlated inheritance of botanical characters in barley and the manner reaction to *Helminthosporium sativum*. J. A. Agric. Res. 30: 915-933.

Grigoryev, M.F., M.V. Lukyanova, N.A. Kabalkina and A.A. Sidorov, 1988. Sources of resistance in barley to *Fusarium* and *Helminthosporium* root rots. Sb. Nauchn. Tr. Prikl. Bot., Genet. i Sel. 122: 87-91. (In Russian).

Gu, X., 1989. Trials of several malting barley cultivars from German Federal Republic. Zuowu Pinzhong Ziyuan 1: 34-35.

Gustafsson, M., and R. von Bothmer, 1994. Present research projects in barley at the Institute of Plant Breeding, Svalöv. J. Swed. Seed Ass. 104: 211-216.

Habgood, R.M., and J.D. Hayes, 1971. The inheritance of resistance to *Rhynchosporium secalis* in barley. Heredity 27: 25-37.

Hahn, S.K., 1968. Resistance of barley (*Hordeum vulgare* L. emend. Lam.) to cereal leaf beetle (*Oulema melanopus* L.). Crop Sci. 8: 461-464.

Hänsel, H., and J. Zakovsky, 1956. Röntgeninduzierte Mutanten der Vollkorngerste. I. Bestrahlung und Auslese auf Mehltauresistenz. Bodenkultur 9: 50-64.

Hansen, L.R., and H.A. Magnus, 1973. Virulence spectrum of *Rhynchosporium secalis* in Norway and sources of resistance in barley. Phytopathol. Z. 76: 303-313.

Harder, D.E, and W.G. Legge, 1996. Stem rust resistance in barley - an overview. In: Proc. 9th European and Mediterranean Cereal Rusts and Powdery Mildew Conference, Lunteren, Netherlands, pp. 182-185.

Harlan, J.R., and D. Zohary, 1966. Distribution of wild wheats and barley. Science 153: 1074-1080.

Hartmann, S., 1993. Differenzierung und Vererbung von Resistenz bei Gerste (*Hordeum* L. sp.) gegen die Netzfleckenkrankheit (*Pyrenophora teres* f. *teres* und *Pyrenophora teres* f. *maculata*). Diss. TU München, 169 p.

Hayashi, J., and H. Heta, 1985. Association of a mildew resistance gene *JMlh* in Hanna CI 906 with chromosome 6. Barley Genet. Newsl. 15: 46-47.

Hayes, H.K., and E.C. Stakman, 1921. Resistance of barley to *Helminthosporium sativum* P.K. and B. Phytopathology 11: 405-411.

Hayes, H.K., E.C. Stakman, F. Griffee and J.J. Christensen, 1923. Reaction of barley varieties to *Helminthosporium sativum*. I. Varietal resistance, II. Inheritance studies in a cross between Lion and Manchuria. Minn. Agric. Exp. Stn. Tech. Bull. 21, 47 p.

Hayes, P., D. Prehn, H. Vivar, T. Blake, A. Comeau, J. Henry, M. Johnston, B. Jones, B.J. Steffenson and C.A. St. Pierre, 1996. Multiple disease resistance loci and their relationship to agronomic and quality loci in a spring barley population. http//www.css.orst.edu/barley/JQTL/JQTL2.HTM.

Hentrich, W., 1977. Test for the selection of mildew-resistant mutants in spring barley. In: Induced Mutations against Plant Diseases, IAEA, Vienna, pp. 333-341.

Heschele, E.E., 1928. The response of barleys to parasitic fungi *Helminthosporium teres* (Sacc.). Byul. Prikl. Bot., Genet. i Sel. 19: 371-384. (In Russian).

Heun, M., 1984. Localization of induced resistance genes in barley for resistance against powdery mildew. Z. Pflanzenzüchtg. 93: 158-168.

Heun, M., 1986. Quantitative differences in powdery mildew resistance among spring barley cultivars. J. Phytopathol 115: 222-228.

Hill, J.P., M.R. Johnston, V.R. Velasco and W.M. Brown, 1995. Responses of selected barley lines to barley stripe rust in Bolivia, Ecuador and Germany. Phytopathology 85: 1040.

Hiura, U., 1960. Studies on the disease resistance in barley. IV. Genetics of resistance to powdery mildew. Ber. Ohara Inst. Landwirtsch. Biol. Okayama 11: 235-300.

Hoffmann, W., and H. Kuckuck, 1938. Versuche zur Züchtung spelzenfreier, eiweißreicher und mehltauwiderstandsfähiger Gersten. Z. Züchtung A 22: 271-302.

Hoffmann, W., and I. Nover, 1959. Ausgangsmaterial für die Züchtung mehltauresistenter Gersten. Z. Pflanzenzüchtg. 42: 68-78.

Hollins, T.W., and P.R. Scott, 1986. Eyespot, take-all, sharp eyespot. In: Ann. Rep. Plant Breed. Inst. Cambridge, pp. 100-103.

Holloway, P.J., and R. Heath, 1992. Identification of polypeptide markers of barley yellow dwarf virus resistance and susceptibility genes in non-infected barley (*Hordeum vulgare*) plants. Theor. Appl. Genet. 85: 346-352.

Honecker L., 1931. Beiträge zum Mehltauproblem bei der Gerste mit besonderer Berücksichtigung der züchterischen Seite. Pflanzenbau 8: 78-84.

Honecker, L., 1938. Über die physiologische Spezialisierung des Gerstenmehltaus als Grundlage für die Immunitätszüchtung. Züchter 10: 169-181.

Hossain, M.A., and D.H.B. Sparrow, 1991. Resistance to powdery mildew (*Erysiphe graminis* f. sp. *hordei*) in the barley cultivar galleon. I. Relationship with known genes for resistance. Euphytica 52: 1-9.

Huth, W., 1989. Ein weiterer Stamm des barley yellow mosaic virus aufgefunden. Nachrichtenbl. Dtsch. Pflanzenschutzdienst 41: 6-7.

Huth, W., and M. Adams, 1990. Barley yellow mosaic virus (BaYMV) and BaYMV-M, two different viruses. Intervirology 31: 38-42.

Iida, Y.T., T. Ban and T. Konishi, 1999. Linkage analysis of the *rym6* resistance gene to Japanese strain II of barley yellow mosaic virus (BaYMV-II) in barley. Barley Genet. Newsl. 29: 31-32.

Isenbeck, K., 1930. Untersuchungen über *Helminthosporium gramineum* Rabh. im Rahmen der Immunitätszüchtung. Phytopathol. Z. 11: 503-555.

Ivandic, V., U. Walther and A. Graner, 1998. Molecular mapping of a new gene in wild barley conferring complete resistance to leaf rust (*Puccinia hordei* Otth). Theor. Appl. Genet. 97: 1235-1239.

Jackson, L.F., and R.K. Webster, 1976. Race differentiation, distribution, and frequency of *Rhynchosporium secalis* in California. Phytopathol. 66: 719-725.

Jahoor, A., and G. Fischbeck, 1987a. Genetical studies of resistance of powdery mildew in barley lines derived from *Hordeum spontaneum* collected from Israel. Plant Breed. 99: 265-273.

Jahoor, A., and G. Fischbeck, 1987b. Sources of resistance to powdery mildew in barley lines derived from *Hordeum spontaneum* collected in Israel. Plant Breed. 99: 274-281.

Jahoor, A., and G. Fischbeck, 1993. Identification of new genes for mildew resistance of barley at the *Mla* locus in lines derived from *Hordeum spontaneum*. Plant Breed. 110: 116-122.

Jahoor, A., A. Ludwig and G. Fischbeck, 1989. New genes for powdery mildew resistance in *Hordeum spontaneum* derived barley lines allelic or closely linked to the *Mlp* locus. Barley Genet. Newsl. 19: 23-26.

Jahoor, A., U. Stephan and G. Fischbeck, 1990. Study of powdery mildew resistance gene from 'Engledow India'. Barley Genet. Newsl. 20: 41-42.

Jana, S., and K.L. Bailey, 1995. Responses of wild and cultivated barley from West Asia to net blotch and spot blotch. Crop Sci. 35: 242-246.

Jedel, P.E., D.R. Metcalf and J.W. Martens, 1989. Assessment of barley accessions PI 382313, PI 382474, PI 382915 and PI 382976 to stem rust resistance. Crop Sci. 29: 1473-1477.

Jensen, H.P., E. Christensen and J.H. Jørgensen, 1992. Powdery mildew resistance genes in 127 Northwest European spring barley varieties. Plant Breed. 108: 210-228.

Jensen, H.P., and J.H. Jørgensen, 1976a. Screening of barley varieties for resistance to the take-all fungus, *Gaeumannomyces graminis*. Z. Pflanzenzüchtg. 76: 152-162.

Jensen, H.P., and J.H. Jørgensen, 1976b. Screening of *Hordeum* species for resistance to the take-all fungus, *Gaeumannomyces graminis*. Z. Pflanzenzüchtg. 76: 200-203.

Jensen., J., 1979. Report from Coordinators. Chromosome 5. Barley Genet. Newsl. 9: 118-120.

Jensen, K.B., and G.D. Griffin, 1994. Resistance of diploid Triticeae species and accessions to the Columbia root-knot nematode, *Meloidogyne chitwoodi*. Suppl. J. Nematol. 26(4S): 635-639.

Jin, Y., B.J. Cui, B.J. Steffenson and J.D. Franckowiak, 1996. New leaf rust resistance genes in barley and their allelic and linkage relationship with other Rph-genes. Phytopathology 86: 887-890.

Jin, Y., G.D. Statler, J.D. Franckowiak and B.J. Steffenson, 1993a. Linkage between leaf rust resistance genes and morphological markers in barley. Phytopathology 83: 230-233.

Jin, Y., and B.J. Steffenson, 1994. Inheritance of resistance to *Puccinia hordei* in cultivated and wild barley. J. Heredity 85: 451-454.

Jin, Y., B.J. Steffenson and H.E. Bockelman, 1995. Evaluation of cultivated and wild barley for resistance to pathotypes of *Puccinia hordei* with wide virulence. Gen. Res. Crop Evol. 42: 1-6.

Jin, Y., B.J. Steffenson and J.D. Franckowiak, 1993b. Linkage between the Rpgl gene for stem rust resistance and the F5-locus on barley chromosome 1. Crop Sci. 33: 642-643.

Jin, Y., B.J. Steffenson and D.J. Miller, 1994. Inheritance of resistance to pathotypes QCC and MCC of Puccinia graminis f. sp. tritici in barley line Q 21861 and temperature effects on the expression of resistance. Phytopathology 84: 452-455.

Jones, A.T., and P.L. Catherall, 1970. The relationship between growth rate and the expression of the tolerance to barley yellow dwarf virus in barley. Ann. Appl. Biol. 65: 137-145.

Jørgensen, J.H., 1975. Identification of powdery mildew resistant barley mutants and their allelic relationship. In: Barley Genetics III. Proc. 3rd Int. Barley Genet. Symp., Thiemig, München, pp. 446-455.

Jørgensen, J.H., 1988. Coordinator report disease and pest resistance. Barley Genet. Newsl. 18: 65-69.

Jørgensen, J.H., 1991. Powdery mildew resistance genes *Mla22* and *Mla23* in barley. In: Barley Genetics VI. Proc. 6th Int. Barley Genet. Symp., Munksgaard, Copenhagen, pp. 596-598.

Jørgensen, J.H., 1992. Sources and genetics of resistance to fungal pathogens. In: P.R. Shewry (ed), Barley. Genetics, Biochemistry, Molecular Biology and Biotechnology, CAB International, pp. 441-457.

Jørgensen, J.H., 1993. Genetics of powdery mildew resistance. Crit. Rev. Plant Sci. 13: 97-119.

Jørgensen, H.J., and V. Smedegaard-Petersen, 1995. Pathogenic variation of *Rhynchosporium secalis* in Denmark and sources of resistance in barley. Plant Dis. 79: 297-301.

Kaiser, R., and W. Friedt, 1989. Chromosomal location of resistance to barley yellow mosaic virus in German winter barley identified by trisomic analysis. Theor. Appl. Genet. 77: 241-245.

Kandawa-Schulz, M.A., 1996. Untersuchungen an Artbastarden von Wild- und Kulturgersten (*Hordeum bulbosum* × *H. vulgare*) zur Introgression züchterisch wichtiger Resistenzeigenschaften. Diss. Univ. Rostock, Mathem.-Naturwiss. Fak.

Kashiwazaki, S., K. Ogawa, T. Usugi and T. Tsuchizaki, 1989. Characterization of several strains of barley yellow mosaic virus. Ann. Phytopathol. Soc. Japan 55: 16-25.

Keeling, B.L., and E.E. Banttari, 1975. Factors associated with the resistance of barley to *Helminthosporium teres*. Phytopathology 65: 464-467.

Khan, T.N., 1982. Changes in pathogenicity of *Drechslera teres* relating to changes in barley cultivars grown in Western Australia. Plant Dis. 66: 655-656.

Khan, T.N., and W.J.R. Boyd, 1969a. Inheritance of resistance to net blotch of barley. II. Genes conditioning resistance against race W.A.-2. Can. J. Genet. Cytol. 11: 592-597.

Khan, T.N., and W.J.R. Boyd, 1969b. Physiologic specialization in *Drechslera teres*. Austral. J. Biol. Sci. 22: 1229-1235.

Khan, T.N., and A. Tekauz, 1982. Occurrence and pathogenicity of *Drechslera teres* isolates causing spot-type symptoms on barley in Western Australia. Plant Dis. 66: 423-425.

Khatskevitch, L.K., and A.A. Benken, 1990. Root and stem rots of cereals. Zashchita rastenii, Moskva 9: 14-15.

Khokhlova, A.P., 1987. Genetic analysis of resistance to loose smut in Ethiopian barleys. Sb. Nauchn. Tr. Prikl. Bot., Genet. i Sel. 110: 45-49. (In Russian).

Khokhlova, A.P., 1990. Resistance of barley to loose smut and the gene complex controlling this character. Thesis Abstract, Cand. Agric. Sci., Leningrad. (In Russian).

Khokhlova, A.P., V.A. Vershinina, M.V. Lukyanova and I.A. Terentyeva, 1989. Catalogue: Characteristics of susceptibility to loose smut, leaf rust and powdery mildew. VIR, St. Petersburg, 30 p. (In Russian).

Kicherer, S., G. Backes, A. Jahoor, U. Walther and G. Fischbeck, 1996. Characterisation of a DH-population with regard to resistance against leaf rust by means of QTL-analysis. In: Proc. 9th European and Mediterranean Cereal Rusts & Powdery Mildew Conf., Lunteren, Netherlands, p. 231.

Kindler, S.D., K.B. Jensen and T.L. Springer, 1993. An overview: resistance to the Russian wheat aphid (Homoptera. Aphididae) within the perennial Triticeae. J. Econ. Entomol. 86: 1609-1618.

Kindler, S.D., and T.L. Springer, 1991. Resistance to Russian wheat aphid in wild *Hordeum* species. Crop Sci. 31: 94-97.

Kintzios, S., A. Jahoor and G. Fischbeck, 1995. Powdery-mildew-resistance genes *Mla29* and *Mla32* in *H. spontaneum* derived winter barley lines. Plant Breed. 1143: 265-266.

Kirdoglo, E.K., 1990. Breeding barley for resistance to smut and leaf/stem diseases. Vestn. Selskokhoz. Nauki, Moskva, 10: 98-104.

Kmecl, A., F. Mauch, H. Winzeler and R. Dudler, 1995. Quantitative field resistance of wheat to powdery mildew and defense reactions at the seedling stage-identification of a potential marker. Physiol. Mol. Plant Pathol. 47: 185-199.

Kølster, P., L. Munck, O. Stølen and J. Løhde, 1986. Near isogenic barley lines with genes for resistance to powdery mildew. Crop Sci. 26: 903-907.

Kozera, W., and A. Roszko, 1995. Inheritance of resistance to *Ustilago nuda* (Jens.) Rostr. in some spring barley cultivars. J. Appl. Genet. 36: 119-127.

Kretschmer, J.M., K.J. Chalmers, S. Manning, A. Karakousis, A.R. Barr, A.K.M.R. Islam, S.J. Logue, Y.W. Choe, S.J. Barker, R.C.M. Lance and P. Langridge, 1997. RFLP mapping of the *Ha2* cereal cyst nematode resistance gene in barley. Theor. Appl. Genet. 94: 1060-1064.

Krishnamurthy, K., N. Venugopal, M.K. Jagannath, R. Ananthanarajan and B.R. Hegge, 1972. Comparison of American, European and Indian barley varieties for yield performance. Indian J. Agric. Res. 6: 1-7.

Krivchenko, V.J., 1984. Resistance of cereals to the pathogens causing smut diseases. Kolos, Moscow. (In Russian).

Lehmann, L., 1988. Isolation and transfer of disease resistance from *Hordeum spontaneum*. Nordisk Jordbrugsforskning 70: 530-532.

Lehmann, L., and R. von Bothmer, 1988. *Hordeum spontaneum* and landraces as a gene resource for barley breeding. In: M.L. Jorna and L.A.J. Slootmaker (eds), Cereal breeding related to integrated cereal production. Proc. Conf. Cereal Sect. EUCARPIA, Wageningen, Netherlands, pp. 190-194.

Leur, J.A.G. van, 1989. Barley pathology. In: Cereal Improvement Program Ann. Rep., 1988, pp. 122-130.

Leur, J.A.G. van, S. Ceccarelli and S. Grando, 1989. Diversity for disease resistance in barley landraces from Syria and Jordan. Plant Breed. 103: 324-335.

Lister, M.R., and R. Ranieri, 1995. Distribution and economic importance of Barley Yellow Dwarf. In: C.J. D'Arcy and P.A. Burnett (eds), Barley Yellow Dwarf - 40 Years of Progress. APS Press, St. Paul, pp. 29-53.

Luig, N.H., 1985. Epidemiology in Australia and New Zealand. In: A.P. Roelfs and W.R. Bushnell (eds), The Cereal Rusts. Volume II. Diseases, distribution, epidemiology, and control, pp. 302-328.

Lukyanova, M.V., 1990. New sources and donors for barley breeding. Byul. VIR 201: 26-30. (In Russian).

Lukyanova, M.V., and I.A. Terentyeva, 1997: Catalogue of genetic collection of barley with identified genes of resistance to powdery mildew. VIR, St. Petersburg, Russia. (In Russian).

Luthra, J.K., 1988. Identification of additional genes conferring resistance to *Puccinia striiformis* in *Yr* lines of barley. Indian J. Genet. Plant Breed. 48: 53-57.

Luthra, J.K., and M.V. Rao, 1973. Inheritance of seedling resistance to *Helminthosporium* blight in barley. Indian J. Genet. Plant Breed. 33: 378-379.

Luthra, J.K., R.S. Verma and K.V. Prabhu, 1992. Development of isogenic lines in barley with reference to *Puccinia striiformis* West. In: Barley Genetics VI. Proc. 6th Int. Barley Genetics Symp. Munksgaard, Copenhagen, Denmark, pp. 622-625.

Magnus, H.A., 1974. Sources of resistance to *Septoria nodorum* in barley. Meldinger Norges Landbrukshøgskole 53, No. 7, 7 p.

Manisterski, J., L. Treeful, J.R. Tomerin, Y. Anikster, J.A. Moseman, I. Wahl and D.E. Wilcoxson, 1986. Resistance of wild barley accessions from Israel to leaf rust collected in USA and Israel. Crop Sci. 26: 727-730.

Marrewijk, G.A.M., and F.L. Dieleman, 1980. Resistance to aphids in barley and wheat. In: A.K. Minks and P. Gruys (eds), Integrated Control of Insect Pests in the Netherlands, Pudoc, Wageningen, pp. 165-167.

Martín, A., and J.L. Cubero, 1981. The use of *Hordeum chilense* in cereal breeding. Cer. Res. Comm. 9: 317-323.

Mathre, D.E., 1982. Compendium of barley diseases. American Phytopathological Society, St. Paul, Minn., 78 p.

Mathur, A.K., and B.S.V. Siradhana, 1990. Reaction of progenies to loose smut in Rajastan. Ind. J. Mycol. Plant Pathol. 20: 165-166.

Mayo, M.A., and C.J. D'Arcy, 1999. Family Luteoviridae. A reclassification of luteoviruses. In: H.G. Smith and H. Barker (eds), The Luteoviridae. CABI Publishing, pp. 15-22.

Mc Donald, B.A., J. Zhan and J.J. Burdon, 1999. Genetic structure of *Rhynchosporium secalis* in Australia. Phytopathology 89: 639-645.

McMullen, M., R. Jones and D. Gallenberg, 1997. Scab of wheat and barley: a re-emerging disease of devastating impact. Plant Dis. 81: 1340-1448.

Meadway M.H., R.A. Bayles and P.L. Stigwood, 1991. Yellow rust of barley. In: Annual report 1990, UK Cereal Pathogen Virulence Survey Committee, pp. 35-36.

Metcalfe, D.R., A.W. Chiko, J.W. Martens and A. Tekauz, 1977. Reaction of barley from the Middle East to Canadian pathogens. Can. J. Plant Sci. 57: 995-999.

Metcalfe, D.R., A.W. Chiko, J.W. Martens and A. Tekauz, 1978. Reaction of Ethiopian barley to Canadian barley pathogens. Can. J. Plant Sci. 58: 885-890.

Mielke, H., 1980. Investigations of the susceptibility of different wild barley species to *Typhula incarnata*. Nachrichtenbl. Deutsch. Pflschutzdienst 32: 39.

Mode, C.J., and C.W. Schaller, 1958. Two additional factors for resistance to net blotch in barley. Agron. J. 50: 15-18.

Moharramipour, S., H. Tsumuki, K. Sato and H. Yoshida, 1997. Mapping resistance to cereal aphids in barley. Theor. Appl. Genet. 94: 592-596.

Moharramipour, S., H. Tsumuki, K. Takeda, K. Sato and H. Yoshida, 1996. Inheritance of resistance to cereal aphid populations in barley. Proc. Jap. Soc. Appl. Entomol. Zool. 56: 134.

Moseman, J.G., 1955. Sources of resistance to powdery mildew of barley. Plant Dis. Rep. 39: 967-972.

Moseman, J.G., 1971. Studies of inheritance in barley to pathogenic organisms, 1963-1969. In: Barley Genetics II. Proc. 2nd Int. Barley Genet. Symp., Pullman, WA, USA, pp. 535-541.

Moseman, J.G., P.S. Baenziger and R.A. Kilpatrick, 1981. Genes conditioning resistance of *Hordeum spontaneum* to *Erysiphe graminis* f. sp. *hordei*. Crop Sci. 21: 229-233.

Moseman, J.G., and J.C. Craddock, 1976. Genetic basis for barley germ-plasm conservation. In: Barley Genetics III. Proc. 3rd Int. Barley Genet. Symp., Thieming, Munich, pp. 51-57.

Moseman, J.G., and J.H. Jørgensen, 1971. Identification of genes at the *Mla* locus in barley for resistance to *Erysiphe graminis* f. sp. *hordei*. Crop Sci. 11: 547-550.

Moseman, J.G., R.C.F. Macer and L.W. Greeley, 1965. Genetic studies with cultures of *Erysiphe graminis* f. sp. *hordei* virulent on *Hordeum spontaneum*. Trans. Brit. Mycol. Soc. 48: 479-489.

Moseman, J.G., E. Nevo and D. Zohary, 1983. Resistance of *Hordeum spontaneum* collected in Israel to infection with *Erysiphe graminis hordei*. Crop Sci. 23: 1115-1119.

Moseman, J.G., and C.W. Schaller, 1960. Genetics of the allelic series at the *Mla* locus in barley and cultures of *Erysiphe graminis* f. sp. *hordei* that differentiate these alleles. Phytopathology 50: 736-741.

Moseman, J.G., and R.T. Smith, 1976. Breeding for multiple pest resistance. In: Barley Genetics III. Proc. 3rd Int. Barley Genet. Symp., Thiemig, München, pp. 421-425.

Munck, L., H. Østergaard, J.B. Pedersen, B. Eriksen and M. Hovmøller, 1999. Disease resistance in cereals. Present state and possibilities. In: 16th Danish Plant Protection Conference. Crop protection in organic farming. Danmarks Jordbrugsforskning, Tjele, Denmark, pp. 107-117. (In Danish).

Narang, S., and J.S. Rana, 1999. Screening of barley genotypes against corn leaf aphid, *Rhopalosiphum maidis* (Fitch.). Cer. Res. Comm. 27: 131-138.

Negassa, M., 1985a. Geographic distribution and genotypic diversity of resistance to powdery mildew of barley in Ethiopia. Hereditas 102: 113-121.

Negassa, M., 1985b. Genetics of resistance to powdery mildew in some Ethiopian barleys. Hereditas 102: 123-138.

Nelson, K.E., and L.W. Burgess, 1994. Reaction of Australian cultivars of oats and barley to infection by *Fusarium graminearum* Group 1. Austral. J. Exp. Agric. 34: 655-658.

Nelson, R.R., 1984. Strategy of breeding for disease resistance. In: Crop Breeding. Pergamon Press, pp. 32-50.

Nettevich, E.D., and N.M. Vlasenko, 1985. Results of evaluating the barley collection in the central nonchernozem zone. Sb. Nauchn. Tr. Prikl. Bot., Genet. i Sel. 95: 28-31. (In Russian).

Newton, A.C., 1990. Detection of components of partial resistance to mildew (*Erysiphe graminis* f. sp. *hordei*) incorporated into advanced breeding lines of barley using measurement of fungal cell wall sterol. Plant Pathol. 39: 598-602.

Nieto-Lopez, R.M., and T.K. Blake, 1994. Russian wheat aphid resistance in barley: inheritance and linked molecular markers. Crop Sci. 34: 655-659.

Nilan, R.A., 1964. The cytology and genetics of barley 1951-1962. Monographic supplement 3 (32), Washington State University Press, 278 p.

Nomura, K., S. Kashiwazaki, H. Hibino, T. Inoue, E. Nakata, Y. Tsuzaki and S. Okuyama, 1996. Biological and serological properties of strains of barley mild mosaic virus. J. Phytopathol. 144: 103-107.

Nover, I., 1972. Untersuchungen mit einer für den Resistenzträger 'Lyallpur 3645' virulenten Rasse von *Erysiphe graminis* DC. f. sp. *hordei* (Marchal). Arch. Pflschutz. 8: 439-445.

Nover, I., C.O. Lehmann and A. Seidenfaden, 1976. Resistenzeigenschaften im Gersten- und Weizensortiment Gatersleben. 20. Prüfung des Verhaltens von Gersten gegen FlugbrandResistance properties of the Gatersleben barley and wheat collection. 20. Testing barley for their reaction to loose smut, *Ustilago nuda* (Jens.) Rostr. Kulturpflanze 24: 237-247.

Nover, I., and R. Mansfeld, 1955. Resistenzeigenschaften im Gersten- und Weizensortiment Gatersleben. Kulturpflanze 3: 105-113.

Nover, I., and R. Mansfeld, 1956. Resistenzeigenschaften im Gersten- und Weizensortiment Gatersleben II. Kulturpflanze 4: 341-349.

Oerke, E.-C., and H.-W. Dehne, 1997. Global crop production and the efficacy of crop protection - current situation and future trends. Eur. J. Plant Pathol. 103: 203-215.

Oerke, E.-C., H.-W. Dehne, F. Schönbeck and A. Weber, 1994. Crop Production and Crop Protection. Estimated Losses in Major Food and Cash Crops. Elsevier, Amsterdam.

Okunowski, I., 1990. Untersuchungen von Gersten aus dem Gaterslebener Weltsortiment auf Resistenz gegen Gelbrost *Puccinia striiformis* West. unter besonderer Berücksichtigung der Feldresistenz. Diss. Akad. Landwirtschaftswissenschaften, Berlin.

Olembo, J.R., F.L. Patterson and R.L. Gallun, 1966. Genetic analyses of the resistance to *Mayetiola destructor* (Say) in *Hordeum vulgare* L. Crop Sci. 6: 563-566.

Oniskova, M.G., 1987. Immunity characteristics in Asian barley in northern Kazakhstan. In: Breeding and genetic research in the production of new cultivars of field crops for northern Kazakhstan. Tselinograd, Kazakhstan, pp. 89-92. (In Russian).

Ordon, F., A. Schiemann and W. Friedt, 1997. Assessment of the genetic relatedness of barley accessions (*H. vulgare* L. s.l.) resistant to soil-borne mosaic inducing viruses (BaMMV, BaYMV, BaYMV-2) using RAPDs. Theor. Appl. Genet. 94: 325-330.

Ordon, F., A. Schiemann, B. Pellio, V. Dauck, E. Bauer, S. Streng, W. Friedt and A. Graner, 1999. Application of molecular markers in breeding for resistance to the barley yellow mosaic virus complex. Z. Pflkrankh. Pflschutz 106: 256-264.

Oswald, J.W., and B.R. Houston, 1951. A new virus disease of cereals, transmissible by aphids. Plant Dis. Rep. 11: 471-475.

Patterson, F.L., and R.G. Shands, 1957. Independent inheritance of four characters of barley. Agron. J. 49: 218-219.

Pecchioni, N., H. Faccioli, H. Toubia-Rahme, G. Vale and V. Terzi, 1996. Quantitative resistance to barley leaf stripe (*Pyrenophora graminea*) is dominated by one major locus. Theor. Appl. Genet. 93: 97-101.

Perkowski, J., I. Kiecana and J. Chelkowski, 1995. Susceptibility of barley cultivars and lines to *Fusarium* infection and mycotoxin accumulation in kernels. J. Phytopathology 143: 547-551.

Perkowski, J., I. Kiecana, U. Schumacher, H.M. Muller, J. Chelkowsi and P. Golinski, 1997. Head infection and accumulation of *Fusarium* toxin in kernels of 12 barley genotypes inoculated with *Fusarium graminearum* isolates of two chemotypes. Europ. J. Plant Pathol. 103: 85-90.

Person-Dedryver, F., J. Jahier and T.E. Miller, 1990. Assessing the resistance to cereal root-knot nematode, *Meloidogyne naasi*, in a wheat line with the added chromosome arm $1H^{ch}S$ of *Hordeum chilense*. J. Genet. Breed. 44: 291-296.

Pickering, R.A., and P.W. Morgan, 1983. Plant regeneration from cultured embryos derived from *Hordeum vulgare* L. pollinated with *H. bulbosum* L. Euphytica 32: 585-591.

Pickering, R.A., W.F. Rennie and M.G. Cromey, 1997. Disease resistant material available from the wide hybridization programme at DSIR. Barley Newsl. 31: 248-259.

Pike, K.S., 1990. A review of barley yellow dwarf virus grain yield losses. In: P.A. Burnett (ed), World Perspectives on Barley Yellow Dwarf. CIMMYT, Mexico, pp. 356-361.

Pluck, D.J., R.L. Evans and C.L. Flegg, 1981. Resistance in barley to *Selenophoma donacis*. Trans. Brit. Mycol. Soc. 77: 509-518.

Prochnow, J., 1997. Selektion eines 500 Sippen umfassenden Sortimentes *Hordeum spontaneum* C. Koch auf qualitative und quantitative Resistenz gegen Zwergrost (*Puccinia hordei* Otth) unter Berücksichtigung weiterer pilzlicher Pathogene. Diss. Univ. Halle/S. Landwirtsch. Fak., 108 p.

Proeseler, G., H. Hartleb, D. Kopahnke and C.O. Lehmann, 1989. Resistenzeigenschaften im Gersten- und Weizensortiment Gatersleben. 29. Prüfung von Gersten auf ihr Verhalten gegenüber dem Milden Gerstenmosaik-Virus (barley mild mosaic virus, BaMMV), Gerstengelbmosaik-Virus (barley yellow mosaic virus, BaYMV), *Drechslera teres* (Sacc.) Shoem. und *Puccinia hordei* Otth. Kulturpflanze 37: 145-154.

Qui, X.Q., R.E. Niks, P. Stam and P. Lindhout, 1998. Identification of QTLs for partial resistance to leaf rust (*Puccinia hordei*) in barley. Theor. Appl. Genet. 96: 1205-1215.

Rautapää, J., 1970. Preference of cereal aphids for various cereal varieties and species of Gramineae, Juncaceae and Cyperaceae. Ann. Agric. Fenn. 9: 267-277.

Reinhold, M., and E.L. Sharp, 1986. Resistance sources in barley to *Puccinia hordei*. Cereal Rusts Bull. 14: 75-83.

Richter, K., J. Schondelmaier and C. Jung, 1998. Mapping of quantitative trait loci affecting *Drechslera teres* resistance in barley with molecular markers. Theor. Appl. Genet. 97: 1225-1234.

Rigina, S.I., 1966. Study of resistance of barley to infectious diseases. Thesis Cand. agric. sci., Leningrad, 196 pp. (In Russian).

Röbbelen, G., and M. Heun, 1991. Genetic analysis of partial resistance against powdery mildew in induced mutants of barley. In: Plant mutation breeding for crop improvement, IAEA, Vienna, pp. 93-111.

Robinson, J., F. Delgado, H.E. Vivar and P.A. Burnett, 1992. Inheritance of resistance to Russian wheat aphid in barley. Euphytica 62: 213-217.

Rochev, M.V., and M.M. Levitin, 1986. Resistance of barley varieties to *Bipolaris sorokiniana* (Sacc.) Shoemaker in the central Urals. Sibir. Vestn. Selskokhoz. Nauk 6: 18-21, 112. (In Russian).

Rochow, W.F., 1969. Biological properties of four isolates of barley yellow dwarf virus. Phytopathology 59: 1580-1589.

Rochow, W.F., and I. Muller, 1971. A fifth variant of barley yellow dwarf virus in New York. Plant Dis. 55: 874-877.

Rodina, N.A., and E.A. Soshnikova, 1986. Evaluation of barley breeding material for smut resistance. In: Breeding of cereal crops for resistance to diseases and abiotic factors in the Volga-Vyatka region, pp. 27-32. (In Russian).

Rodriguez, E., 1973. Barley protection. In: R.C. Anderson (ed.), Wheat, triticale and barley seminar. El Batan, Mexico, Intern. Maize and Wheat Improvem. Center, pp. 152-154.

Rubiales, D., C.H.A. Snijders, P. Nicholson and A. Martín, 1996. Reaction of *Tritordeum* to *Fusarium culmorum* and *Septoria nodorum*. Euphytica 88: 165-174.

Ruge, B., M. Michel, R. Pickering, G. Proeseler and P. Wehling, 2000. Gene introgressions from *H. bulbosum* into cultivated barley cause resistance to different pathogens. In: Plant & Animal Genome VIII, San Diego, p. 228.

Saeki, K., C. Miyazaki, N. Hirota, A. Saito, K. Io and T. Konishi, 1999. RFLP mapping of BaYMV resistance gene *rym3* in barley (*Hordeum vulgare*). Theor. Appl. Genet. 99: 727-732.

Sakti, J., and K.L. Bailey, 1995. Responses of wild and cultivated barley from west Asia to net blotch and spot blotch. Crop Sci. 35: 242-246.

Sandoval-Islas, J.S., L.H.M. Broers, H. Vivar and K.S. Osada, 1998. Evaluation of quantitative resistance to yellow rust (*Puccinia striiformis* f. sp. *hordei*) in the ICARDA/CIMMYT barley breeding programme. Plant Breed. 117: 127-130.

Sato, K., and K. Takeda, 1991. Studies on the conidia formation of *Pyrenophora teres* Drechs. II. Effects of day-length, medium and temperature under near-ultraviolet radiation. Nogaku Kenkyu 62: 165-176.

Sato, K., and K. Takeda, 1992. An establishment of seedling test and a search for resistant varieties to net blotch in barley. Bull. Res. Inst. Biores., Okayama Univ. 1: 75-90.

Sato, K., and K. Takeda, 1993. Pathogenic variation of *Pyrenophora teres* isolates collected from Japanese and Canadian spring barley. Bull. Res. Inst. Biores., Okayama Univ. 1: 147-158.

Sato, K., and K. Takeda, 1994. Sources of resistance to net blotch in barley germplasm. Bull. Res. Inst. Biores., Okayama Univ. 2: 91-102.

Sato, K., and K. Takeda, 1997. Net blotch resistance in wild species of *Hordeum*. Euphytica 95: 179-185.

Schaller, C.W., 1955. Inheritance of resistance to net blotch of barley. Phytopathology 45: 174-176.

Schaller, C.W., 1984. The genetics of resistance to barley yellow dwarf virus in barley. In: P.A. Burnett (ed), Barley Yellow Dwarf, a Proceeding of the Workshop. CIMMYT, Mexico, pp. 93-99.

Schaller, C.W., and F.N. Briggs, 1955. Inheritance of resistance to mildew *Erysiphe graminis hordei* in the barley variety, Black Russian. Genetics 40: 421-428.

Schaller, C.W., C.O. Qualset and J.N. Rutger, 1964. Inheritance and linkage of the *Yd2* gene conditioning resistance to barley yellow dwarf virus disease in barley. Crop Sci. 4: 544-548.

Schaller, C.W., D.C. Rasmusson and C.O. Qualset, 1963. Sources of resistance to the yellow dwarf virus in barley. Crop Sci. 3: 342-344.

Schaller, C.W., and G.A. Wiebe, 1952. Sources of resistance to net blotch of barley. Agronomy J. 44: 334-336.

Schiemann, A., 1999. Entwicklung PCR-basierter Marker für Resistenzgene gegen Gelbmosaikviren der Gerste und Erstellung einer hochauflösenden Kartierungspopulation für das Resistenzgen *ym5*. Diss. Univ. Giessen, Verlag Shaker, Aachen, 135 p.

Scholz, F., and I. Nover, 1967. Genetische Untersuchungen mit einer vollständig mehltauresistenten Gerstenlinie. Kulturpflanze 15: 243-254.

Schönfeld, R.M., 1997. Identifizierung und molekulare Lokalisierung neuer Resistenzgenloci der Wildgerste (*Hordeum vulgare* ssp. *spontaneum*) gegen Mehltau (*Erysiphe graminis* f. sp. *hordei*). Diss., Technische Universität München, 220 p.

Schönfeld, R.M., A. Ragni and G. Fischbeck, 1996. RFLP-mapping of three new loci for resistance genes to powdery mildew (*Erysiphe graminis* f. sp. *hordei*) in barley. Theor. Appl. Genet. 93: 48-56.

Schwarzbach, E., 1967. Recessive total resistance of barley to mildew (*Erysiphe graminis* DC f. sp. *hordei* Marchal) as a mutation induced by ethylmethansulfonate. Genetika a Šlechtení 3: 159-162.

Schwarzbach, E., and G. Fischbeck, 1981. Die Mehltauresistenzfaktoren von Sommer- und Wintergerstensorten in der Bundesrepublik Deutschland. Z. Pflanzenzüchtg. 87: 309-318.

Segal, A., K.H. Dörr, D. Fischbeck, D. Zohary and I. Wahl, 1987. Genotypic composition and mildew resistance in natural population of wild barley. Plant Breed. 99: 118-127.

Semeane, Y., 1995. Importance and control of barley leaf blights in Ethiopia. Rachis 14: 83-89.

Sharma, H.S.S., and A.E. Brown, 1983. The assessment of susceptibility or resistance of spring and winter barley cultivars to *Septoria nodorum*. In: Record Agric. Res., Dept. Agric. Northern Ireland, 31: 55-57.

Shchelko, L.G., 1969. Study of initial material of barley for resistance to loose smut. In: Proc. V^{th} All-Union Symp. on Immunity, Kiev, No. 5, pp. 7-12. (In Russian).

Skou, J.P., and V. Haahr, 1985. The barleys in Nordic Gene Bank screened for resistance against barley leaf stripe (*Drechslera graminea*). Nordisk Jordbrugsforskning 67: 262-263.

Skou, J.P., B.J. Nielsen and V. Haahr, 1992. The effectivity of Vada resistance against leaf stripe in barley varieties. Nordisk Jordbrugsforskning 74: 34.

Skou, J.P., B.J. Nielsen and V. Haahr, 1994. Evaluation and importance of genetic resistance to leaf stripe in western European barleys. Acta Agric. Scand. Sect. B, Soil and Plant Sci., 44: 98-106.

Smedegaard-Petersen, V., 1971. *Pyrenophora teres* f. *maculata* f. nov. and *Pyrenophora teres* f. *teres* on barley in Denmark. In: Yearbook Roy. Vet. Agric. Univ., Copenhagen, pp. 124-144.

Smirnova, Z.G. and A.Ya. Trofimovskaya, 1985. Sources of resistance of barley to net blotch. Sb. Nauchn. Tr. Prikl. Bot., Genet. i Sel. 95: 52-56. (In Russian).

Smith, L., 1951. Cytology and genetics of barley. Bot. Rev. 17: 1-51, 133-202, 285-355.

Smith, O.D., A.M. Schlehuber and B.C. Curtis, 1962. Inheritance studies of greenbug (*Toxoptera graminum* Rond.) resistance in four varieties of winter barley. Crop Sci. 2: 489-491.

Søgaard, B., and P. von Wettstein-Knowles, 1987. Barley: genes and chromosomes. Carlsberg Res. Comm. 52: 123-196.

Spaner, D., L.P. Shugar, T.M. Choo, I. Falak, K.G. Briggs, W.G. Legge, D.E. Falk, S.E. Ullrich, N.A. Tinker, B.J. Steffenson and D.E. Mather, 1998. Mapping of disease resistance loci in barley on the basis of visual assessment of naturally occurring symptoms. Crop Sci. 38: 843-850.

Stanford, E.H., and F.N. Briggs, 1940. Two additional factors for resistance to mildew in barley. J. Agric. Res. 61: 231-236.

Starks, K.J., and J.A. Webster, 1985. Insects and related pests. In: D.C. Rasmusson (ed), Barley. Amer. Soc. Agron., Madison, Wisconsin, No. 26, pp. 337-365.

Steffenson, B.J., 1992. Analysis of durable resistance to stem rust in barley. Euphytica 63: 153-167.

Steffenson, B.J., 1999. Coordinator's report. Disease and pest resistance genes. Barley Genet. Newsl. 29: 62.

Steffenson, B.J., P.M. Hayes and A. Kleinhofs, 1996. Genetics of seedling and adult plant resistance to net blotch (*Pyrenophora teres* f. *teres*) and spot blotch (*Cochliobolus sativus*) in barley. Theor. Appl. Genet. 92: 552-558.

Steffenson, B.J., Y. Jin and C.A. Griffey, 1993. Pathotypes of *Puccinia hordei* with virulence for the barley leaf rust resistance gene *Rph7* in the United States. Plant Disease 77: 867-869.

Steffenson, B.J., and R.K. Webster, 1992. Quantitative resistance to *Pyrenophora teres* in barley. Phytopathology 82: 407-411.

Steffenson, B.J., R.K. Webster and L.F. Jackson, 1991. Reduction in yield loss using incomplete resistance to *Pyrenophora teres* f. *teres* in barley. Plant Dis. 75: 96-100.

Sterling, J.D.E., and R.B. Maclaren, 1975. Kinkora: a jointworm-resistant barley. Can. J. Plant Sci. 55: 1075-1076.

Stubbs, R.W., 1985. The Cereal Rusts. Vol. 2. Disease, Epidemiology and Control. A. Roelfs, Bushnell, W.R., Orlando, Academic Press, pp. 61-101.

Su, L., A. Maric, B. Kostic and K. Mikic, 1989. Some properties of *Helminthosporium gramineum* and barley resistance to the pathogen. Zastita Bilja 40: 151-164.

Subrahmanyam, N.C., and R. von Bothmer, 1987. Interspecific hybridization with *Hordeum bulbosum* and development of hybrids and haploids. Hereditas 106: 119-127.

Sun, Y., B.J. Steffenson and Y. Jin, 1996. Genetics of *Puccinia graminis* f. sp. *secalis* in barley line Q 21681. Phytopathology 86: 1299-1302.

Suneson, C.A., 1955. Breeding for resistance to yellow dwarf virus in barley. Agron. J. 47: 283.

Surin, N.A., 1989. Source material and problems of breeding spring barley in the eastern Siberia. Sb. Nauchn. Tr. Prikl. Bot., Genet. i Sel. 129: 37-41. (In Russian).

Surin, N.A., and N.E. Lyakhova, 1993. Barley breeding in Siberia. Novosibirsk, pp. 33-62, 107-219. (In Russian).

Takeda, K., 1992. Current topics on the scab disease resistance in barley and wheat. Proc. Japan. Assoc. Mycotoxicol. 36: 13-17.

Takeda, K., and H. Heta, 1989. Establishing the testing method and a search for the resistant varieties to *Fusarium* head blight in barley. Japan. J. Breed. 39: 203-216.

Tekauz, A., 1978. Incidence and severity of net blotch of barley and distribution of *Pyrenophora teres* biotypes in the Canadian Prairies 1976. Can. Plant Dis. Surv. 58: 9-11.

Tekauz, A., 1983. Reaction of Canadian barley cultivars to *Pyrenophora graminea*, the incident of leaf stripe. Can. J. Plant Pathol. 5: 294-301.

Tekauz, A., 1991. Pathogenic variation in *Rhynchosporium secalis* on barley in Canada. Can. J. Plant Pathol. 13: 298-304.

Tekauz, A., and K.W. Buchannon, 1977. Distribution of and sources of resistance of new biotypes of *Pyrenophora teres* in Western Canada. Can. J. Plant Sci. 57: 389-395.

Tekauz, A., and J.T. Millis, 1974. New types of virulence in *Pyrenophora teres* in Canada. Can. J. Plant Sci. 54: 731-734.

Thomson, S.B., H.P. Jensen, J. Jensen, J.P. Skou and J.H. Jørgensen 1997. Localisation of a resistance gene and identification of sources of resistance to barley leaf stripe. Plant Breed. 116: 455-459.

Toojinda, T., E. Baird, A. Booth, L. Broers, P. Hayes, P. Powell, W. Thomas, H. Vivar and G. Young, 1998. Introgression of quantitative trait loci (QTLs) determining stripe rust resistance in barley. An example of marker-assisted line development. Theor. Appl. Genet. 96: 123-131.

Torp, J., H.P. Jensen and J.H. Jørgensen, 1978. Powdery mildew resistance genes in 106 Northwest European spring barley varieties. In: Kgl. Vet.-. og Landbohøjsk. Årsskr. 1978, pp. 75-102.

USDA-ARS, National Genetic Resources Program. Germplasm Resources Information Network - (GRIN). [Online Database] National Germplasm Resources Laboratory, Beltsville, Maryland. Available: www.ars-grin.gov/cgi-bin/npgs/html/acchtml.pl 1111522 (30 December 1999) [Mianwali].

Vavilov, N.I., 1926. Centres of origin of cultivated plants. Trudy po Prikl. Bot. 16: 139-245. (In Russian).

Vavilov, N.I., 1951. Phytogeographic basis of plant breeding. The origin, variation, immunity and breeding of cultivated plants. Chronica Bot. 13: 1-366.

Velibekova, E.L., 1981. Susceptibility of barley to root rots. Selektsiya i Semenovodstvo, USSR 10: 20-21. (In Russian).

Vincent, J.R., R.M. Lister and B.A. Larkins, 1991. Nucleotide sequence analysis and genomic organization of the NY-RPV isolate of barley yellow dwarf virus. J. Gen. Virol. 72: 2347-2355.

Wahl, I., A. Eshed, N. Segal and Z. Sobel, 1978. Significance of wild relatives of small grains and other wild grasses in cereal powdery mildews. In: D.M. Spencer (ed), Powdery Mildews, Acad. Press London, pp. 83-100.

Wahl, I., Y. Anikster and J. Manisterski, 1988. Evolution of host-parasite relations in the *Puccinia hordei* - *Hordeum* spp. system at the center of origin. In: Proc. 7th Cereal Rusts Conf., Vienna, pp. 122-123.

Walther, U., 1996. Development of pathogen populations of *Puccinia hordei* Otth in Germany (1974-1994) and in the neighbouring countries (since 1992) under consideration of biotic and abiotic influences. In: E. Limpert Finkh and M. Wolfe (eds), COST 817, Proc. Integrated Control of Cereal Mildew and Rusts. Towards coordination of research across Europe. European Commission, ECSC-EC-EAEC, Brussels, pp. 119-124.

Walther, U., D. Kopahnke, A. Habekuß, E. Schliephake and G. Proeseler, 1999. Aussichten der Nutzung von Wild- und Primitivformen der Gerste in der Resistenzzüchtung. Vortr. Pflanzenzücht. 46: 30-44.

Walther, U., H. Rapke, G. Proeseler and G. Szigat, 2000. *Hordeum bulbosum* – a new source of disease resistance-transfer of resistance to leaf rust and mosaic viruses from *H. bulbosum* in winter barley. Plant Breed. 119: 215-218.

Wan, Y.F., C. Chen, J.J. Yang and F.Q. Liu, 1997. Evaluation of *Roegneria* for resistance to head scab caused by *Fusarium graminearum* Schwabe. Gen. Res. Crop Evol. 44: 211-215.

Webster, J.A., C.A. Baker and D.R. Porter, 1991. Detection and mechanisms of Russian wheat aphid (Homoptera: Aphididae) resistance in barley. J. Econ. Entomol. 84: 669-673.

Weibull, J., 1987. Screening for resistance against *Rhopalosiphum padi* (L.). 2. *Hordeum* species and interspecific hybrids. Euphytica 36: 571-576.

Weibull, J., 1994. Resistance to *Rhopalosiphum padi* (Homoptera. Aphididae) in *Hordeum vulgare* subsp. *spontaneum* and in hybrids with *H. vulgare* subsp. *vulgare*. Euphytica 78: 97-101.

Weibull, J., and R. Hanson, 1986. Possibilities for resistance breeding against aphids in cereals. In: 27th Weed & Plant Prot. Conf., Swed. Univ. Agric. Sci., pp. 39-44.

Wells, S.A., and W.P. Skoropad, 1963. Inheritance of reaction to *Rhynchosporium secalis* in barley. Can. J. Plant Sci. 43: 184-187.

Weltzien, E., 1988. Evaluation of barley (*Hordeum vulgare* L.) landrace populations originating from different growing regions in the Near East. Plant Breed. 101: 95-106.

Wiberg, A., 1973. Mutants of barley with induced resistance to powdery mildew. Hereditas 75: 83-100.

Wiberg, A., 1974a. Genetical studies of spontaneous sources of resistance to powdery mildew in barley. Hereditas 77: 89-148.

Wiberg, A., 1974b. Sources of resistance to powdery mildew in barley. Hereditas 78: 1-40.

Wicke, H., and H.C. Weltzien, 1986. The resistance situation of German spring barley varieties with respect to loose smut *Ustilago nuda* (Jens.) Rostr. Mitt. Biol. Bundesanst. Land- u. Forstwirtschaft 232: 159.

Wilcoxson, R.D., D.C. Rasmusson and M.R. Miles, 1990. Development of barley resistant to spot blotch and genetic of resistance. Plant Dis. 74: 207-210.

Williams, K.J., A. Lichon, P. Gianquitto, J.M. Kretschmer, A. Karkousis, S. Manning, P. Langridge and H. Wallwork, 1999. Identification and mapping of a gene conferring resistance to the spot form of net blotch (*Pyrenophora teres* f. *maculata*) in barley. Theor. Appl. Genet. 99: 323-327.

Williams, R.J., and H. Owen, 1973. Physiologic races of *Rhynchosporium secalis* on barley in Britain. Can. J. Plant Pathol. 8: 380-386.

Wolfe, M.S., and E. Schwarzbach, 1978. Patterns of race changes in powdery mildews. Ann. Rev. Phytopathol 16: 159-180.

Wright, A.J., and J.B. Heale, 1984. Adult plant resistance to powdery mildew (*Erysiphe graminis*) in three barley cultivars. Plant Pathol. 33: 225-231.

Xu J., and J.W. Snape, 1989. The resistance of *Hordeum bulbosum* and its hybrids with *H. vulgare* to common fungal pathogens. Euphytica 41: 273-276.

Yahyaou, A.H., E.L. Sharp and M. Reinhold, 1988. New sources of resistance to *Puccinia hordei* in barley landrace cultivars. Phytopathology 78: 905-908.

Yamaguchi, I., and A. Yamashita, 1979. Induced mutation of two-rowed barley resistant to powdery mildew, *Erysiphe graminis* f. sp. *hordei*. I. Comparison of effects of gamma rays and ethyleneimine induction of resistant mutation. Japan. J. Breed. 29: 217-227.

Yasuda, S., and K. Rikiishi, 1997. Screening of the World Barley Collection for resistance to barley yellow mosaic virus. Barley Genet. Newsl. 28: 64-66.

Yitbarek, S., L. Berhane, A. Fikadu, J.A.G. van Leur, S. Grando and S. Ceccarelli, 1998. Variation in Ethiopian barley landrace populations for resistance to barley leaf scald and net blotch. Plant Breed. 117: 419-423.

Zeven, A.C., and P.M. Zhukovky, 1975. Dictionary of cultivated plants and their centers of diversity. Pudoc, Wageningen.

Diversity in Barley *(Hordeum Vulgare)*
Roland von Bothmer et al (Editors).

Chapter 9

Diversity in abiotic stress tolerances

A. Michele Stanca[a], Ignacio Romagosa[b], Kazuyoshi Takeda[c], Tomas Lundborg[d], Valeria Terzi[a] and Luigi Cattivelli[a]

[a]Experimental Institute for Cereal Research, Section of Fiorenzuola d'Arda, Via S. Protaso, 302 I-29017 Fiorenzuola d'Arda (PC), Italy
[b]Centre UdL-IRTA – Universitat de Lleida, Alcade Rovira Roure 177–25198 Lleida, Spain
[c]Barley Germplasm Center, Research Institute for Bioresources, Okayama University, Kurashiki 710-0046, Japan
[d]Department of Crop Science, Swedish University of Agricultural Sciences, SE-230 53 Alnarp, Sweden

Introduction

Abiotic stress is responsible for significant yield losses in barley on a worldwide scale, and yet under severe stress conditions, barley is one of the most important sources of energy for human food and animal feed.

It is abiotic stress, which has driven the evolution, the distribution and ecology of the genus *Hordeum*, whose species are widespread in temperate, subtropical and arctic areas, from sea level to heights of more than 4,500 m in the Andes and Himalayas (Bothmer *et al.*, 1995). Overall, the genus *Hordeum* shows a high degree of adaptation to different stressful environments, realised through morphological, physiological and reproductive variants. *Hordeum* species are annual or perennial: the annual habit is related to environments in which the climatic conditions do not permit a permanent vegetation cover, but even the perennial species when subjected to stressful factors show a short lifespan, similar to an annual life cycle (Bothmer *et al.*, 1995). It follows that the evaluation of biodiversity for abiotic stress resistance or tolerance, now supported by molecular tools, is one of the most important aims, to provide new resistance genes for breeding work.

Drought tolerance

Drought is a multidimensional stress, which is difficult to characterise (Belhassen, 1996). It is a function of the genotype, rainfall, temperature, and soil water holding capacity. Its occurrence and severity varies at any site from year to year. Drought is the most common abiotic constraint for stable barley production in rainfed areas. Under Mediterranean conditions, water stress is particularly common at the end of barley life cycle as grain filling usually occurs under relatively high evapotranspirative demands and low precipitation. Drought affects plant processes in a wide

---

Stanca, A.M., I. Romagosa, K. Takeda, T. Lundborg, V. Terzi and L. Cattivelli, 2003. Diversity in abiotic stress tolerances. In: R. von Bothmer, Th. van Hintum, H. Knüpffer and K. Sato (eds), Diversity in Barley (*Hordeum vulgare*), pp. 179-199. Elsevier Science B.V., Amsterdam, The Netherlands.

range of time scales, from a few minutes to weeks or months, causing an array of physiological effects from minor stomatal adjustments to significant yield losses (Passiuora, 1996).

Compared to other cereals, barley is well adapted to arid environments. Ecological studies have shown that the root system of the immediate progenitor of cultivated barley *H. vulgare* ssp. *spontaneum* penetrates deeply into the warm steppes and deserts (Nevo, 1992; Zohary and Hopf, 1998). Ecotypes characterised by average plant height and many tillers per plant were identified in desert locations in Jordan (Jaradat *et al.*, 1996). Higher genetic diversity, measured by protein polymorphisms (Nevo *et al.*, 1994, 1997) and higher phenotypic variation (Volis *et al.*, 1998) have been associated with drought stressed microsites, supporting the hypothesis that polymorphic populations are better adapted to environmental changes. Genotypic diversity, studied by molecular markers in six sub-populations of *H. vulgare* ssp. *spontaneum* sampled from the "Evolution Canyon" microsite, parallels allozyme diversity in the same populations, confirming that this genetic variability driven by natural selection is related to plant climatic adaptation (Owuor *et al.*, 1997). The correlation between polymorphic loci, assessed as molecular marker band frequencies, and environmental characteristics were evaluated in a collection of 21 populations of ssp. *spontaneum* sampled in Israel, Turkey and Iran. A higher genetic diversity index was found associated with stressful environments suggesting that a high diversity at molecular level is required for adaptation to stress conditions (Nevo *et al.*, 1998). In the work of Papa *et al.* (1998) on barley landraces from Sardinia (Italy), a differentiation between southern and northern populations, measured by molecular and morphophysiological traits, was found in relation to different environmental characteristics.

Ample genetic variability in the way cultivated barley genotypes cope with drought has been demonstrated empirically. Studies have particularly been carried out at the International Center for Agricultural Research in the Dry Areas (ICARDA) in Aleppo, Syria by Ceccarelli and co-workers (Ceccarelli and Grando, 1996). They have consistently found that the best performing cultivars under very low- (below 3 t ha^{-1}) and high-yielding (above 5 t ha^{-1}) conditions were different, indicating the presence of genetic diversity for drought tolerance as measured by grain yield. Conversely, under mild stress conditions (yield potential from 3 to 5 t ha^{-1}), high-yielding cultivars selected in fertile environments perform better than any other genotypes, suggesting that selection for moderate stress conditions could be carried out successfully also in high-yielding conditions (Cattivelli *et al.*, 1994). Voltas *et al.* (1999) clearly reported differential responses of genotypes to rainfall regimes. Several studies have experimentally found contrasting yield responses across the environment for different germplasm groups, such as local landraces *vs.* improved cultivars (Ceccarelli and Grando, 1996); and old *vs.* modern cultivars (Muñoz *et al.*, 1998). This is probably reflecting the consequences of differences in the rainfall pattern.

Variation in heading date is the primary cause for yield differences at different water regimes during grain filling (Ludlow and Muchow, 1990; Passiuora, 1996; Richards, 1996; Slafer and Araus, 1998). Thus, the earliest cultivars generally perform better in rainfed low-yielding environments (Oosterom *et al.*, 1993; Jackson *et al.*, 1994; Abay and Cahalan, 1995), escaping the harshest conditions at the end of the growing season. There is a great deal of genetic diversity for traits related to the heading date. A number of well-defined loci are known to control the flowering time following the interaction with environmental signals (temperature and day-length). Three different genetic components are known: photoperiod response (depending from day-length), vernalisation response (depending from temperature), and "earliness *per se*" largely independent from both day-length and low temperature.

Grain yield in water-limited environments is directly related to the plant ability to capture water (T, reflected by the amount of water transpired); water use efficiency (WUE, the ratio of above ground biomass produced by unit water transpired); and harvest index (HI; Passiuora, 1977, 1996). Little is known about how much genetic diversity is present for T or WUE. For barley, the results of genetic gains in total plant biomass (the product of these two components) with time are contradictory. Some studies suggest that there is not much genetic variation for this trait since biomass production has varied little in recent decades (Riggs *et al.*, 1981; Jedel and Helm, 1994). However, other studies have found an increase in total biomass with time (Martiniello *et al.*, 1987; Boukerrou and Rasmusson, 1990). Increasing the amount of water extracted by the crop from the soil is meaningful if provided soil water is still present at maturity. In rice, genetic factors controlling the root characteristics were found to be associated with field drought tolerance (Champoux *et al.*, 1995). Genetic variability for root characteristics has also been shown for barley (Grando and Ceccarelli, 1995; Gorny, 1996). Early vigour, which to some extent depends on phenological development (Slafer and Araus, 1998), has been repeatedly proposed as a secondary trait associated with total water extracted from the soil and drought tolerance (López Castañeda *et al.*, 1996; Richards, 1996). Early vigour improves T by increasing ground coverage and, thus, reducing losses due to direct evaporation from the soil. There is also large genetic variation for this trait among cereals (López Castañeda *et al.*, 1996).

Induced osmotic adjustment may allow plants to extract water from the soil under drought, and therefore may be an important component of drought resistance in barley (Blum, 1989). Differences in osmoregulation in response to water deficits have been found in both *H. vulgare* ssp. *vulgare* and ssp. *spontaneum* genotypes (Gunasekera *et al.*, 1994). In this context, increased accumulation of an osmoprotectant such as glycine-betaine has been associated with drought stress tolerance (Ishitani *et al.*, 1995).

Arnau *et al.* (1997) studied the effect of water stress on plant water status and net photosynthetic gas exchange (P-N) in six barley cultivars and landraces differing in productivity and drought tolerance. Variability for some parameters like osmotic adjustment capacity and P-N was observed among drought tolerant genotypes and susceptible ones. The ratios of the isotope pairs ^{15}N to ^{14}N and ^{13}C to ^{12}C were measured in 30 genotypes of *H. vulgare* ssp. *spontaneum* collected in ecologically diverse sites and subjected to mild, short-term drought. Among the potentially most productive genotypes, the most stress-tolerant ones had the most negative whole-plant ratio ^{15}N (Robinson *et al.*, 2000).

All data presented above describe the phenotypic diversity for drought tolerance-related traits and this diversity is the consequence of a corresponding DNA diversity. The development of molecular markers and QTL (quantitative trait loci) analysis has allowed the identification of the genetic bases of drought tolerance, and different alleles (with either positive or negative effects) have been identified at the drought tolerance loci. Teulat *et al.* (1997; 1998) identified two regions on the chromosome 6H and 7H where most of the loci controlling drought related traits are located. Several drought-responsive genes have been cloned in barley (reviewed by Cattivelli *et al.*, 2001), the majority belonging to the class 2 LEA (Late Embryogenesis Abundant) gene family. The proteins encoded by these genes (known as dehydrins) are characterised by one or more conserved lysine-rich 15-amino acid sequences near the C-terminus and, in most cases, by a stretch of serine residues. Their hydrophilic characteristics suggest an osmoprotective role during cell dehydration (Grossi *et al.*, 1995; Close, 1997; Choi *et al.*, 1999). Molecular diversity for drought-induced proteins has also been found in barley cultivars with different drought-tolerance (Grossi *et al.*, 1992). Other drought-responsive genes cloned in barley are known to encode for proteins with enzymatic activities, such as PG22-69 (Bartels *et al.*, 1991), homologous to aldose reductase, or BADH (Ishitani *et al.*, 1995) coding for betaine aldehyde dehydrogenase. Betaine and sorbitol are considered osmoprotective molecules.

A recurrent feature of a harsh environment is the concomitance of drought stress and high temperatures. A sudden increase from 22 °C to a level above 34-37 °C determines several modifications in the protein synthesis system and the induction of specific heat shock proteins. Temperature stress in barley genotypes results in specific protein synthesis with a significant degree of genotype dependency (Marmiroli *et al.*, 1989; Terzi *et al.*, 1990). In barley, three cDNA clones encoding for 17 kDa, 18 kDa (Marmiroli *et al.*, 1995) and 90 kDa (Wolthen-Larsen *et al.*, 1993) HSPs have been characterised. The latter is also involved in the response to pathogen attack.

Winter hardiness

Winter hardiness can be defined as the ability to survive throughout the winter. By the virtue of the wide range of stressful conditions that a plant may experience during the cold season, winter hardiness is a complex trait. Freezing temperature is the most relevant stress factor, although other stress situations, such as anoxia due to excess water or to ice encasement and photoinhibition due to the combination of light and low temperature, may also occur.

The adaptation to cold climate can be achieved either by the development of a powerful frost tolerance ability or by limiting the life cycle to the short summer season. Winter barley cultivars are generally less hardy than winter wheat, rye and triticale. Nevertheless, barley is grown beyond the Polar Circle because early maturing spring cultivars are able to run their life cycle in the short summer season. Plant growth habit and heading date can, therefore, be considered as the basic traits involved in barley adaptation to environments since they allow synchronisation of the plant life cycle with seasonal changes.

In temperate areas, winter cultivars are preferred, whenever possible, as they are higher-yielding than spring forms, and the identification of new gene sources for frost tolerance is still an important task for barley improvement (Cattivelli *et al.*, 1994). A systematic evaluation of frost tolerance carried out with a large number of barley cultivars by several authors (Jenkins and Roffey, 1974; Fowler and Limin, 1987; Kolar *et al.*, 1991; Hömmö, 1994; Rizza *et al.*, 1994; Pulli *et al.*, 1996) led to three main conclusions:

1) Winter barley is generally less hardy than the corresponding wheat or rye cultivars;
2) Winter cultivars are more frost tolerant than spring ones;
3) Diversity for frost tolerance can be found among winter cultivars.

The German winter barley 'Borwina' was found to be the hardiest cultivar in a Finnish winter field trial network (Hömmö, 1994). Similarly, the Italian winter barley 'Onice' was found to be more frost tolerant in a collection of about 30 cultivars commonly grown in south Europe (Rizza *et al.*, 1994).

Despite the fact that genetic diversity has also been observed for freezing tolerance of non-acclimated barley plants (Bravo *et al.*, 1998), the ability of overwintering plants to withstand cold is mainly based on an adaptive response, known as cold acclimation or hardening, activated during growth at low non-freezing temperatures. Frost resistance can be assessed through field evaluation methods, although this strategy is rather inefficient because of the irregular occurrence of natural conditions that satisfactorily differentiated genotypes. As a consequence, researchers have developed artificial freezing tests such as percentage of post-stress survival (Doll *et al.*, 1989), LT_{50} (temperature at which 50% of the population is killed) (Gullord *et al.*, 1975; Fowler and Carles, 1979), percentage of crown moisture (Brule-Babel and Fowler, 1989), integrity of cell membranes after freezing (Dexter *et al.*, 1932; Jenkins and Roffey, 1974), etc. A number of physiological traits such as proline accumulation, ABA and crown fructan content have been associated with the development of frost tolerance. Generally, the amount of these metabolites increases during acclimation, being higher in frost-tolerant genotypes *vs.* frost-susceptible ones (Livingston *et al.*, 1989; Murelli *et al.*, 1995; Bravo *et al.*, 1998). A comparison of a number of frost resistance evaluation methods in different crop species was carried out by Pulli *et al.* (1996). The value of the laboratory tests appears to vary with the plant species, although for cereals frost resistance evaluation methods based on freezing assays provide the highest correlation with field survival.

Genetic factors controlling frost tolerance

Growth habit is the best-known genetic trait affecting the winter hardiness of barley cultivars, the spring type being frost-sensitive and the winter type frost-tolerant. Winter habit depends on the presence of the dominant allele at locus *Sgh* and of the recessive alleles at loci *sgh2* and *sgh3*. All other possible allele combinations of these three genes are found in spring genotypes. The loci *Sgh*, *Sgh2* and *Sgh3* are located on chromosomes 4H, 5H and 1H, respectively. The homozygous genotype *sghsgh* is epistatic with respect to the recessive alleles *sgh2* and *sgh3*, and evinces a facultative behaviour towards the spring habit when sown in spring. The *Sgh3* allele is epistatic with respect to alleles *Sgh* and *sgh2*. Without vernalisation and in long-day conditions, all the *Sgh3Sgh3* cultivars are essentially spring types. *Sgh2*, which is epistatic *vis à vis* alleles *Sgh* and *Sgh3*, has a series of multiple alleles, which induce several spring-to-winter variants (Cattivelli *et al.*, 1994).

Winter cultivars are generally hardier than their spring counterparts, either due to pleiotropic effects of the growth habit loci or to genetic linkage between the *Sgh* loci and the loci controlling frost tolerance. Genetic studies of traits associated with winter hardiness in barley (field survival and crown fructan content) have found that a major QTL effect for winter survival is located on the barley chromosome 5H, in association with the *Sgh2* locus (Hayes *et al.*, 1993) and with a QTL for heading date and vernalisation response under long-day conditions (Pan *et al.*, 1994). These results are due to genetic linkage rather than pleiotropic effects since recombinants between vernalisation requirement and winter survival traits have been described (Doll *et al.*, 1989). RFLP analysis performed on the homoeologous 5A chromosome of wheat (highly syntenic to the barley chromosome 5H) has proved that vernalisation and frost resistance are controlled by two different, but tightly linked loci. *Vrn1* (hortologous to *Sgh2*) and *Fr1*, respectively (Galiba *et al.*, 1995).

Molecular diversity

Molecular analysis of the cold acclimation process has led to the isolation of many cold-regulated (*cor*) genes; in barley more than 20 clones whose expression is affected by low temperatures have been isolated and associated with the development of frost tolerance (Cattivelli *et al.*, 2001). The level of hardiness among barley plants grown in different temperature environments was found to be strongly correlated with the expression level of *cor* genes (Pearce *et al.*, 1996).

Although cold acclimation in barley involves the expression of many cold-regulated genes, genetic studies have proved that only few chromosome regions (mainly on chromosome 5H) carry loci that play an important role in frost tolerance. By using the highly syntenic wheat genome it has been demonstrated that on the homoeologous group 5, in genetic linkage with the growth habit locus, there are two loci which control the expression of several *cor* genes

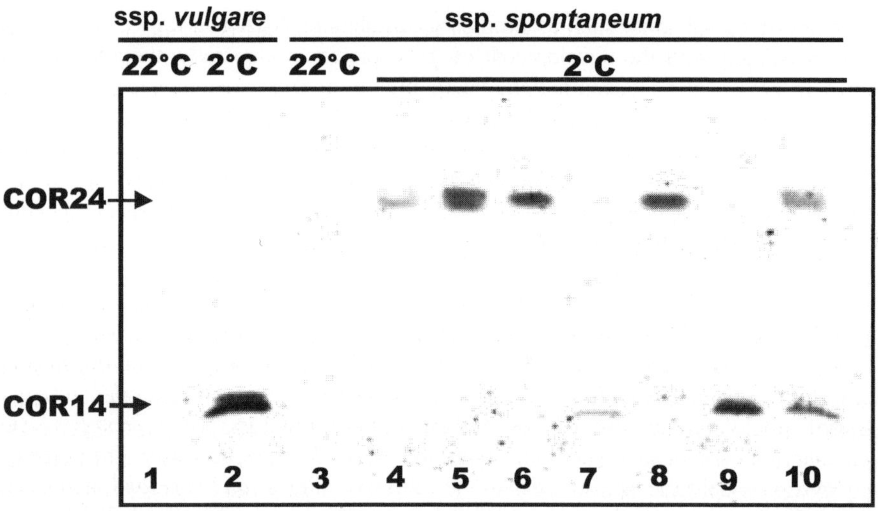

Figure 9.1. Molecular variation for accumulation of the cold-regulated proteins COR14 and COR24. Two *H. vulgare* ssp. *vulgare* samples, either grown at 22 °C (lane 1) or acclimated for seven days at 2 °C (lane 2), are compared with a collection of ssp. *spontaneum* accessions grown in the same contrasting conditions (lanes 3 and 4 to 10, respectively). Proteins were detected after western analysis through a COR14-specific antibody. COR14 and COR24 are indicated.

(Vagujfalvi *et al.*, 2000). This work demonstrates the association with growth habit, frost resistance and the expression of *cor* genes.

A large diversity for frost tolerance has been described in barley, although the molecular bases of this diversity is only in part understood. Barley cultivars are rarely polymorphic for the *cor* gene sequences, i.e., the *cor* gene *tmc-ap3* was found to be identical in spring and winter barley (Mastrangelo *et al.*, 2000), while a number of polymorphisms were found looking at the expression of *cor* genes. Susceptible and resistant genotypes have often the same *cor* gene sequences, but they express the genes at different level. The cold-regulated expression of the gene *tmc-ap3*, a sequence coding for a chloroplastic amino acid selective channel protein, is higher in frost-tolerant cultivars than in frost-sensitive ones (Baldi *et al.*, 1999). The cold-regulated gene *cor14b* (formerly *pt59*) encodes for a chloroplast localised protein of 14 kDa (Crosatti *et al.*, 1995, 1999) accumulated in higher amount in winter cultivars than in spring ones, particularly when plants are acclimated at +8 °C (Crosatti *et al.*, 1996). These results suggest that frost-tolerant genotypes have a higher induction-temperature threshold of *cor14b* than frost-susceptible ones (Giorni *et al.*, 1999).

Natural populations of *H. vulgare* ssp. *spontaneum* are known to represent a rich source of genetic variability for resistance to physiological stresses (Nevo, 1992) and this was also true when the molecular response to low temperature was assessed. A typical example of molecular diversity is the polymorphism of the cold-regulated proteins COR14/COR24. In *H. vulgare* ssp. *vulgare* the COR14 antibody produced against the protein encoded by the *cor14b* gene cross-reacts with a polypeptide with a relative molecular weight of about 14 kDa (COR14; Figure 9.1, lane 2) induced by low temperatures (and therefore absent in the control sample; Figure 9.1, lane 1). Within a collection of *H. vulgare* ssp. *spontaneum* a clear polymorphism was found for the corresponding COR proteins. While some accessions showed the same COR pattern as cultivated barley (Figure 9.1, lane 9), in other ssp. *spontaneum* accessions the COR14 antibody cross-reacted with an additional cold-induced protein with a relative molecular weight of about 24 kDa (COR24). The accumulation of COR24 was often associated with the absence of COR14b (Figure 9.1, lanes 4, 5, 6 and 8). Although in some *H. vulgare* ssp. *spontaneum* accessions both proteins were detected (Figure 9.1, lane 10) possibly due to heterozygosity at the corresponding locus. A large variation in the amount of COR14 and COR24 accumulated by *H. vulgare* ssp. *spontaneum* plants under low temperatures has also been detected. Most accessions accumulated COR proteins at high level (Figure 9.1, lanes 5, 6, 8 and 9), but there were also accessions expressing the same proteins at very low level (Figure 9.1, lanes 4 and 7). The COR14/24 polymorphism was detected in *H. vulgare* ssp. *spontaneum* populations collected in highland regions occasionally or normally exposed to frost. By contrast, COR24 was found to be monomorphic and COR14 was absent in warm desert and mild, usually frost-free coastal plain populations. Thus, the polymorphism of COR24 and COR14 may be climatically adaptive. Several *H. vulgare* ssp. *spontaneum* genotypes were also tested for the expression of another *cor* gene named *blt14* and a large variability was found. When the same genotypes were tested for frost resistance, the most hardy ssp. *spontaneum* genotypes showed a high expression of the *blt14* gene family, although this molecular trait *per se* is not sufficient to confer frost tolerance (Grossi *et al.*, 1998).

Flooding tolerance

Barley cultivars in East Asia have differentiated in a particular way. They have been exposed to water damage due to the Asian monsoon climate where annual precipitation exceeds 2,000 mm. The crop is often cultivated in paddy fields with a rice-barley double cropping system. Barley cultivars in East Asia are, therefore, highly tolerant to water damage and pre-harvest sprouting.

Excess of water in the soils often induces hypoxic conditions. When plants are subjected to anaerobic stress, there is a shift in carbohydrate metabolism from the oxidative to the fermentative pathway, with the induction of enzymes, mainly alcohol dehydrogenase (ADH), associated with the flow of carbon into glycolysis and alcoholic fermentation (Dennis *et al.*, 1992). In barley, three ADH loci are known: *Adh1* and *Adh2* are closely linked on chromosome 4H; whereas the third locus, *Adh3*, is located on chromosome 6H (Trick *et al.*, 1988). *Adh1* is constitutively expressed in developing seeds, while all isoenzymes are strongly induced in root cells under anoxia conditions (Mayne and Lea, 1984). The sequence of *Adh1* has been analysed in *H. vulgare* ssp. *vulgare* and ssp. *spontaneum* accessions in screening for molecular diversity, but only a few polymorphisms, restricted to the non-coding part of the gene, were found (Petersen and Seberg, 1998). Besides *Adh* lactate dehydrogenase (LDH) and alanine amino-transferase (AlaAT) was also found to be enhanced in barley roots under hypoxic conditions (Hondred and Hanson, 1990; Muench and Good, 1994). AlaAT is the enzyme that catalyses the production of alanine and indeed an increase in alanine content under hypoxia has been found.

Water sensitivity

Water sensitivity is defined as the reduction in germinability due to excess water conditions. Takeda and Fukuyama (1983) evaluated the water sensitivity of 2,212 Asian barley cultivars by germination tests. Water sensitivity showed very large genotypic variation, 0-100%. All 16 hulled isogenic lines examined were more sensitive than the hull-less counterparts (Figure 9.2). However, hull-removed grains showed a similar water sensitivity to hull-less ones, suggesting

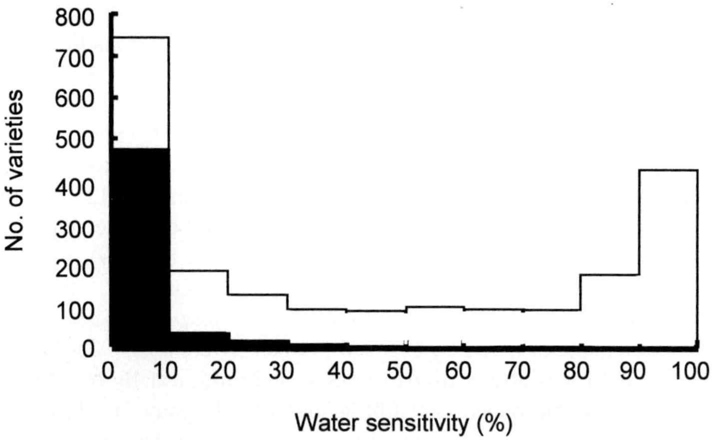

Figure 9.2. Variation for water sensitivity in Asian barley cultivars. Open: hulled varieties (n=1,604; mean=61.4), solid: hull-less varieties (n=608; mean=6.7).

that the presence of hulls and probably the glue between the hull and caryopsis prohibited gas exchange and reduced the germinability in excess water conditions. Since the water sensitivity is affected by the hull and genotypic variation of the water sensitivity in hull-less cultivars is rather small (Figure 9.2), true genetic variation of the trait might be small.

Tolerance to pre-germination flooding

Among the cereal crops, only rice can germinate under anoxia conditions (Perata et al., 1998). Takeda (unpublished data) tested thousands of barley accessions for germinability under water-soaking conditions to find that none of them could germinate in the water. However, after a certain period of soaking there was a large variation in germinability, i.e., some of the cultivars were completely killed while others survived.

Germinability of barley seeds decreased according to time of soaking and temperature of the water. Arney and Leben (1955) soaked barley seeds in water at 22-25 °C for 56 hours to control loose smut and found that germinability was reduced by 19% in an average of seven cultivars examined, with a range from 40% reduction in the most sensitive 'Kindred' to only 5% in the most tolerant 'Wisconsin Barbless'.

Takeda and Fukuyama (1987) examined the germinability of ca. 3,400 accessions after four days' soaking at 25 °C. Genotypic variation showed a bimodal distribution (Figure 9.3): cultivars which originated in the western part of India were generally sensitive to soaking, while genotypes derived from the eastern region of Nepal were tolerant. Barleys from China, many of them from the Long River basin, were most tolerant, and those from Southwest Asia and Ethiopia were most sensitive. This geographical differentiation may reflect the climatic condition of the regions. Some 90% of the most tolerant cultivars germinated after seven days of soaking at 25 °C. The sensitive barley seeds fermented but the seeds of tolerant cultivars appeared dormant in the water.

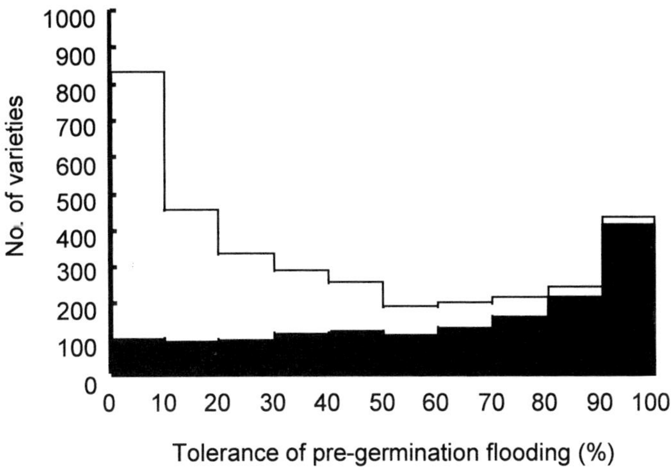

Figure 9.3. Variation for tolerance to pre-germination flooding. Open: varieties in India and west of India (n=1,864; mean=20.3), solid: varieties in Nepal and east of Nepal (n=1,563; mean=61.3).

Flooding tolerance after germination

Flooding during the period after the germination to maturity often occurs in Asian monsoon areas where barley is grown in paddy fields. Even in the uplands, topographical and climate conditions including snow-melting water, cause occasional flooding.

Towscy *et al.* (1983) tested 25 barley cultivars and reported no marked differences. Oram and Driscoll (1983) evaluated breeding materials and found that 'Dampier', 'Ilineckii 43', 'Murasaki-moch' and 'Arimont' were useful sources of tolerance to water-logging. 'Murasaki-moch' is a Japanese local waxy cultivar. In China, Qiu and Ke (1991) examined 4,572 accessions. Only 0.4% of the germplasm showed a high level of tolerance. Suh (1982) tested 57 barley cultivars in Korea and found that 'Molyang 12' and 'Molyang 16' were tolerant. Sasaki (1984) evaluated flooding tolerance of a total of 198 cultivars at the internode elongation stage. 'Minorimugi', 'Saikaikawa 17', 'Akashinriki', 'Izumiwasehadaka' and 'Swannech' were least affected by excess soil moisture. The former four are Japanese cultivars.

Takeda (1989) evaluated 4,096 genotypes and the variation showed a normal distribution. The 360 most tolerant forms were selected and repeatedly tested. These genotypes normally ripened in paddy field conditions. Tolerant germplasm was found mainly in Japan, Korea, China and Ethiopia. Tolerant genotypes abundantly developed adventitious roots and aerenchyma in the roots. Alcohol dehydrogenase activity was not very high in tolerant genotypes.

Okubo and Takeda (1991) also evaluated 3,165 accessions for flooding tolerance. Genotypes from each region showed a wide range of variation. The six-rowed forms were more tolerant than two-rowed, but the difference between the hulled and hull-less groups was statistically non-significant.

Tolerance to pre-harvest sprouting

Pre-harvest sprouting reduces the starch quality of barley grains (Ringlund, 1980), and a rate of more than 5% pre-harvest germination is considered unsuitable for malting (Brookes, 1980). Seed dormancy is the strongest factor preventing pre-harvest sprouting.

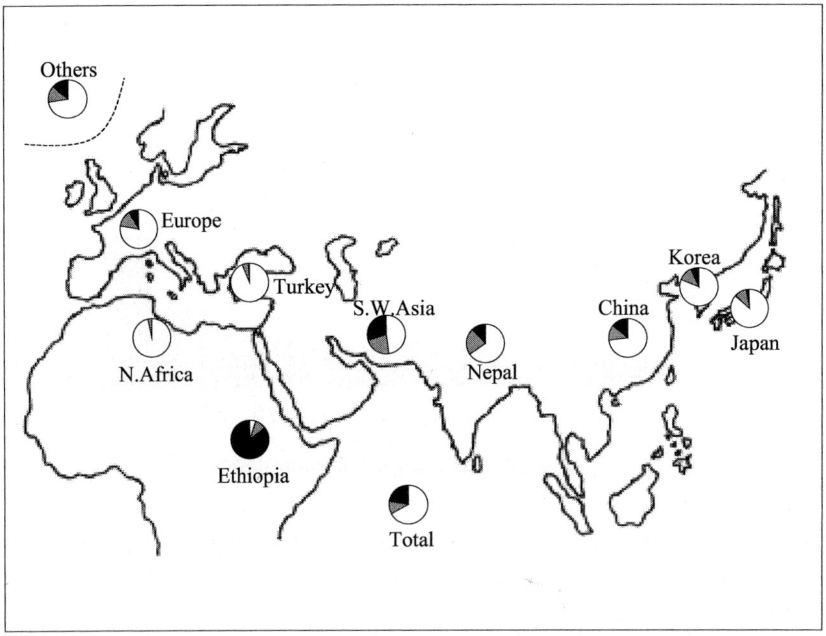

Figure 9.4. Variation for seed dormancy (n=4,422). Black: none; grey: medium; white: deep.

It was previously proposed that dormancy of barley grains is determined by at least three factors: (1) synthesis of the plant hormone abscisic acid (ABA); (2) breakdown and/or removal of ABA; and (3) sensitivity to ABA (Wang, 1997). Gibberellic acid stimulated germination of dormant seeds and ABA severely inhibited the response and prevented germination (Dunwell, 1981). In 'Kristia', which showed a short dormancy period and low resistance to sprouting, ABA content was lower and the decrease during drying was more pronounced than in 'Oriol' (Goldbach and Michael, 1976). Molina-Cano *et al.* (1999) reported that a less dormant mutant induced from 'Triumph' by sodium azide treatment was ABA-insensitive and showed a high level of alpha-amylase activity.

Hulled conditions have been shown to influence seed dormancy (LaBerge, 1983; Lenoir *et al.*, 1983; Benech-Arnold *et al.*, 1999) and removal of the palea and lemma strongly promoted seed germination (Dunwell, 1981).

The ABA content of grains was highest at physiological maturity and decreased as the grain dried (Goldbach and Michael, 1976). The rate of dormancy loss or the germinability curves followed a positive cumulative normal distribution with storage time that allowed the application of probit analysis techniques to the data (Favier and Woods, 1993). Therefore, the degree of dormancy can be numerically expressed by germination percentage after a certain period of storage in controlled conditions.

Takeda (1995) evaluated the dormancy of more than 4,000 cultivars and 177 wild (*Hordeum vulgare* ssp. *spontaneum*) accessions. The germinability showed sigmoid curves as also reported by Favier and Woods (1993). Germination percentage at harvest time showed the largest variation among cultivars. It showed a typical bimodal distribution pattern, i.e., about half of the entries germinated less than 10% and about 20% of the accessions germinated more than 90%.

Very few accessions showed 20-80% germination. This indicates that these accessions consist of two groups of genotypes, dormant and non-dormant, at the yellow ripeness stage. On average, germination percentage was 5% higher in hull-less accessions than in hulled ones.

It is clear from Figure 9.4 that most accessions from Ethiopia have no seed dormancy whereas most of the accessions from East Asia, Turkey and North Africa (except Ethiopia), have highly dormant seed. Many of the accessions from Nepal, Europe and US were also dormant. Accessions from Southwest Asia, i.e., Iran, Iraq, Afghanistan, Pakistan and India, showed a wide range of seed dormancy. All wild accessions tested were highly dormant.

For further comparisons of levels of seed dormancy, the accessions were grouped according to agro-morphological characteristics such as hulled/hull-less, two-rowed/six-rowed, etc. When all the accessions were grouped no clear difference in seed dormancy was found between the types. However, if the accessions from Southwest Asia alone were analysed, it appeared that accessions with hulled, two-rowed, short-haired rachilla, blue aleurone or black hull tended to be more dormant than their counterparts. This indicates that either the genes controlling these traits or the genetic backgrounds of these varietal groups affect seed dormancy.

Deep-seeding tolerance

When the barley growing season precedes the rainy season, the soil surface is generally too dry for germination and the farmers have to seed in depth to utilise the water stored in deeper soil. Under these conditions deep-seeding tolerance represents an important trait for a successful crop. Takahashi and Takeda (1999) examined the percentage of emergence of a total of 5,082 accessions from a world collection preserved at the Barley Germplasm Center, Okayama University. Deep-seeding tolerance showed a wide range of variation, 0-100% emerging from a depth of 12 cm, and the genotypes from North Africa and Nepal were more tolerant than those from Europe and China. A semi-dwarf '*uzu*' type was decidedly weak (Figure 9.5). The most tolerant genotype can normally emerge from 18 cm.

Deep-seeding tolerance correlated positively with coleoptile length, the first internode length

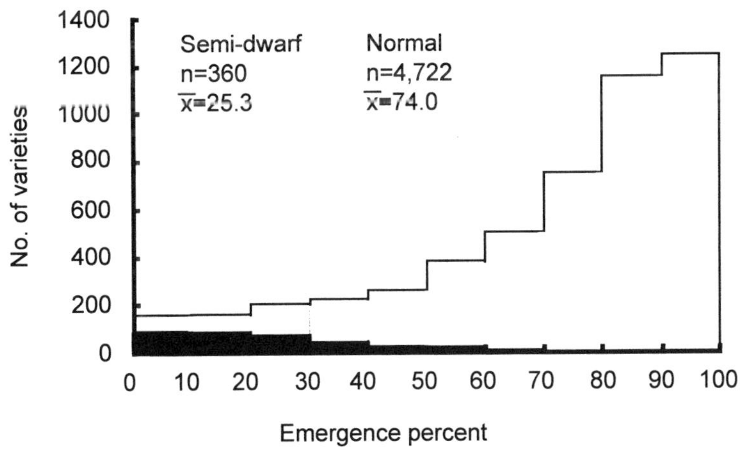

Figure 9.5. Variation for emergence percent from 12 cm seedling depth in semi-dwarf, *uzu* (solid) and normal (open) cultivars.

and seed weight. Hulled isogenic lines were more tolerant than the hull-less counterparts. Deep-seeding tolerance of six-rowed central grains were comparable to two-rowed isogenic counterparts, but six-rowed lateral grains were weaker then the central grains.

Salinity tolerance

Soil salinity and soil sodicity are common problems in arid and semi-arid areas. Barley genotypes grown in these marginal areas have to be tolerant to soil salinity. Nair and Khulbe (1990) reported that barley was more salt-tolerant than wheat.

In plants there is no specific carrier for Na^+ uptake; however, Na^+ enters by competition with other cations, particularly K^+. Regulation of cellular Na^+ is achieved by effluxing Na^+ through a Na^+/H^+ antiporter, driven by the electrochemical H^+ gradient across the plasmalemma. Intracellular compartmentalisation can also occur due to the work of the vacuolar Na^+/H^+ antiporter driven by the electrochemical H^+ gradient across the tonoplast (Schachtman and Liu, 1999). Salt tolerance in cereals is known to be associated with the control of shoot Na^+ content; tolerant lines have more efficient systems to exclude sodium from their cells. Loci involved in salt tolerance have been identified on chromosomes 4H and 5H of *H. vulgare* and $1H^{ch}$, $4H^{ch}$ and $5H^{ch}$ of *H. chilense* (Forster *et al.*, 1990). The genepool of *H. vulgare* ssp. *spontaneum* may also represent an interesting source of new loci for salt tolerance and several QTLs were detected on chromosomes 7H, 4H, 1H and 6H (Ellis *et al.*, 1997).

Many studies have been dedicated to the evaluation of salinity tolerance and a vast number of barley accessions have been tested for their ability to growth in high salt conditions. For instance, Omara *et al.* (1987) tested 488 Egyptian landraces and selected 37 tolerant genotypes that germinated more than 80% in 1.2% NaCl solution. Towsey *et al.* (1983) evaluated 600 genotypes and identified a group of accessions (among them 'California Mariout') with high salt tolerance. Abo-Elenin *et al.* (1981) tested 1,163 entries in the field and 777 in lysimeters. 'Abyssinia' was the most tolerant germplasm. Slavich *et al.* (1990) grew 38 genotypes in saline soil conditions. 'Forrest' and 'O'Connor' ranked more tolerant than 'California Mariout', a tolerant standard.

Mano *et al.* (1996) evaluated 6,712 accessions of a world collection for salt tolerance at germination. Cultivar variation showed a normal distribution; the most sensitive genotypes failed to germinate in 1% NaCl solution and the most tolerant ones could germinate in sea water. Tight correlation (r=0.789) between germinability in NaCl solution and in polyethylene glycol solution with comparative osmotic pressure suggested that salt inhibits seed germination primarily by osmotic effect as Bliss *et al.* (1986) indicated. Geographical differentiation was not clear. But six-rowed genotypes were more tolerant than two-rowed, hull-less than hulled, normal than semi-dwarf '*uzu*', winter than spring. Germinability in saline conditions showed a negative correlation with seed weight (r=-0.300) indicating that accessions with smaller seeds were more tolerant.

Mano and Takeda (1995) evaluated 5,182 barley cultivars for salt tolerance at seedling stage. Geographical differentiation was not clear. Comparing isogenic pairs for major genes it was demonstrated that six-rowed types were more tolerant than two-rowed, semi-dwarf '*uzu*' than normal, while hulled was equal to hull-less. There was no correlation between salt tolerance at germination and at seedling stage (r=-0.061), suggesting that the mechanisms of tolerance are different at the different stages. As mentioned before, salinity sensitivity at germination is primarily due to an osmotic effect, and the sensitivity at seedling stage may be due to sodium toxicity. In a further study Mano and Takeda (1998) examined 340 wild accessions including various species for salt tolerance at germination and seedling stages. Salt tolerance of wild species at germination was at the same level as that of cultivars, while most of the wild accessions showed better tolerance to salinity than cultivated species at the seedling stages. Nevo *et al.* (1993) also reported that superior genotypes of *H. vulgare* ssp. *spontaneum* ripened in 60% sea water.

Acid and alkaline soils and tolerance to heavy metals

Barley, like other plant species, suffers when grown in soils with non-neutral pH conditions. Both alkaline and acid soils can produce unfavourable conditions. In some cases a pure pH effect is present, related to hydrogen concentration, but mostly the effects are indirect and depend on the availability of other ions in the soil.

When plants are grown in alkaline soils, a manganese (Mn) deficiency may occur. Manganese, although present in these soils, is usually in a complex form not available to plants. Among cereals, rye absorbs about five times as much Mn as wheat or barley (Gladstones and Loneragan, 1970). On the other hand, bread wheat or barley is more tolerant to manganese deficiency than durum wheat (Saberi *et al.*, 1999). In barley, a genetic analysis of tolerance to Mn-deficiency has revealed the presence of an Mn-efficiency locus (*MelI*) located on chromosome 4H (Jefferies *et al.*, 2000). The interaction between Mn and nitrogen availability was studied using barley genotypes with different tolerance to manganese deficiency. At high nitrate fertilisation and low rates of Mn fertilisation, the Mn-efficient genotype 'Weeah' (tolerant to Mn deficiency) showed better growth of shoot and root than the Mn inefficient 'Galleon' (sensitive to Mn deficiency; Tong *et al.*, 1997).

Acid soils are common in several areas worldwide. Lower soil pH affects soil structure, microflora, and the availability of mineral nutrients. Situations of mineral deficiency and mineral toxicity are both present in acid soils. Generally, barley is regarded as a species tolerating a very wide range of soil pH values. However, according to Bona *et al.* (1991) barley was found to be less tolerant to acid conditions than rye, oat, millet and common wheat.

A low soil pH also increases the availability of toxic heavy metals to the plant roots. In a screening of different grass species for phytoremediation purposes, barley, together with oat, was found to be able to tolerate relatively high concentrations of copper, cadmium and zinc. These results indicated the potential role of barley in the phytoremediation of contaminated soils (Ebbs and Kochian, 1998).

Upon exposure to heavy metals, barley cells activated the expression of a set of stress-responsive genes. Tamás *et al.* (1997) showed that copper, cadmium and cobalt, as well as aluminium, induced extracellular accumulation of several polypeptides originally described as PR (pathogenesis-related) proteins. The apoplastic PR-protein concentration was higher in response to nickel than to cadmium or zinc treatment (Blinda *et al.*, 1997). A genetic diversity in the molecular response to heavy metal stress was also found between resistant and sensitive cultivars. After treatment with copper, cadmium or cobalt, the aluminium-sensitive barley 'Alfor' was found to accumulate two cytoplasmic polypeptides absent in the aluminium resistant 'Bavaria' (Tamás *et al.*, 2000).

In acid soils with pH values below 5.5, the presence of available aluminium probably represents the most important growth-limiting factor. Among winter cereals barley is the most sensitive to the excess of soluble or exchangeable aluminium and a limited genetic diversity for aluminium tolerance has been found (Minella and Sorrells, 1992; Ma *et al.*, 1997; Gallardo *et al.*, 1999). The inheritance of aluminium tolerance in barley is reported to be controlled by a single major locus located on chromosome 4H (Minella and Sorrells, 1997; Jeffries *et al.*, 2000). The limited genetic variability found so far suggests that new and still uncharacterised germplasm and related species should be investigated in the future in an attempt to improve aluminium tolerance in barley.

Conclusions and outlook

All evidence indicates that abiotic stress tolerance in barley is a complex trait largely under polygenic control. Further, more thorough investigations are necessary to elucidate the basic metabolic processes involved in the plant response and their genetic control when insulted by single or multiple stresses. The knowledge derived from this research represents a milestone for developing new strategies for a better use of barley biodiversity to improve this crop species and to use it as a model plant. For this purpose, new barley variability, and 'new mutants', are required to amplify the spectrum of specific morphological and physiological traits. For all these reasons, it is essential to improve the evaluation for abiotic stress resistance of genetic resources conserved *ex situ*, and to maintain and promote the *in situ* conservation to ensure the germplasm can cope with stress conditions.

From the breeding point of view, a never-ending debate between germplasm *vs.* methodology exists among breeders and physiologists (Weltzien and Fischbeck, 1990; Rajaram and Ginkel, 1996). Whereas intensive work has been carried out continuously by physiologists in the area of tolerance to abiotic stresses, few barley breeders have routinely used the developed physiological criteria in their mainstream-breeding programme. Of those who do, the majority consider development and assessment of new criteria a sideline and many do not routinely apply their own published methodology. Nonetheless, many breeders develop a profound understanding of their environments and adaptation of their genetic materials. However, we believe that physiological assessment of adaptation to the environment is needed to complement breeders' impressions particularly in the first and last stages of a breeding programme: selection of parents and assessment of adaptation of new advanced lines. Most barley breeders focus on just the elite genepool, reflecting decades of crossing, selection and recombination (Rasmusson, 1996).

This classic strategy depends on the fact that adding genetic diversity has had limited success compared with programmes that develop new cultivars using already existing elite germplasm. Even if this process indicates that each cycle within narrow genepools (good × good crosses) is expected to lead to reduced variability, Rasmusson and Phillips (1997) showed that by using this breeding procedure *de novo* variation is generated. These authors posed the hypothesis that this newly generated variation makes an important contribution to further genetic gain. However, to avoid a drastic reduction in genetic variability, more emphasis should be placed on the use of new genetic variability, particularly by genetically building new parents for crosses, incorporating desired traits into the genepool after a series of pre-breeding activities. By using molecular markers associated with single genes or QTLs, in the near future marker-assisted selection will be a powerful tool to be introduced during all stages of pre-breeding – breeding processes with the final aim of identifying and introducing useful genetic traits into targeted breeding material more effectively than can be reached by traditional breeding methodologies.

Acknowledgements

Part of the evaluation work reported in this paper was supported by the EU project GENRES CT-98-104 and by the Italian Ministry of Agriculture (MIPAF) special project "Risorse genetiche."

References

Abay, F., and C. Cahalan, 1995. Evaluation of response of some barley landraces in drought prone sites of Tigray (Northern Ethiopia). Crop Improv. 22: 125-132.

Abo-Elenin, R.A., M.S. Heakal, A.S. Gomaa and J.G. Moseman, 1981. Studies on salt tolerance in barley and wheat. Source of tolerance in barley germplasm. In: Barley Genetics IV. Proc. 4th Int. Barley Genet. Symp. Edinburgh, UK, pp. 402-409.

Arnau, G., P. Monneveaux, D. This and L. Alegre, 1997. Photosynthesis of six barley genotypes as affected by water stress. Photosynthetica 34: 67-76.

Arney, D.C., and C. Leben, 1955. The effect of the water-soak seed treatment on the germination of certain barley varieties grown at different locations. Phytopathology 45: 518-519.

Baldi, P., M. Grossi, N. Pecchioni, G. Valè and L. Cattivelli, 1999. High expression level of a gene coding for a chloroplastic amino acid selective channel protein is correlated to cold acclimation in cereals. Plant Mol. Biol. 41: 233-243.

Bartels, D., K. Engelhardt, R. Roncarati, K. Schneider, A.M. Rotter and F. Salamini, 1991. An ABA and GA modulated gene expressed in the barley embryo encodes an aldose reductase related protein. EMBO J. 10: 1037-1043.

Belhassen, E. (ed), 1996. Drought Tolerance in Higher Plants: Genetical, Physiological and Molecular Biology Analysis. Plant Growth Reg. 20: 79-103.

Benech-Arnold, R.L., M.C. Giallorenzi, J. Frank and V. Rodriguez, 1999. Termination of hull-imposed dormancy in developing barley grains is correlated with changes in embryonic ABC levels and sensitivity. Seed Sci. Res. 9: 39-47.

Blinda, A., B. Koch, S. Ramanjulu and K.J. Dietz, 1997. *De novo* synthesis and accumulation of apoplastic proteins in leaves of heavy metal-exposed barley seedlings. Plant Cell Environ. 20: 969-981.

Bliss, R.D., K.A. Platt-Aloia and W.W. Thomson, 1986. Osmotic sensitivity in relation to salt sensitivity in germinating barley seeds. Plant Cell Environ. 9: 721-725.

Blum, A., 1989. Osmotic adjustment and growth of barley genotypes under drought stress. Crop Sci. 29: 230-233.

Bona, L., R.J. Wright and V.C. Baligar, 1991. A rapid method for screening cereals for acid soil tolerance. Cereal Res. Comm. 19: 465-468.

Bothmer, R. von, N. Jacobsen, C. Baden, R.B. Jørgensen and I. Linde-Laursen, 1995. An ecogeographical study of the genus *Hordeum*. Systematic and Ecogeographic Studies on Crop Genepools, 7. IPGRI, Rome, 2nd ed., 129 p.

Boukerrou, L., and C. Rasmusson, 1990. Breeding for high biomass yield in a spring barley. Crop Sci. 30: 31-35.

Bravo, L.A., G.E. Zuniga, M. Alberdi and L.J. Corcuera, 1998. The role of ABA in freezing tolerance and cold acclimation in barley. Physiol. Plant. 103: 17-23.

Brookes, P.A., 1980. The significance of pre-harvest sprouting of barley in malting and brewing. Cereal Res. Comm. 8: 29-38.

Brule-Babel, A.L., and D.B. Fowler, 1989. Use of controlled environments for winter cereal cold hardiness evaluation: controlled freeze tests and tissue water content as prediction tests. Can. J. Plant Sci. 69: 355-366.

Cattivelli, L., P. Baldi, C. Crosatti, M. Grossi, G. Valè and A.M. Stanca, 2001. Genetic bases of barley physiological response to stressful conditions. In: G.A. Slafer, J.L. Molina-Cano, J.L. Araus, R. Savin and I. Ramagosa (eds), Barley Science: Recent Advances from Molecular Biology to Agronomy of Yield and Quality. Food Product Press, New York, USA, pp. 307-360.

Cattivelli, L., G. Delogu, V. Terzi and A.M. Stanca, 1994. Progress in barley breeding. In: G.A. Slafer (ed), Genetic Improvement of Field Crops. Marcel Dekker Inc., New York, pp. 95-181.

Ceccarelli, S., and S. Grando, 1996. Drought as a challenge for the plant breeders. Plant Growth Regul. 20: 149-155.

Champoux, M.C., G. Wang, S. Sarkarung, D.J. Mackill, J.C. O'Toole, N. Huang and S.R. McCouch, 1995. Locating genes associated with root morphology and drought avoidance in rice via linkage to molecular markers. Theor. Appl. Genet. 90: 969-981.

Choi, D.W., B. Zhu and T.J. Close, 1999. The barley (*Hordeum vulgare* L.) dehydrin multigene family: sequences, allele types, chromosome assignments and expression characteristics of 11 *Dhn* genes of cv. Dicktoo. Theor. Appl. Genet. 98: 1234-1247.

Close, T.J., 1997. Dehydrins: a commonality in the response of plants to dehydration and low temperature. Physiol. Plant. 100: 291-296.

Crosatti, C., E. Nevo, A.M. Stanca and L. Cattivelli, 1996. Genetic analysis of the COR14 proteins accumulation in wild (*Hordeum spontaneum*) and cultivated (*Hordeum vulgare*) barley. Theor. Appl. Genet. 93: 975-981.

Crosatti, C., P. Polverino de Laureto, R. Bassi and L. Cattivelli, 1999. The interaction between cold and light controls the expression of the cold-regulated barley gene *cor14b* and the accumulation of the corresponding protein. Plant Physiol. 119: 671-680.

Crosatti, C., C. Soncini, A.M. Stanca and L. Cattivelli, 1995. The accumulation of a cold-regulated chloroplastic protein is light-dependent. Planta 196: 458-463.

Dennis, E.S., M. Olive, R. Dolferus, A. Millar, W.J. Peacock and T.L. Setter, 1992. Biochemistry and molecular biology of the anaerobic response. In: J.L. Wray (ed), Inducible Plant Proteins, Cambridge Univ. Press, pp. 231-245.

Dexter, S.T., W.E. Tottingham and L.F. Graber, 1932. Investigation of the hardiness of plants by measurement of electrical conductivity. Plant Physiol. 7: 63-78.

Doll, H., V. Haahr and B. Søgaard, 1989. Relationship between vernalisation requirement and winter hardiness in doubled haploids of barley. Euphytica 42: 209-213.

Dunwell, J.M., 1981. Dormancy and germination in embryos of *Hordeum vulgare* L. effect of dissection, incubation temperature and hormone application. Ann. Bot. 48: 203-213.

Ebbs, S.D., and L.V. Kochian, 1998. Phytoextraction of zink by oat (*Avena sativa*), barley (*Hordeum vulgare*) and Indian mustard (*Brassica juncea*). Environ. Sci. Technol. 32: 802-806.

Ellis, R., B.P. Forster, R. Waugh, N. Bonar, L.L. Handley, D. Robinson, D.C. Gordon and W. Powell, 1997. Mapping physiological traits in barley. New Phytol. 137: 149-157.

Favier, J.F., and J.L. Woods, 1993. The quantification of dormancy loss in barley (*Hordeum vulgare* L.). Seed Sci. Technol. 21: 653-674.

Forster, B.P., M.S. Phillips, T.E. Miller, E. Baird and W. Powell, 1990. Chromosome location of genes controlling tolerance to salt (NaCl) and vigour in *Hordeum vulgare* and *H. chilense*. Heredity 65: 99-107.

Fowler, D.B., and R.J. Carles, 1979. Growth, development and cold tolerance of fall-acclimated cereal grains. Crop Sci. 19: 915-922.

Fowler, D.B., and A.E. Limin, 1987. Exploitable genetic variability for cold tolerance in commercially grown cereals. Can. J. Plant Sci. 67: 278.

Galiba, G., S.A. Quarrie, J. Sutka, A. Morgounov and J.W. Snape, 1995. RFLP mapping of the vernalisation (*Vrn1*) and frost resistance (*Fr1*) genes on chromosome 5A of wheat. Theor. Appl. Genet. 90: 1174-1179.

Gallardo, F., F. Borie, M. Alvear and E. von Baer, 1999. Evaluation of aluminium tolerance of three barley cultivars by two short-term screening methods and field experiments. Soil Sci. Plant Nutr. 45: 713-719.

Giorni, E., C. Crosatti, P. Baldi, M. Grossi, C. Marè, A.M. Stanca and L. Cattivelli, 1999. Cold-regulated gene expression during winter in frost tolerant and frost susceptible barley cultivars grown under field conditions. Euphytica 106: 149-157.

Gladstones, J.S., and J.F. Loneragan, 1970. Nutrient elements in herbage plants, in relation to soil adaptation and animal nutrition. Proc. XIth Int. Grassld. Congr. Brisbane, pp. 350-354.

Goldbach, H., and G. Michael, 1976. Abscisic acid content of barley grains during ripening as affected by temperature and variety. Crop Sci. 16: 797-799.

Gorny, A.G., 1996. Evaluation of the response to limited water, nitrogen and phosphorus supply in spring barley genotypes selected for vigorous seminal roots. J. Appl. Genet. 37: 11-27.

Grando, S. and S. Ceccarelli, 1995. Seminal root morphology and coleoptile length in wild and cultivated barley. Euphytica 86: 73-80.

Grossi, M., L. Cattivelli, V. Terzi and A.M. Stanca, 1992. Modification of gene expression induced by ABA, in relation to drought and cold stress in barley shoots. Plant Physiol. Biochem. 30: 97-103.

Grossi, M., E. Giorni, F. Rizza, A.M. Stanca and L. Cattivelli, 1998. Wild and cultivated barleys show differences in the expression pattern of a cold-regulated gene family under different light and temperature conditions. Plant Mol. Biol. 38: 1061-1069.

Grossi, M., M. Gulli, A.M. Stanca and L. Cattivelli, 1995. Characterization of two barley genes that respond rapidly to dehydration stress. Plant Sci. 105: 71-80.

Gullord, M., C.R. Olien and E.H. Everson, 1975. Evaluation of freezing hardiness in winter wheat. Crop Sci. 15: 153-157.

Gunasekera, M., Santakumari, Z. Glinka and G.A. Berkowitz, 1994. Wild and cultivated genotypes demonstrate varying ability to plant water deficits. Plant Sci. 99: 125-134.

Hayes, P.M., T. Blake, T.H.H. Chen, S. Tragoonrung, F. Chen, A. Pan and B. Liu, 1993. Quantitative trait loci on barley (*Hordeum vulgare* L.) chromosome-7 associated with components of winterhardiness. Genome 36: 66-71.

Hömmö, L.M., 1994. Hardening of some winter wheat (*Triticum aestivum* L.), rye (*Secale cereale* L.), triticale (*Triticosecale* Wittmack) and winter barley (*Hordeum vulgare* L.) cultivars during autumn and final winter survival in Finland. Plant Breed. 112: 285-293.

Hondred, D., and A.D. Hanson, 1990. Hypoxycally inducible barley lactate dehydrogenase: cDNA cloning and molecular analysis. Proc. Natl. Acad. Sci. USA 87: 7300-7304.

Ishitani, M., T. Nakamura, S.Y. Han and T. Takabe, 1995. Expression of the betaine aldehyde dehydrogenase in barley in response to osmotic stress and abscisic acid. Plant Mol. Biol. 27: 307-315.

Jackson, P.A., D.E. Byth, K.S. Fischer and R.P. Johnston, 1994. Genotype × environment interactions in progeny from a barley cross: II. Variation in grain yield, yield components and dry matter production among lines with similar times to anthesis. Field Crops Res. 37: 11-23.

Jaradat, A.A., M.M. Ajlouni and M.A. Duwayri, 1996. Genetic resources of cereals and their wild relatives in Jordan: revisited. In: A.A. Jaradat (ed), Plant Genetic Resources of Jordan. IPGRI Regional Office for West Asia and North Africa, Aleppo, Syria, pp. 55-75.

Jedel, P.E., and J.H. Helm, 1994. Assessment of western Canadian barleys of historical interest. I. Yield and agronomic traits. Crop Sci. 34: 922-927.

Jefferies, S.P., H. Raman, M.A. Pallotta and A.R. Barr, 2000. Genetic approaches to ameliorating mineral stresses in barley. In: Barley Genetics VIII. Proc. 8th Int. Barley Genet. Symp., Adelaide, Australia, pp. 253-259.

Jenkins, G., and P. Roffey, 1974. A method of estimating the cold hardiness of cereals by measuring electrical conductance after freezing. J. Agr. Sci. Cambridge 83: 87-92.

Kolar, S.C., P.M. Hayes, T.H.H. Chen and R.G. Linderman, 1991. Genotypic variation for cold tolerance in winter and facultative barley. Crop Sci. 31: 1149-1152.

LaBerge, D.E., 1983. Sprouting resistance in barley. In: J.E. Kruger and D.E. LaBerge (eds), Proc. IIIth Int. Symp. Pre-harvest Sprouting in Cereals. Westview Press, Boulder, Colorado, pp. 239-243.

Lenoir, C., F. Corbineau and D. Come, 1983. Role of paleae in dormancy of barley seeds. Physiol. Vegetal 21: 633-643.

Livingston, D.P., C.R. Olien and R.D. Freed, 1989. Sugar composition and freezing tolerance in barley crowns at varying carbohydrate levels. Crop Sci. 29: 1266-1270.

López Castañeda, C., R.A. Richards, G.D. Farquhar and R.E. Williamson, 1996. Seed and seedling characteristics contributing to variation in seedling vigour among temperate cereals. Crop Sci. 36: 1257-1266.

Ludlow, M.M., and R.C. Muchow, 1990. A critical evaluation of traits for improving crop yields in water-limited environments. Adv. Agron. 43: 107-153.

Ma, J.F., S.J. Zheng, X.F. Li, K. Takeda and H. Matsomoto, 1997. A rapid hydroponic screening for aluminium tolerance in barley. Plant and Soil 191: 133-137.

Mano, Y., H. Nakazumi and K. Takeda, 1996. Varietal variation in and effect of some major genes on salt tolerance at the germination stage in barley. Breed. Sci. 46: 227-233.

Mano, Y., and K. Takeda, 1995. Varietal variation and effects of some major genes on salt tolerance in barley seedlings. Bull. Res. Inst. Biores., Okayama University, 3: 71-81.

Mano, Y., and K. Takeda, 1998. Genetic resources of salt tolerance in wild *Hordeum* species. Euphytica 103: 137-141.

Marmiroli, N., C. Lorenzoni, L. Cattivelli, A.M. Stanca and V. Terzi, 1989. Induction of heat shock proteins and acquisition of thermotolerance in barley (*Hordeum vulgare* L.). Variations associated with growth habit and plant development. J. Plant Physiol. 135: 267-273.

Marmiroli, N., A. Pavesi, G. Di Cola, H. Hartings, G. Raho, M.R. Conte and C. Perrotta, 1995. Identification, characterisation and analysis of cDNA and genomic sequences encoding two different small heat shock proteins in *Hordeum vulgare*. Genome 36: 1111-1118.

Martiniello, P., G. Delogu, M. Odoardi, G. Boggini and A.M. Stanca, 1987. Breeding progress in grain yield and selected agronomic characteristics of winter barley over the last quarter of a century. Plant Breed. 99: 289-294.

Mastrangelo, A.M., P. Baldi, C. Marè, V. Terzi, G. Galiba, L. Cattivelli and N. Di Fonzo, 2000. The cold dependent accumulation of COR TMC-AP3 in cereals with contrasting frost tolerance is regulated by different mRNA expression and protein stability. Plant Sci. 156: 47-54.

Mayne, R.G., and P.J. Lea, 1984. Alcohol dehydrogenase in *Hordeum vulgare*: changes in isoenzyme level under hypoxia. Plant Sci. Letter 37: 73-78.

Minella, E., and M.E. Sorrells, 1992. Aluminium tolerance in barley: genetic relationships among genotypes of diverse origin. Crop Sci. 32: 593-598.

Minella, E., and M.E. Sorrells, 1997. Inheritance and chromosome location of *Alp*, a gene controlling aluminium tolerance in 'Dayton' barley. Plant Breed. 116: 465-469.

Molina-Cano, J.L., A. Spoena, J.S. Swanston, A.M. Casas, M.A. Moralejo, A. Ubieto, I. Lara, A.M. Perez-Vendrell and I. Romagosa, 1999. A mutant induced in the malting barley cv. Triumph with reduced dormancy and ABA response. Theor. Appl. Genet. 98: 347-355.

Muench, D.G., and A.G. Good, 1994. Hypoxycally inducible barley alanine aminotransferase: cDNa cloning and expression analysis. Plant Mol. Biol. 24: 417-427.

Muñoz, P., J. Voltas, J.L. Araus, E. Igartua and I. Romagosa, 1998. Changes over time in the adaptation of barley releases in North-eastern Spain. Plant Breed. 117: 531-535.

Murelli, C., F. Rizza, F. Marinone Albini, A. Dulio, V. Terzi and L. Cattivelli, 1995. Metabolic changes associated with cold-acclimation in contrasting cultivars of barley. Physiol. Plant. 94: 87-93.

Nair, K.P.P., and N.C. Khulbe, 1990. Differential response of wheat and barley genotypes to substrate-induced salinity under North Indian conditions. Exp. Agr. 26: 221-225.

Nevo, E., 1992. Origin, evolution, population genetics and resources for breeding of wild barley, *Hordeum spontaneum*, in the Fertile Crescent. In: P.R. Shewry (ed), Barley: Genetics, Biochemistry, Molecular Biology and Biotechnology. CAB International, pp. 19-45.

Nevo, E., I. Apelbaum Elkaher, J. Garty and A. Beiles, 1997. Natural selection causes microscale allozyme diversity in wild barley and a lichen at "Evolution Canyon", Mt Carmel, Israel. Heredity 78: 373-382.

Nevo, E., B. Baum, A. Beiles and D.A. Johnson, 1998. Ecological correlates of RAPD DNA diversity of wild barley, *Hordeum spontaneum*, in the Fertile Crescent. Genet. Res. Crop Evol. 45: 151-159.

Nevo, E., T. Krugman and A. Beiles, 1993. Genetic resources for salt tolerance in the wild progenitors of wheat (*Triticum dicoccoides*) and barley (*Hordeum spontaneum*) in Israel. Plant Breed. 110: 338-341.

Nevo, E., T. Krugman and A. Beiles, 1994. Edaphic natural selection of allozyme polymorphisms in *Aegilops peregrina* at a Galilee microsite in Israel. Heredity 72: 109-112.

Okubo, K., and K. Takeda, 1991. Varietal variation of flooding tolerance in barley seedlings. Japanese J. Breed. 41 (suppl. 2): 396-397.

Omara, M.K., K.A. Abdal-Rahman and M.Y. Hussein, 1987. Selection for salt stress tolerance in barley. Assiut J. Agric. Sci. 18: 199-218.

Oosterom, E.J. van, D. Kleijn, S. Ceccarelli and M.M. Nachit, 1993. Genotype-by-environment interactions of barley in the Mediterranean region. Crop Sci. 33: 669-674.

Oram, R.N., and C.J. Driscoll, 1983. Breeding barley tolerant of soil acidity and waterlogging. Proc. Austr. Plant Breed. Conf., pp. 71-73.

Owuor, E.D., T. Fahima, A. Beiles, A. Korol and E. Nevo, 1997. Population genetic response to microsite ecological stress in wild barley, *Hordeum spontaneum*. Mol. Ecol. 6: 1177-1187.

Pan, A., P.M. Hayes, F. Chen, T.H.H. Chen, T. Blake, S. Wright, I. Karsai and Z. Bedő, 1994. Genetic analysis of the components of winterhardiness in barley (*Hordeum vulgare* L.). Theor. Appl. Genet. 89: 900-910.

Papa, R., G. Attene, G. Barcaccia, A. Ohgata and T. Konishi, 1998. Genetic diversity in landraces populations of *Hordeum vulgare* L. from Sardinia, Italy, as revealed by RAPDs, isozymes and morphophenological traits. Plant Breed. 117: 523-530.

Passiuora, J.B., 1977. Grain yield, harvest index and water use of wheat. J. Aust. Inst. Agric. Sci. 43: 117-120.

Passiuora, J.B., 1996. Drought and drought tolerance. Plant Growth Reg. 20: 79-83.

Pearce, R.S., M.A. Dunn, J. Rixon, P. Harrison and M.A. Hughes, 1996. Expression of cold-inducible genes and frost hardiness in the crown meristem of young barley (*Hordeum vulgare* L. cv. Igri) plants grown in different environments. Plant Cell Environ. 19: 275-290.

Perata, P., E. Loreti, L. Guglielminetti and A. Alpi, 1998. Carbohydrate metabolism and anoxia tolerance in cereal grains. Acta Bot. Neerlandica 47: 269-283.

Petersen, G., and O. Seberg, 1998. Molecular characterization and sequence polymorphism of the alcohol dehydrogenase 1 gene in *Hordeum vulgare* L. Euphytica 102: 57-63.

Pulli, S., K. Hjortsholm, A. Larsen, B. Gudleifsson, S. Larsson, B. Kristiansson, L. Hömmö, A.M. Tronsmo, P. Ruuth and C. Kristensson, 1996. Development and evaluation of laboratory testing methods for winterhardiness breeding. Nordic Gene Bank.

Qiu, J.D., and Y. Ke, 1991. A study of determination of wet tolerance of 4572 barley germplasm resources. Acta Agr. Shanghai 7: 27-32.

Rajaram, S., and M. van Ginkel, 1996. Yield potential debate: germplasm *vs.* methodology, or both. In: M.P. Reynolds, D. Rajaram and M. McNab (eds), Increasing Yield Potential in Wheat: Breaking the Barriers. Mexico, DF, CIMMYT, pp. 11-18.

Rasmusson, D.C., 1996. Germplasm is paramount. In M.P. Reynolds, D. Rajaram and M. McNab (eds), Increasing Yield Potential in Wheat: Breaking the Barriers. Mexico, DF, CIMMYT, pp. 28-35.

Rasmusson, D.C., and R.L. Phillips, 1997. Plant breeding progress and genetic diversity from *de novo* variation and elevated epistasis. Crop Sci. 37: 303-310.

Richards, R.A., 1996. Defining selection criteria to improve yield under drought. Plant Growth Reg. 20: 157-166.

Riggs, T.J., P.R. Hanson, N.D. Start, D.M. Miles, C.L. Morgan and M.A. Ford, 1981. Comparison of spring barley varieties grown in England and Wales between 1880 and 1980. J. Agric. Sci. 97: 599-610.

Ringlund, K., 1980. Starch quality in wheat and barley at different maturity stages in relation to seed dormancy. Cereal Res. Comm. 8: 193-197.

Rizza, F., C. Crosatti, A.M. Stanca and L. Cattivelli, 1994. Studies for assessing the influence of hardening on cold tolerance of barley genotypes. Euphytica 75: 131-138.

Robinson, L.L., Handley, C.M. Scrimgeour, D.C. Gordon, B.P. Forster and R.P. Ellis, 2000. Using stable isotope natural abundances (delta N-15 and delta C-13) to integrate the stress responses of wild barley (*Hordeum spontaneum* C. Koch) genotypes. J. Exp. Bot. 51: 41-50.

Saberi, H.K., R.D. Graham and A.J. Rathjen, 1999. Inheritance of manganese efficiency in durum wheat. J. Plant Nutr. 22: 11-21.

Sasaki, A., 1984. Varietal differences in response to excess soil moisture at the internode elongation stage in barley. Japanese J. Breed. 34: 79-86.

Schachtman, D., and W. Liu, 1999. Molecular pieces to the puzzle of the interaction between potassium and sodium uptake in plants. Trends Plant Sci. 4: 281-287.

Slafer, G.A., and J.L. Araus, 1998. Improving wheat responses to abiotic stresses. In: A.E. Slinkard (ed), Proc. IXth Int. Wheat Genet. Symp. Saskatoon. Univ. Saskatchewan Extension Press, vol. I, pp. 201-213.

Slavich, P.G., B.J. Read and B.R. Cullis, 1990. Yield response of barley germplasm to field variation in salinity quantified using the EM-38. Aust. J. Exp. Agr. 30: 551-556.

Suh, H.S., 1982. Studies on the breeding and characteristics of barley varieties resistant to flooding in Korea. Res. Reports Office Rural Develop. Korea 24: 128-167.

Takahashi, H., and K. Takeda, 1999. Varietal variation for deep-seeding tolerance in barley and wheat. Breed. Res. 1: 1-8.

Takeda, K., 1989. Varietal variation of flooding tolerance in barley seedlings, and its diallel analysis. Japanese J. Breed. 39 (Suppl. 1): 174-175.

Takeda, K., 1995. Varietal variation and inheritance of seed dormancy in barley. In: K. Noda and D.J. Mares (eds), VIIth Int. Symp. Pre-harvest Sprouting in Cereals. Center for Academic Societies, Osaka, Japan, pp. 205-212.

Takeda, K., and T. Fukuyama, 1983. Studies on the water sensitivity of barley seeds. Varietal variation and geographical distribution in Asia. Japanese J. Breed. 33 (Suppl. 1): 226-227.

Takeda, K., and T. Fukuyama, 1987. Tolerance to pre-germination flooding in the world collection of barley varieties. In: Barley Genetics V. Proc. 5th Int. Barley Genet. Symp., Okayama, Japan, pp. 735-740.

Tamás, L., J. Huttová, I. Mistrík and L. Hajasová, 2000. Accumulation of two cytoplasmic polypeptides in roots of Al-sensitive and Al-resistant barley cultivars during Al, pH and metal stress. Biologia 55: 259-265.

Tamás, L., J. Huttová and Z. Žigová, 1997. Accumulation of stress-proteins in intercellular spaces of barley leaves induced by biotic and abiotic factors. Biol. Plant. 39: 387-394.

Terzi, V., L. Cattivelli, A. Gatti, A.M. Stanca, C. Lorenzoni and N. Marmiroli, 1990. Physiological and molecular studies of heat and drought tolerance in barley. Bull. Soc. Bot. Fr. 137, Actual. Bot. 1: 73-80.

Teulat, B., P. Monneveux, J. Wery, C. Borries, I. Souyris, A. Charrier and D. This, 1997. Relationship between relative water content and growth parameters under water stress in barley: a QTL study. New Phytol. 137: 99-107.

Teulat, B., D. This, M. Khairall, C. Borries, C. Ragot, P. Sourdille, P. Leroy, P. Monneveux and A. Charrier, 1998. Several QTLs involved in osmotic-adjustment trait variation in barley. Theor. Appl. Genet. 96: 688-698.

Tong, Y.P., Z. Rengel and R.D. Graham, 1997. Interactions between nitrogen and manganese nutrition of barley genotypes differing in manganese efficiency. Ann. Bot. 79 53-58.

Towsey, M.W., J.R. Boyd and C.J. Driscoll, 1983. The screening of barley germplasm for tolerance to soil salinity and waterlogging. Proc. Austr. Plant Breed. Conf., pp. 321-322.

Trick, M., E.S. Dennis, K.J.R. Edwards and N.J. Peacock, 1988. Molecular analysis of the alcohol dehydrogenase gene family of barley. Plant Mol. Biol. 11: 147-160.

Vagujfalvi, A., C. Crosatti, G. Galiba, J. Dubcovsky and L. Cattivelli, 2000. Two loci on wheat chromosome 5A regulate the differential cold-dependent expression of the *cor14b* gene in frost tolerant and sensitive genotypes. Mol. Gen. Genet. 263: 194-200.

Volis, S., S. Mendlinger, L. Olsvig-Whittaker, U.N. Safriel and N. Orlovsky, 1998. Phenotypic variation and stress resistance in core and peripheral populations of *Hordeum spontaneum*. Biodivers. Conserv. 7: 799-813.

Voltas, J., I. Romagosa, A. Lafarga, A.P. Armesto, A. Sombrero and J.L. Araus, 1999. Genotype by environment interaction for grain yield and carbon isotope discrimination of barley in Mediterranean Spain. Aust. J. Agric. Res. 50: 1263-1271.

Wang, M., 1997. The role of abscisic acid in the regulation of barley grain germination. Seed Sci. Technol. 25: 67-74.

Weltzien, E., and G. Fischbeck, 1990. Performance and variability of local barley landraces in near-eastern environments. Plant Breed. 104: 58-67.

Wolthen-Larsen H., J. Brandt, D.B. Collinge and H. Thordal-Christensen, 1993. A pathogen induced gene of barley encodes a HSP90 homologue showing striking similarity to vertebrate forms resident in the endoplasmic reticulum. Plant Mol. Biol. 21: 1097-1108.

Zohary, D., and M. Hopf, 1988. Domestication of Plants in the Old World. Clarendon, Oxford.

Diversity in Barley *(Hordeum Vulgare)*
Roland von Bothmer et al (Editors).

Chapter 10

Genetic diversity for quantitatively inherited agronomic and malting quality traits

Patrick M. Hayes[a], Ariel Castro[a], Luis Marquez-Cedillo[a], Ann Corey[a], Cynthia Henson[b,c], Berne L. Jones[b,c], Jennifer Kling[a], Diane Mather[d], Ivan Matus[a], Carlos Rossi[a] and Kazuhiro Sato[e]

[a]Department of Crop and Soil Science, Oregon State University, Corvallis, OR, 97331, USA
[b]Cereal Crops Research Unit, USDA-ARS, 501 Walnut St., Madison, WI 53705, USA
[c]Department of Agronomy, University of Wisconsin, Madison, WI 53706, USA
[d]Department of Plant Science, McGill University, Ste Anne de Bellevue, QC H9X 3V9 Canada
[e]Research Institute for Bioresources, Okayama University, Kurashiki 710-0046, Japan

Introduction

Agronomic and quality traits were undoubtedly key issues for the domesticators of barley. According to the archeological record, these early farmers used both wild and cultivated (non-brittle rachis) forms of barley (Harlan, 1995). Crop productivity would clearly have been an attribute of key interest, and the selection of shattering-resistant mutants probably led to a quantum leap in yield. Because barley has been used both as a food and as a principal ingredient of fermented beverages from the earliest times, there may well have been conscious selection for end-use properties. The selection of hull-less mutants in areas of the world where barley was a principal foodstuff underscores the importance of end-use properties in domestication. The malting and brewing properties of wild barley accessions and landraces have not been well described and are, in fact, extremely difficult to measure. There are no absolute definitions of malting and brewing quality, due to differences in malting and brewing practices and consumer preferences. Both direct consumers of barley and the first brewers would have selected for uniform, plump kernels. However, only the latter would have selected for the higher levels of enzymatic activity necessary for satisfactory malting and brewing performance.

Plant breeding efforts are directed primarily at traits exhibiting quantitative variation. However, diversity and quantitative traits are not terms that are often encountered together in the plant breeding and genetics literature. This is certainly true for barley, a crop with a rich legacy of genetics research and an impressive record of cultivar improvement. Genetics was limited to models where phenotypes allowed for Mendelian analysis. Genetic tools, when applied to cultivar development, were of a statistical nature. In the mid-1980s, the availability of abundant DNA-level markers, accompanied by the development of computer hardware and software,

---

Hayes, P.M., A. Castro, L. Marquez-Cedillo, A. Corey, C. Henson, B.L. Jones, J. Kling, D. Mather, I. Matus, C. Rossi and K. Sato, 2003. Genetic diversity for quantitatively inherited agronomic and malting quality traits. In: R. von Bothmer, Th. van Hintum, H. Knüpffer and K. Sato (eds), Diversity in Barley (*Hordeum vulgare*), pp. 201-226. Elsevier Science B.V., Amsterdam, The Netherlands.

caused these parallel lines of investigation to change course and head toward intersection. Breeders and geneticists were now able to collaborate in developing and testing hypotheses regarding the number, location, effect, and interactions of genes influencing quantitative traits. These genes were termed "Quantitative Trait Loci" (QTL). In this chapter, we will review (1) diversity in agronomic traits; (2) diversity in malting quality traits; and (3) the current status of QTL analysis in barley and the application of QTL tools to the analysis of genetic diversity in barley and crop improvement.

Agronomic traits

Barley is one of the most widely adapted cereal crops. It is grown in a range of extreme environments that vary from northern Scandinavia to the Himalayan mountains to monsoon paddies. No barley genotype is adapted to all environments and, in fact, very different genepools have evolved in the major barley production areas of the world. The genepools may be defined by essential physiological parameters that determine adaptation to a production environment – such as vernalization and/or photoperiod response – or they may be defined by evolutionary bottlenecks and the accidents of history, such as regional preferences for two-rowed or six-rowed cultivars. Within these genepools, agronomic performance will be determined by the entire allelic architecture of each genotype. The actual number of genes in barley (with a genome size of 5.3 billion base pairs) is not known at this point. There are at least 26,588 genes in the euchromatic regions of the human genome (2.91 billion base pairs) and there is some evidence for an additional 12,000 genes (Venter *et al.*, 2001). The first higher plant genome – *Arabidopsis thaliana*, with an estimated genome size of 125 million base pairs – was sequenced and the number of genes was estimated at 25,800 (Pennisi, 2000). Surely, most of the genes in barley are involved in one way or another with yield – the "ultimate" fitness trait. Likewise, a large number of them must be involved in malting quality, which is the end result of the fundamental processes of seed carbohydrate deposition and hydrolysis.

In terms of agronomic traits, growth habit genes may be major "specious" yield genes (see also Chapter 2). For example, highly significant genotype × environment interaction will occur when the progeny derived from a cross of spring × winter habit parents are evaluated under spring and fall-sown conditions. A genetic analysis of such germplasm would likely reveal vernalization requirement, photoperiod reaction, and low temperature tolerance genes as primary determinants of yield. This was indeed the case when Quantitative Trait Locus (QTL) tools were applied to a population derived from the cross of North American winter and spring genotypes (Oziel *et al.,* 1996). Likewise, resistance to biotic stresses can be the primary determinant of yield, if resistance to the disease in question is endemic to a production zone. A "cost", or lower yield potential of resistant *vs.* disease-susceptible germplasm in disease-free environments, could be due to linkage of resistance and "yield-determining" genes or pleiotropic effects of the resistance genes (Hayes *et al.,* 1996).

Effects of inflorescence type on yield

The role of inflorescence type in determining yield is complex. In barley, each rachis node has three spikelets, each of them bearing one floret. The two-rowed and six-rowed germplasm groups are defined by the number of fertile florets per rachis node. In two-rowed barley, there is one fertile floret per rachis node, whereas in six-rowed barley all three florets are fertile (Hitchcock, 1971; see also Chapter 2). Most barley cultivars of commercial importance are six-rowed or two-rowed inbred lines. This simple genetic system defines the two principal germplasm groups of barley, and these germplasm groups have historically defined end uses. Two-rowed barleys are favored for malting throughout most of the world, except for the USA and Mexico, where six-rowed barleys are used extensively for this purpose (Riggs and Kirby, 1978).

Two-rowed cultivars usually have a higher number of tillers per plant and larger, heavier seed than six-rowed ones. Six-rowed cultivars on the other hand, usually have more seeds per inflorescence. Thus, the compensatory effects of yield components lead to similar levels of yield potential. However, historical patterns of geographic distribution and end-use of the two-rowed and six-rowed germplasm groups have led to the idea that the two germplasm groups carry different alleles at other loci in addition to those determining lateral floret fertility (Takahashi *et al.,* 1975). Accordingly, crosses between the two germplasm groups might be expected to produce positive transgressive segregants for economically important phenotypes. The experience of plant breeders, however, has generally been that two-rowed × six-rowed crosses are not suitable for cultivar development (Harlan, 1957; Kjær and Jensen, 1996). Allard (1988) concluded that "Evidently the two-row/six-row locus affects developmental processes in ways that leave few quantitative characters untouched" and "this locus had large effects on survival and adaptedness". Large pleiotropic effects on multiple phenotypes have been attributed to alleles at the *vrs1* locus, based on studies of the progeny of two-rowed × six-rowed crosses (Kjær and Jensen, 1996; Jui *et al.*, 1997). However, it is possible that the correlated phenotypes are due to linkage rather than pleiotropy. It is difficult to distinguish between these phenomena in the case of the *vrs1* locus, which is located near the centromere on linkage maps. On the physical map of Künzel *et al.* (2000), however, the *vrs1* locus is located in one of the higher recombination regions on the long (minus) arm of chromosome 2H. Powell *et al.* (1990), in a comparative analysis of two types of progeny from a six-rowed × two-rowed cross, concluded that some associations between quantitative phenotypes and the *vrs1* locus were due to linkage rather than pleiotropy. Marquez-Cedillo *et al.* (2001) used QTL tools to analyze a two-rowed × six-rowed mapping population for agronomic traits. The major QTL related with agronomic traits coincided with the loci that determined inflorescence type.

Two-rowed barley in Japan provides an example of the role of growth habit and maturity genes within a germplasm class. Yasuda *et al.* (1992) analyzed a collection of two-rowed barleys from Japan, Turkey, Ethiopia and Europe and found that Japanese two-rowed malting cultivars, which were introduced from Europe less than 130 years ago, were quite different from the other germplasm groups. Adaptation to Japanese conditions was achieved within a relatively short period, and a major distinguishing attribute of the Japanese germplasm is early maturity. In Japan, barley should be harvested before the rainy season, when the fields are converted to rice paddies. Yasuda (1964) used hybrid populations derived from winter/spring barley crosses to demonstrate selection for adaptation to Japanese conditions and later showed the effect of four spring growth habit genes on yield performance (Yasuda, 1977; see also Chapter 2).

Identifying factors that affect yield

Once the more obvious "yield-limiting genes" are accounted for, what determines yield in barley, and how much genetic diversity is there at these loci? There have been three principal approaches to the study of "yield genes". Plant physiologists have defined at the cellular, organ, and organism level the fundamental physiological processes that lead to yield in barley (Kirby and Faris, 1972; Gallagher *et al.*, 1975; Lauer and Simmons, 1989; Dreccer *et al.*, 1997). Crop physiologists and plant breeders have defined physiological parameters for production level phenotypes – e.g., yield components, harvest index, and ideotypes – that determine yield (Rasmusson, 1987; Hamblin, 1993). This research has led to a better understanding of fundamental plant processes and better *a posteriori* understanding of crop performance, but in many cases the two approaches have not led to predictive tools that are useful for barley improvement (Beldford and Sedgley, 1991; Rasmusson, 1991). A problem has been that, often of necessity, these studies were conducted in agronomically irrelevant genetic stocks and environments. Furthermore, it has been exceptionally difficult, as shown in the examples of growth habit and inflorescence type, to separate the effects of individual genes with large effects from "genetic background".

Because of these challenges, and the fact that most agronomic traits showed inheritance patterns too complex for Mendelian inheritance, barley breeders and crop geneticists approached the issue of the genetics of yield using biometrical tools. However, as stated by Allard (1988), "Most quantitative genetic models require numerous assumptions, many of which are invalid, thus causing estimates of genetic parameters to be imprecise and even to exceed their theoretical limits. Whatever the cause or causes, the laborious biometrical experiments I conducted provided little information about the numbers of alleles per locus, numbers of loci, types of gene action (additive, dominance, and epistatic), the impact of various single locus or multilocus genotypes on fitness, and other genetic factors about which we must learn more if we are to understand the evolution of adaptedness".

For over a decade, molecular approaches have been used to map and describe Quantitative Trait Loci that determine yield in barley. As discussed in a subsequent section of this chapter, our understanding of QTL and their effects on agronomic traits is still incomplete. Nonetheless, developments in molecular biology have provided the necessary tools for determining the functional basis of yield and adaptation in barley.

Malting quality traits

Historically, the biochemistry and genetics of malting quality have been parallel areas of study. The former has focused on the systematic characterization of the deposition and hydrolysis of starch and proteins. This research has provided a more comprehensive understanding of the underlying processes, but has not provided breeders with a better tool kit for improving malting quality. Genetic studies of malting quality, to date, have provided perspectives on allelic diversity at only a few key loci. Consequently, barley breeders must still conduct expensive tests to determine the malting quality of their experimental germplasm and, because of the expense, can only carry out a limited number of assays. This has precluded the use of extensive population-based analyses of malting quality genetics. Furthermore, when such analyses have been conducted, malting quality phenotypes have shown frequency distributions that defy Mendelian analysis. Breeders have therefore relied upon phenotypic selection of malting quality in agronomically relevant germplasm, with occasional attempts to estimate genetic variances and numbers of "effective factors". As is evidenced by the quotation from Allard (1988) cited in the previous section on "Agronomic traits", these procedures have had limited practical applications. Information on allelic diversity in the genes that determine malting quality is limited. Accordingly, we will focus on providing a brief summary of the malting process and an enumeration of the genes known to be involved in these processes. The reader is referred to earlier reviews of the malting process (Burger and LaBerge, 1985; Bamforth and Barclay, 1993) and biochemical aspects of barley seed germination (Briggs, 1992; Fincher and Stone, 1993) for further background information. Expressed Sequence Tag (EST) projects have recently been initiated for malting quality in several laboratories around the world, but results are not yet available. Characterization of the structure and expression of genes expressed at key times and in key tissues relevant to the malting process will provide a foundation for efficient characterization and use of allelic diversity for malting quality traits.

The malting process

Malting is an exercise in applied biochemistry, especially enzymology. The starch, protein and nucleic acid molecules that are stored in barley grains are neither good nutrients for brewing yeast nor do they support the fermentation reactions performed by brewing yeasts. These large and structurally complex compounds must be partially or, in some instances, fully degraded into their component sugars, amino acids, and nucleotides before the yeast can use them. When barley seeds germinate, hydrolytic enzymes are synthesized or converted to active forms that can readily degrade these large compounds.

During "malting", barley seeds are germinated under controlled conditions so that degradative enzymes form and begin to hydrolyze the starch, protein, and nucleic acid molecules into small molecules that are needed at appropriate stages of the brewing process. To arrest the malting process, the green malt is kilned (gently dried, with heat) and the rootlets are removed. By this stage, little of the starch has been converted to sugars, but about 70% of the protein that needs to be solubilized during malting and mashing has already been rendered soluble. There is still some question as to how much free amino nitrogen (FAN) is released during malting. Modification is a collective term that is used to refer to all of the polymer-degrading processes that occur during malting. If malting is allowed to continue too long, the malt obtained will be overmodified and will not produce beers of optimal quality.

The malt is treated with water under appropriate conditions (a process called "mashing") to obtain an extract (wort) that must perform several critical functions. The extract must provide adequate nourishment to the yeast so that fermentation can occur. Secondly, the extract must provide sufficient fermentable sugars to enable the yeast to produce the desired levels of alcohol. A high quality malt will contain the right amount of hydrolytic enzymes and metabolites to fulfill these requirements and will have the right degree of friability to allow many of its components to be readily solubilized during mashing. During malting and mashing, the barley starch should be almost completely degraded into sugars that can be utilized by the brewing yeasts, whereas only about 45% of the barley protein should be solubilized. Too much protein solubilization is thought to result in beers with poor foaming characteristics. When insufficient protein hydrolysis occurs, the remaining proteins may interact with polyphenols to form beer haze precipitates.

The malting process, accordingly, involves a host of interacting genes involved in the fundamental processes of seed germination, growth and development. Domestication and selection have accumulated favorable alleles at multiple loci that determine malting quality. The specific alleles that have been accumulated in the major malting barley germplasm groups may differ, based on regional preferences and genetic drift. A summary of current knowledge of the genes and enzymes that control malting quality, and key citations, are presented in Table 10.1.

Enzymes and genes that control carbohydrate degradation

Four amylolytic enzymes are generally thought to participate in converting the starch in malted barley into fermentable sugars: these are α-amylase, β-amylase, α-glucosidase and limit dextrinase. Alpha-amylase and α-glucosidase can release α-glucans from native (nongelatinized) starch granules (Sun and Henson, 1990; Sissons and MacGregor, 1994). All four enzymes can hydrolyze gelatinized starch and/or glucan fragments. A fifth carbohydrase, isoamylase, has recently been discovered and it may have a role in degrading starch and/or producing fermentable sugars during mashing (Sun *et al.*, 1999), but the precise role of this enzyme is not yet known.

The α-amylases are particularly important in the production of fermentable sugars during

Table 10.1. Genes and gene products that control malting quality in barley.

Enzyme/protein	Function	Gene	Chr.	Citations
Carbohydrate-degrading enzymes and inhibitors				
α-amylase	Converts native starch, gelatinized starch and glucans into sugars	*Amy1* *Amy2*	7H 6H	Henson and Stone, 1988; Bamforth and Barclay, 1993; Ko and Henry, 1994; Ko *et al.*, 1996
β-amylase	Converts gelatinized starch and glucans into sugars; acts as storage protein (globulin)	*Bmy1* *Bmy2*	4H 2H	Bamforth and Berclay, 1993; Eglinton *et al.*, 1998; Erkkila *et al.*, 1998
α-glucosidase	Converts native starch, gelatinized starch and glucans into sugars	*Agl*	7H	Konishi *et al.*, 1994; Henson and Sun, 1995; Im and Henson, 1996
limit dextrinase	Breaks down branched starch and amylopectin molecules	LD	7H4H	MacGregor *et al.*, 1994; Langridge *et al.*, 1996; Burton *et al.*, 1999; Li *et al.*, 1999
isoamylase	May convert starch to sugars, but precise role not known			Sun and Henson, 1999
β-glucanase	Degrades cell wall β-glucans	*Glb1* *Glb2* *Glb31-37*	5H 7H 3H	Loi *et al.*, 1988; MacLeod *et al.*, 1991; Viëtor *et al.*, 1993; Li *et al.*, 1999
barley amylase subtilisin inhibitor (BASI)	Inhibits α-amylase encoded by *Amy2*	*Isa1*	2H	Munck *et al.*, 1985; Cannel *et al.*, 1992
limit dextrinase inhibitor (LDI)	Inactivates limit dextrinase			MacGregor *et al.*, 1995
Protein Z	β-amylase binding protein; acts as storage protein (globulin)	*Paz1*	4H	Cannel *et al.*, 1992
Storage proteins, protein-degrading enzymes and inhibitors				
B hordein	Storage protein (prolamin)	*Hor2*	1H	Shewry, 1993
C hordein	Storage protein (prolamin)	*Hor1*	1H	
D hordein	Storage protein (prolamin)	*Hor3*	1H	
hordenin (GluA, GluB, GluC)	Storage and structural protein (glutelin)	*GluA* *GluB* *Glu.c1*	1H 1H 1H	Kleinhofs *et al.*, 1993; Giese *et al.*, 1994
Carboxypeptidase (exopeptidase)	Hydrolyzes storage protein	*Cxp1* *Cxp3*	3H 6H	Kleinhofs *et al.*, 1993
malt cysteine endoproteinase	Hydrolyzes hordeins	*Cys1* *Cys2*	 3H	Cannel *et al.*, 1992; Mikkonen *et al.*, 1996; Jones, 1999; Jones and Budde, 1999
phytepsin; aspartic class endoproteinase	Hydrolyzes storage protein			Runeberg-Roos *et al.*, 1991
lipid transfer protein (LTP1); aka Probable Amylase Protease Inhibitor (PAPI)	Regulates activities of malt cysteine class endoproteinases	*Ltp1*	5H	Kleinhofs *et al.*, 1993; Sørensen *et al.*, 1993; Lusk *et al.*, 1995 Jones and Marinac, 2000b
lipid transfer protein (LTP2)	Inhibits endoproteinase	*Ltp2*		Jones and Marinac, 2000b
limit dextrinase inhibitor (LDI)	Inactivates limit dextrinase			MacGregor *et al.*, 1995
Protein Z	β-amylase binding protein; also acts as storage protein (globulin)	*Paz1*	4H	Cannel *et al.*, 1992; Evans *et al.*, 1999

mashing because they are the only amylolytic enzymes present that are sufficiently thermostable to retain at least some level of activity for the full duration of mashing (Bamforth and Barclay, 1993). Furthermore, most fermentable sugars are produced during the late stages of mashing when temperatures are quite high. The *Amy1* and *Amy2* loci encode two α-amylase isozymes that differ somewhat in their biochemical and biophysical characteristics. Some allelic variation has been documented in the *Amy2* locus of barley (Ko and Henry, 1994; Ko *et al.*, 1996), although the usefulness of this variation is not clear.

Potentially useful allelic variation in the genes that encode the β-amylases has recently been identified. Erkkila *et al.* (1998) identified three alleles of β-amylase1, including one that might be responsible for very high enzyme activity. Eglinton *et al.* (1998) identified β-amylase1 alleles that encode proteins with enhanced thermostabilities.

Limit dextrinase activity is thought to be necessary for the complete degradation of starch (MacGregor *et al.*, 1994). Li *et al.* (1999) identified five limit dextrinase alleles in cultivated barley and an additional 10 different genes or alleles in non-cultivated barley species. It is not yet known if the alleles identified confer useful characteristics for malting and brewing.

Inhibitors of carbohydrate-degrading enzymes

There are proteins in barley that interact with and inhibit the activities of some of the enzymes that are involved in degradation of starch and arabinoxylans. The best studied of these proteinaceous inhibitors is the barley amylase subtilisin inhibitor (BASI), which specifically inhibits the α-amylase encoded by *Amy2* (Mundy *et al.*, 1986). A portion of the β-amylase in barley and malt is in a bound form that is rendered soluble, and thereby active, via the action of reducing agents or endoproteinases. Genes coding for these β-amylase-solubilizing-proteinases, which have yet to be identified, might affect the amount of β-amylase activity in a malt or mash. One of the proteins that is bound to β-amylase in malt may be "Protein Z". Additional information on, and citations for, BASI and Protein Z are presented in Table 10.1. Proteins have been isolated from wheat that inhibit the activity of barley malt arabinoxylanases and barley has been shown to contain proteins that cause the same effect (Debyser *et al.*, 1999). The presence of these inhibitors in barley could reduce the rate of arabinoxylan degradation during malting and mashing, leading to problems of high viscosity in the mash. The barley arabinoxylanase inhibitor has not yet been purified, and nothing is known about the gene that codes for its synthesis.

Enzymes and genes that control protein degradation

The hordeins (prolamins) and hordenins (glutelins) are the major storage proteins of barley, but Protein Z and β-amylase, which are globulins, also appear to behave as storage proteins (Shewry, 1993). Relatively large amounts of hordenin occur in some cultivars, but since these proteins generally serve structural purposes, they may not play much of a role in determining malting quality. The enzyme systems that reduce the barley storage proteins to 'soluble protein' (proteins, peptides and amino acids that are soluble in warm water) are very complicated, and involve both endoproteinases and exopeptidases (mainly carboxypeptidases). The rate-limiting step for protein degradation is the hydrolysis of the original proteins into soluble protein by the endoproteinases, so it is the activities of these enzymes in malt that will usually determine whether a barley genotype is acceptable for malting. There are at least 40 endoproteolytic activities, including representatives of all four of the common endoprotease classes, in green malt (Zhang and Jones, 1995). It seems likely that the cysteine proteinases play a major role in the degradation of storage proteins during mashing and malting, and that the aspartic and metalloproteinases also contribute significantly to this process (Jones, 1999). A gene coding for one aspartic class endoproteinase called phytepsin has been cloned (Runeberg-Roos *et al.*, 1991). Its map location has not been reported. The small peptides and amino acids that are released by the exopeptidases comprise the majority of the free amino nitrogen (FAN) fraction. FAN concentration is measured to indicate how well the original protein material can be utilized by yeasts during brewing. Three carboxypeptidase (Cxp) genes have been cloned and two have been mapped (see also Table 10.1).

Inhibitors of protein-degrading enzymes

The activities of the cysteine class malt endoproteinases are strongly inhibited by a series of proteinaceous inhibitors that occur in both barley and malt, with malt containing about 2.5 times as much inhibitory activity as barley. These inhibitors appear to play a role in controlling the rate of protein solubilization during mashing (Jones and Marinac, 2000a). Two of these inhibitors are lipid transfer proteins – LTP1 and LTP2. LTP1 was initially purified and described as a Probable Amylase Protease Inhibitor (PAPI) (Mundy and Rogers, 1986) and its locus, *Ltp1,* has been mapped to chromosome 5H (Kleinhofs *et al.*, 1993; see also Table 10.1).

Future improvement in malting quality

By manipulating the carbohydrate and protein-degrading enzymes that occur in malts, as well as their inhibitor proteins, it should be possible to produce even better malting barleys. Because of the large number of enzymes and inhibitors involved, however, it is clear that this will not be an easy process. Similar to the situation for grain yield, malting quality is an economically important phenotype with complex genetic inheritance. Many of the component determinants of malting quality such as malt extract and α-amylase activity are known. However, we are still in the process of developing a comprehensive understanding of how the individual determinants interact to determine the final phenotypes. In terms of genetic diversity, we have yet to develop comprehensive catalogs of the alleles that determine these traits.

Quantitative Trait Loci (QTL)

The QTL concept represents an incremental step forward in understanding traits showing quantitative variation. The "QTL" acronym and the often-used term "quantitative inheritance" may perpetuate the idea that the genes affecting quantitative traits are somehow different from the genes that determine qualitative traits that show Mendelian inheritance. Because alleles at both quantitative and qualitative trait loci can ultimately be reduced to DNA sequences, what are the distinguishing features of QTL? Defining QTL based on assumptions regarding the number of loci determining the target phenotype is not appropriate, since the number of loci that can be characterized in an unbiased fashion is dictated by the terms of a given experiment. These terms include population size, heritability, and the number and types of environments that are sampled. A simpler operative definition may be to consider QTL as genes underlying phenotypes that (i) defy Mendelian classification and (ii) are measured by a quantitative scale. Thus defined, QTL approaches are appropriate for most agronomic and malting quality traits. As elegantly demonstrated in rice and tomato, with sufficient resources and the appropriate genetic stocks, QTL can be "converted" to Mendelian loci and cloned (Yano *et al.*, 1997, 2000; Yamamoto *et al.*, 1998; Frary *et al.*, 2000).

In addition to map-based cloning, a goal of many QTL projects in barley, and other crops, is to discover, dissect, and manipulate the genes that determine quantitatively inherited phenotypes that pose particular challenges for plant breeders. In essence, this involves a systematic characterization of the existing genetic diversity. The first QTL reports were particularly promising in this regard, as complex phenotypes such as yield, malting quality, and quantitative disease resistance were reduced to relatively few loci that showed little evidence of epistasis or interaction with the environment. Based on such results, it seemed likely that the effectiveness of marker-assisted selection for QTL alleles would be limited only by the availability and cost of markers flanking target QTL. However, as more data were generated on more germplasm combinations, and QTL detection and characterization methodologies were improved, QTL mapping has turned out to be nearly as complex as the phenotypes it was meant to simplify. The essentially descriptive genome mapping and QTL detection projects of the late 20th century have been followed by the functional genomics, proteomics, and large-scale sequencing projects of the early 21st century. With our genetics horizons now limited only by our ability to phrase interesting questions, and to afford the technology, the time is right to summarize results from over ten years of barley QTL studies (Hayes *et al.*, 2000).

The successful cloning of QTL in two model systems (rice and tomato) justifies an attempt to systematically dissect quantitative traits in barley. However, investigators need to recognize that the initial estimates of the number, genome location, and effects of QTL from mapping populations are probably crude representations of very complex biological processes. Finer structure analyses employing alternative genetic stocks and more refined measures of target phenotypes will be required for precise and accurate QTL analysis. When alleles at multiple QTL in multiple germplasm accessions are isolated and characterized at the DNA sequence, expression, and protein levels, techniques for describing, quantifying and classifying genetic diversity can be applied to these data. This will permit hypotheses to be developed and tested regarding the evolution, distribution, and function of QTL alleles in barley germplasm. The results of these experiments will provide a rational basis for germplasm conservation and utilization. The current barley QTL data, while extensive, are too fraught with bias and low precision to undertake such formal analyses of QTL allelic diversity. Additional limitations of the available data are: (1) inadequate sampling of the global diversity in barley germplasm; (2) the use of different QTL analysis procedures; and (3) a lack of common markers. Accordingly, we approached the question of QTL diversity in barley from descriptive and enumerative perspectives.

Summary of QTL diversity

We have summarized the available barley QTL data and addressed the levels of diversity in Figure 10.1 using the following criteria: (1) germplasm sampled; (2) phenotypes sampled; (3) distribution of QTL across and within linkage groups; and (4) distribution of QTL on the physical map of barley that was reported by Künzel *et al.* (2000). QTL were assigned to the Bin Map, herein referred to as the "BM", of Kleinhofs and Graner (2000) as follows: if markers flanking a QTL were also in the BM, we were able to unequivocally assign the QTL to a bin or region. Binned QTL were assigned to the physical map of Künzel *et al.* (2000) – referred to as the "PM" – as follows. First, we identified markers that occurred in both the BM and the 'Igri'/'Franka' map, referred to as "IF". The IF map was used in developing both the BM (Kleinhofs and Graner, 2000) and the PM (Künzel *et al.*, 2000). We used markers in common to the BM and IF map as our starting framework. Then, IF markers without BM positions were assigned to the latter using the proportional distances to common flanking markers in the IF and CM maps. This generated an adjusted "Linkage Bin Map" (LBM). We then aligned the LBM

Table 10.2. Summary of QTL reports in barley, sorted by chromosome and phenotype. Phenotype classifications correspond to those used by Hayes *et al.* (2000).

Chromosome	Number of QTL					Total
	Abiotic stress resistance	Agronomic traits	Biotic stress resistance	Quality	Other	
1H	8	35	9	24	1	77
2H	7	81	16	35	6	145
3H	3	70	17	19	4	113
4H	8	52	15	21	1	97
5H	14	56	14	47	4	135
6H	15	36	13	10	1	75
7H	12	59	19	24	1	115
Total	67	389	103	180	18	757

and the PM to generate a "Physical Bin Map" (PBM). The LBM and PBM bins were assigned recombination values, following the nomenclature of Künzel *et al.* (2000), to generate a "Recombination Bin Map" (RBM). We then aligned the RBM with the PM. As shown in Figure 10.1, this allows for approximations of the physical size of the bins to which QTL are assigned and an estimate of the degree of recombination in each bin. Finally, as shown in Figure 10.1, QTL assigned to each bin in the LBM were drawn proportional to the bin size in the PBM.

For the five trait groups – abiotic stress resistance, agronomic traits, biotic stress resistance, quality, and other – 757 QTL were distributed across the seven chromosomes, with the most QTL on chromosomes 2H and 5H and the fewest on chromosomes 1H and 6H (Table 10.2). The high number of QTL on chromosome 2 is attributable to QTL coincident with, or located near, the *vrs1* locus.

There are QTL for all traits on all chromosomes that are located in bins where there is a low physical-to-linkage map ratio (Figure 10.1). These QTL are sufficiently resolved to represent defined targets for marker-assisted selection, finer structure mapping, and cloning. However, there are also QTL mapping to large physical distances, which often correspond to the centromeric regions on linkage maps. Additional resolution will be required before these QTL can be considered as targets for map-based cloning. For the purposes of marker-assisted selection, linkage drag is clearly an issue. However, procedures for the marker-assisted selection of QTL

Table 10.3. Summary of barley QTL reports. See also Hayes *et al.* (2000) for criteria employed in compiling this summary.

Trait	Number of phenotypes measured	Number of QTL[1]	Number of populations[2]	Number of citations[3]
Abiotic stress resistance	26	67	7	9
Agronomic traits	58	389	16	24
Biotic stress resistance	15	103	10	15
Quality traits	27	180	8	22
Other	5	18	3	4

[1]The number of QTL per trait is the number of unique QTL per population. Please refer to Hayes *et al.* (2000) for QTL for the same trait mapping to the same bin in different populations.
[2]The number of populations is unique for each trait, but multiple traits are often measured on the same reference population.
[3]The number of citations per trait is unique, but multiple traits are often described in the same citation.

alleles in such regions are no different from those for QTL in high recombination regions, since the issue is recombination, not physical distance.

The world barley community, particularly in Europe and North America, has been very active in QTL mapping. Of the 757 QTL, the greatest number of reports – 389 – is for agronomic traits, followed by 180 for quality traits (Table 10.3). However, despite the abundance of reports, the wealth of genetic diversity in barley has barely been characterized. Germplasm from important barley production areas of the world, including Africa, Asia, the former USSR, and the Middle East, is underrepresented. Despite the vigorous barley genomics research effort in Australia, there is only one report of these studies (Jefferies *et al.*, 1999). The breadth of the Australian mapping and QTL effort is, however, apparent in the summary posted on GrainGenes (Langridge *et al.*, 1996). Even in the case of the best-characterized germplasm groups, there are only preliminary data that allow for a comprehensive characterization of allelic diversity in any particular set of accessions.

Applications of QTL analyses

Information about the QTL that influence economically important traits is useful, necessary, and valuable. QTL need to be assigned to regions of the genome – either as a platform for map-based cloning or as a way of assigning functional roles to sequences. Genetic and physical map coordinates for the determinants of phenotypic variation will be essential pieces of information for characterizing and using genetic diversity. The QTL summary we have generated is based on a limited sampling of germplasm, but it is a useful model for addressing two of the key questions in the conservation and use of barley germplasm. "How many genes determine a phenotype?" and "How much allelic diversity is there at these loci?" Answers to these questions will allow conservators to develop representative, non-redundant collections and they will allow plant breeders to make informed decisions regarding the accumulation of favorable alleles in single genotypes. In the following section, we have highlighted some of the ways in which QTL research can be of assistance in answering the two fundamental questions we have posed, and we have drawn on the QTL summary to illustrate these applications.

Catalogs of mapped loci for economically and evolutionarily important phenotypes

A goal of germplasm collection programs is to systematically catalog allelic diversity at key loci. To date, this cataloging has focused, of necessity, on simply scorable morphological descriptors and molecular markers. However, these indicators are often of limited biological utility. What is needed is a characterization of allelic diversity at loci that determine phenotypes showing complex inheritance, and QTL analysis is a first step in this direction. An example of a powerful and immediate application of this is developing catalogs of allelic diversity at genes that confer resistance to biotic stresses. A wealth of theoretical and empirical studies demonstrates that, from the standpoint of probable durability, quantitative resistance may be more desirable than qualitative resistance (Vanderplank, 1978; Johnson, 1981; Parlevliet, 1983). QTL mapping allows for an indirect allelism test via the classification of different germplasm accessions showing the same phenotype. This "QTL allelism" data can be used as a basis for constructing resistance gene pyramids or deploying resistance genes. This strategy has been used to catalog genes that confer quantitative resistance to barley stripe rust (incited by *Puccinia striiformis* f. sp. *hordei*) (Chen *et al.*, 1994; Hayes *et al.*, 1996; Toojinda *et al.*, 2000) and to pyramid quantitative resistance alleles in single genotypes (Castro *et al.*, 2000). QTL conferring resistance to six of the most important foliar pathogens of barley are reported on all seven chromosomes and in bins corresponding to low, middle, and high recombination regions (Figure 10.1). The proximity of these disease resistance QTL to QTL for agronomic and malting quality traits can be used as a guide for future selection studies, considering the potential for linkage drag. The relationship between physical and linkage map distances can indicate QTL that would be good candidates for finer structure mapping and isolation.

Understanding correlated responses to selection

QTL analysis can be used to determine, at a gross level, if trait associations are due to linkage or pleiotropy. In the case of linkage, repulsion phase linkages can be broken via marker-assisted selection, or the QTL information can be used to develop populations that will generate the desired recombinant phenotypes for selection. In the case of coincident QTL, breeding objectives may need to be re-defined, pending a finer dissection of the genetic structure of the region. As discussed in the previous section, the genome-wide distribution of disease resistance QTL, and their associations with agronomic and malting quality QTL can assist in defining breeding objectives. As demonstrated by Zhu *et al.* (1999), plant architecture, phenology, and morphology have important implications for breeding for *Fusarium* head blight (FHB) resistance. Finer structure analysis will be required to determine if, in the case of FHB resistance, function follows form (pleiotropy) or if coincident QTL are due to tight linkages. The assignment of QTL to bins that correspond to the degree of recombination on the physical map will assist in prioritizing targets for finer structure mapping and gene isolation. For example, the FHB resistance QTL on chromosome 2H were reported in populations that were derived from crosses between six-rowed genotypes ('Chevron'/M64) and between two-rowed genotypes ('Gobernadora'/CMB). The bin at which QTL are coincident in the two populations contains the *vrs1* locus. In progeny of crosses between two-rowed and six-rowed genotypes, the principal FHB rust resistance QTL maps to this region. This is commonly assumed to be a pleiotropic effect of inflorescence morphology (reviewed by Zhu *et al.*, 1999). However, inflorescence type is not segregating in these populations, which suggests that there is potential for recombination between the genes that determine inflorescence type and FHB resistance. The bin to which the FHB resistance QTL and the *vrs1* locus map lies at the juncture of middle to high recombination regions, and is thus a reasonable target for genetic dissection.

Characterization of blocks of the genome that are necessary for essential phenotypes

Certain complex phenotypes – e.g., quality profiles, resistance to a spectrum of stresses – may be requisites for barley production in some environments. If these regions can be defined via QTL analysis, better efficiencies can be achieved in breeding programs by using a high throughput genotypic screening procedure to identify progeny carrying the target blocks of alleles. Phenotypic and/or genotypic selection can then be carried out for genes that reside outside the target blocks. This strategy has been proposed for malting quality in North American barley germplasm, based on malting quality QTL alleles that are contributed by the reference cultivar 'Morex' in multiple mapping populations (Marquez-Cedillo *et al.*, 2000). The regions, which are targeted in the quality footprint, are apparent in Figure 10.1. The two regions with the most consistent effects on malting quality are the large blocks on chromosomes 7H and 5H that span the centromeric regions. QTL for multiple malting quality traits map to each of these regions.

Assessment of alternative procedures for measuring the same phenotype

Many complex phenotypes are difficult to improve through breeding because phenotyping is laborious or expensive. When new procedures are developed for measuring a given phenotype, the commonality of QTL detected by the new and the standard procedure can be used as a measure of the correspondence of the two assays. For example, Iyamabo and Hayes (1995) used the consistency of QTL detection as a criterion for comparing grain yield in hill plots and row plots and concluded that the former were useful for detecting large-effect QTL. This has important implications for large-scale germplasm assessment and screening, where the costs of full yield plot assessments are prohibitive. Experiments are underway to compare malting quality QTL that are detected by Near Infrared Spectroscopy (NIR) and conventional wet chemistry analyses, and the summary of observed QTL will be useful in establishing relationships between the two approaches. The efficiencies of NIR would make it an attractive tool for characterizing extensive germplasm arrays for phenotypic diversity in malting quality parameters, and the genetic determinants of this diversity can be assigned to map positions with QTL tools.

Introgression of exotic and/or ancestral germplasm

Exotic germplasm can be a source of alternative alleles at known loci or of entirely new genes, in cases where there is no allelic variation at these loci in cultivated germplasm. Tanksley and Nelson (1996) stimulated widespread interest in discovering and using favorable QTL alleles in phenotypically unattractive tomato germplasm, and this research has borne fruit with the recent cloning of a fruit size QTL (Frary *et al.*, 2000). In barley, there are encouraging preliminary data on the systematic introgression of favorable alleles from ssp. *spontaneum* into cultivated barley (Matus *et al.*, 2000). The QTL summary does not yet include data from such introgression projects. When QTL are mapped in these studies, a priori information about other loci that are found in the introgression region will be invaluable for minimizing linkage drag and establishing allelism.

Conclusions and outlook

Despite evidence that there is diversity for traits of economic importance in barley, our understanding of allelic architecture at the many loci determining agronomic performance and malting quality is incomplete. It does not yet provide a basis for systematic classification and utilization of barley germplasm resources in breeding programs.

The focus of future QTL studies could be to thoroughly characterize multiple phenotypes and to develop dynamic profiles of QTL expression in different tissues and at different points in crop development. Most QTL studies have been based on single or end point measures of a phenotype. To obtain a more complete understanding of the genetic basis of complex traits, it will be necessary to decompose complex traits into their components. For example, QTL analyses of quantitative disease resistance revealed that multiple regions of the genome were associated with symptom expression. Some associations may be the consequence of alleles that condition host-pathogen interactions, whereas others may reflect variation in alleles that determine crop architecture, phenology, and morphology. Information that relates QTL to underlying mechanisms of resistance will be invaluable for understanding the genetics of disease resistance and in achieving selection response. The full genetic potential of an organism may not be realized due to temporal and spatial changes in gene expression. Functional genomic tools may be developed to measure these changes in any given tissue or at any point in time. Understanding how developmental expression of alleles determines quantitative traits may reveal complexities, such as changes in favorable allele phase, that explain some of the unexpected results obtained from selection experiments.

There is a pressing need for more comprehensive sampling and reporting of QTL in barley germplasm. Important barley production areas of the world, with much to offer in terms of genetic diversity, are conspicuously underrepresented in the barley QTL literature. This is, in large part, due to economic constraints that face barley researchers outside of Europe, North America, and Oceania. Every effort should be made to stimulate collaborative and mutually rewarding analysis of the world's barley genetic resources.

In general, the trend in crop improvement programs has been to rely on ever-narrower germplasm bases to satisfy increasingly stringent yield and quality expectations of producers and consumers. Barley breeders are faced with the challenge of developing cultivars that satisfy multiple end-use requirements for brewing, animal feed, and human consumption. Malting quality standards for barley impose particularly rigid constraints on breeding programs. As a result, the barley industry tends to be conservative about adopting new cultivars and exploiting genetic diversity. Nonetheless, impressive progress has been made through the use of narrow germplasm bases, although the genetic mechanisms that account for selection response remain a matter of conjecture (Rasmusson and Phillips, 1997). Cultivars developed from the North American six-row barley germplasm have exceeded the expectations of growers and the malting and brewing industries. However, the extreme susceptibility of these cultivars to *Fusarium* head blight clearly demonstrates the risk of reliance on such a narrow germplasm base. In response to recent devastating epidemics of this disease in the Midwest, breeders have screened exotic germplasm and are incorporating diverse sources of resistance into new cultivars adapted to the region. This is eloquent testimony to the need for systematically characterizing diversity in barley and maintaining barley germplasm resources.

Barley possesses enormous genetic diversity that enables it to grow under a wide range of environmental conditions and to tolerate stresses such as drought and salinity. In addition to its utility in malting and brewing, barley can provide food security in regions where other cereal crops cannot be produced. It can be readily crossed with its wild ancestor (ssp. *spontaneum*), which may also represent a tremendous reservoir of variation in disease resistance and stress tolerance that remains to be exploited. Although it is difficult to quantify the variation in agronomic and malting quality traits that is present in barley, due to the complex inheritance of these phenotypes, progress is being made in this regard. Efficient strategies for discovering, characterizing, and using alleles in exotic and ancestral germplasm will help to ensure the sustainability, productivity, and profitability of barley production.

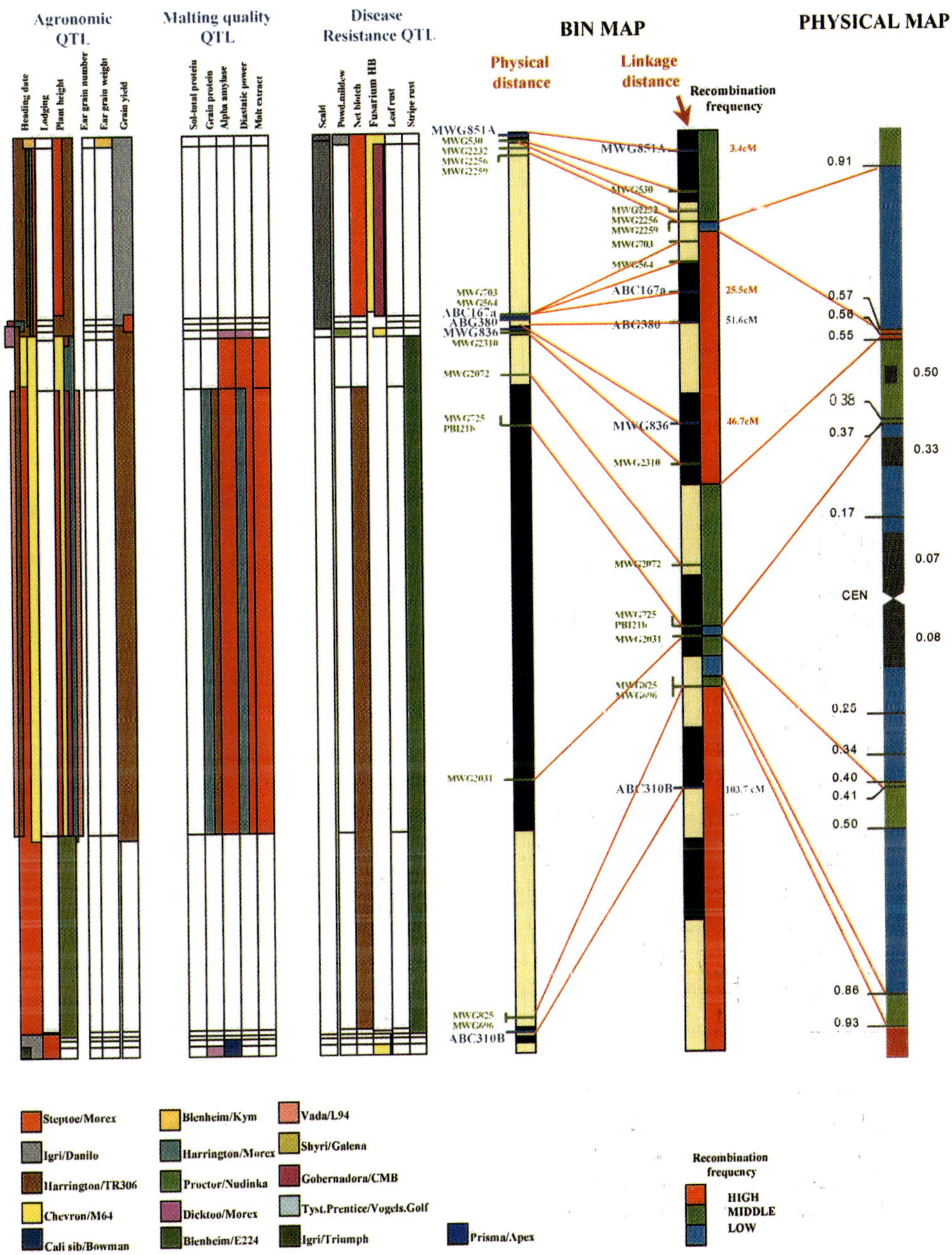

Chromosome 1H

Figure 10.1. Quantitative Trait Loci (QTL) determining economically important traits in barley, assigned to bins on the consensus linkage map (Kleinhofs and Graner, 2000) and the physical map of Künzel *et al.* (2000). The criteria used for QTL assignment to map locations were described by Hayes *et al.* (2000).

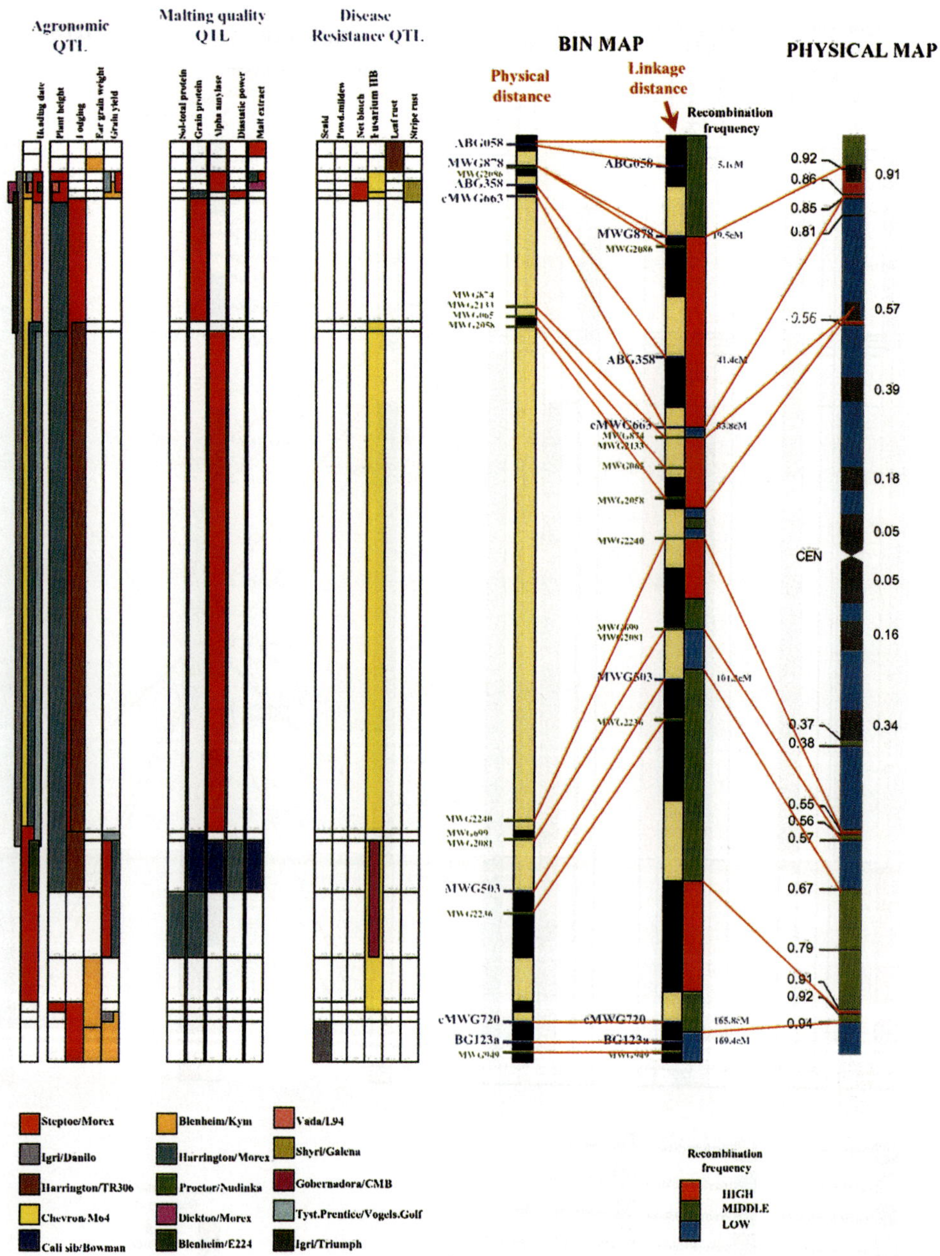

Chromosome 2H

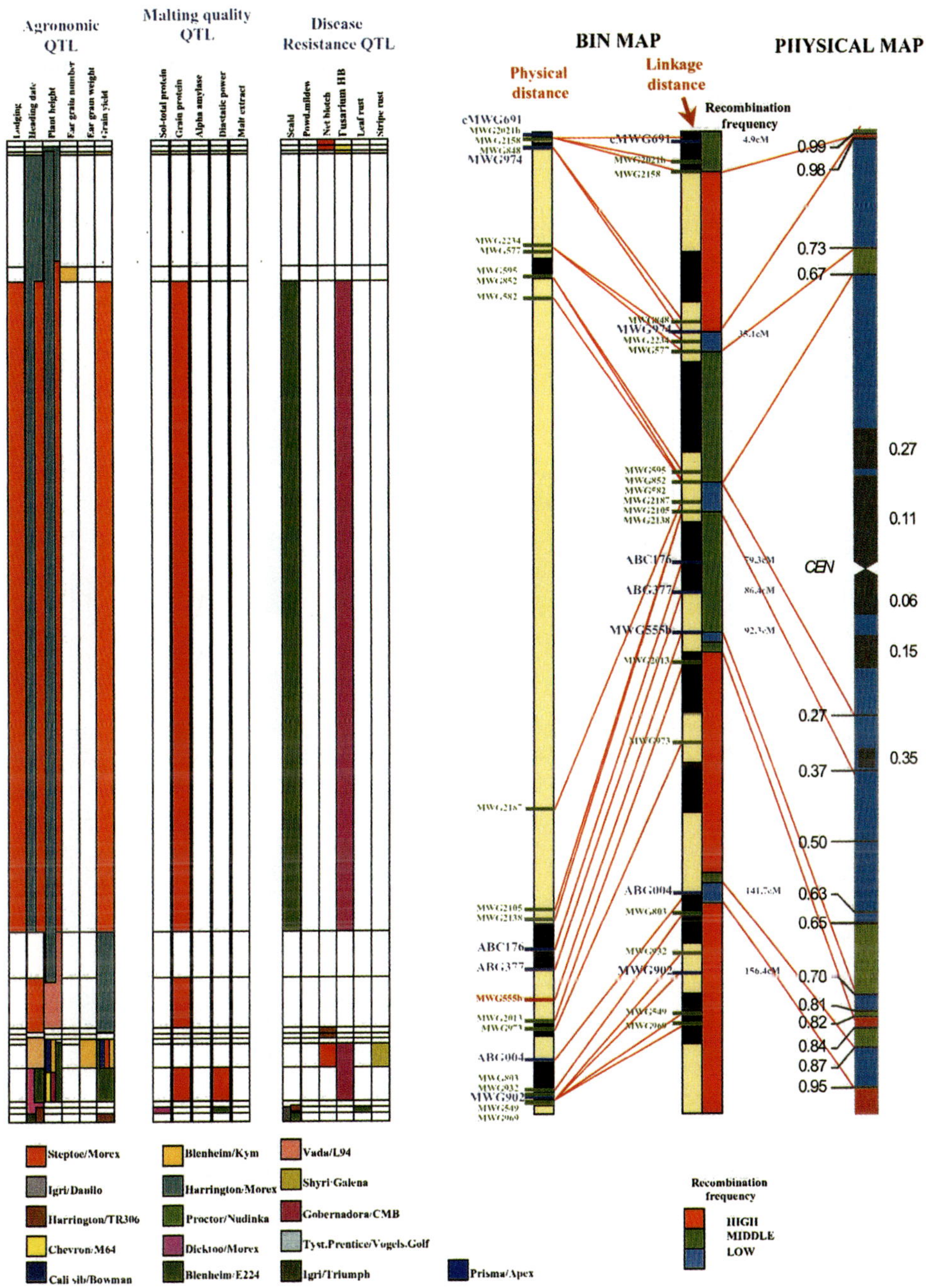

Chromosome 3H

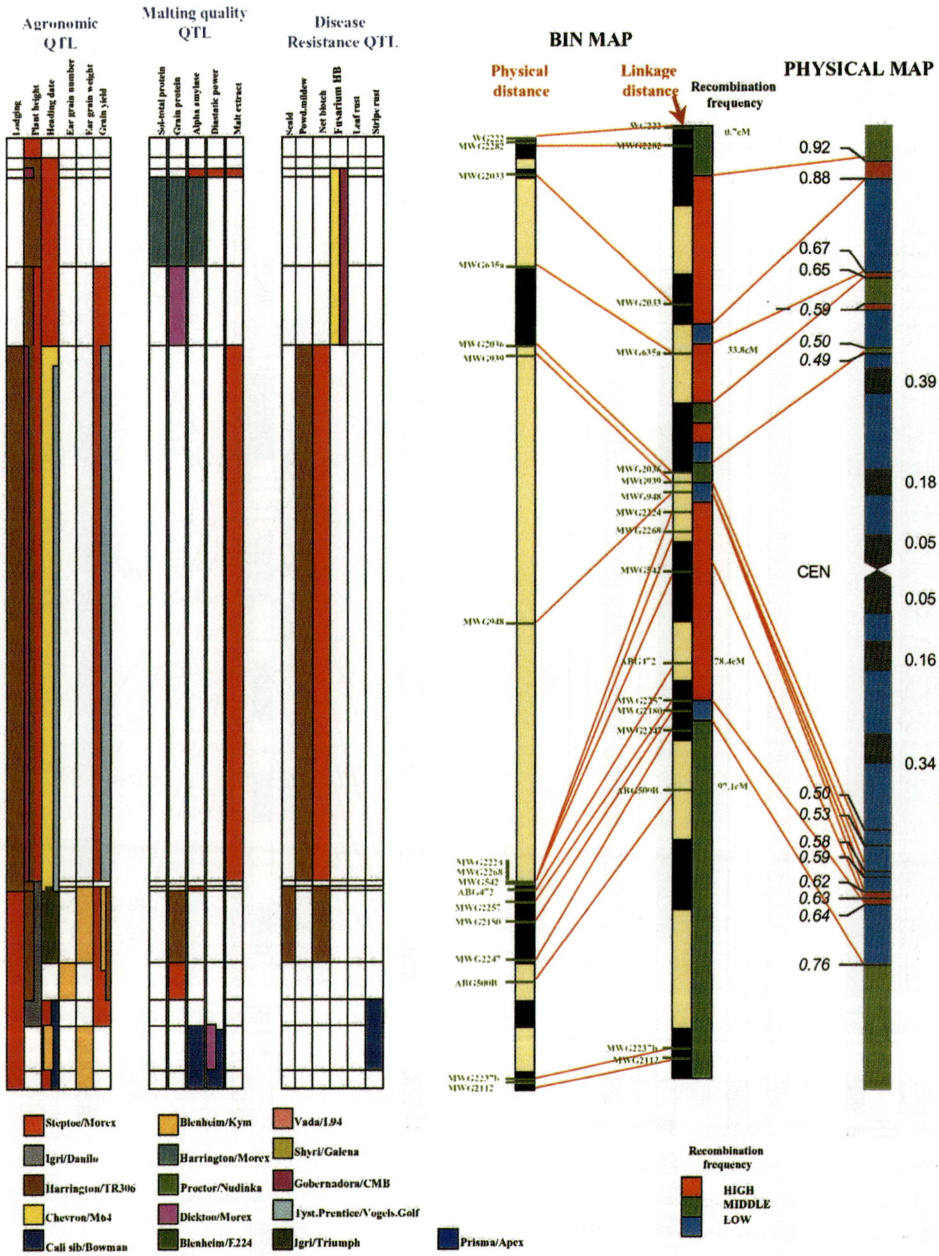

Chromosome 4H

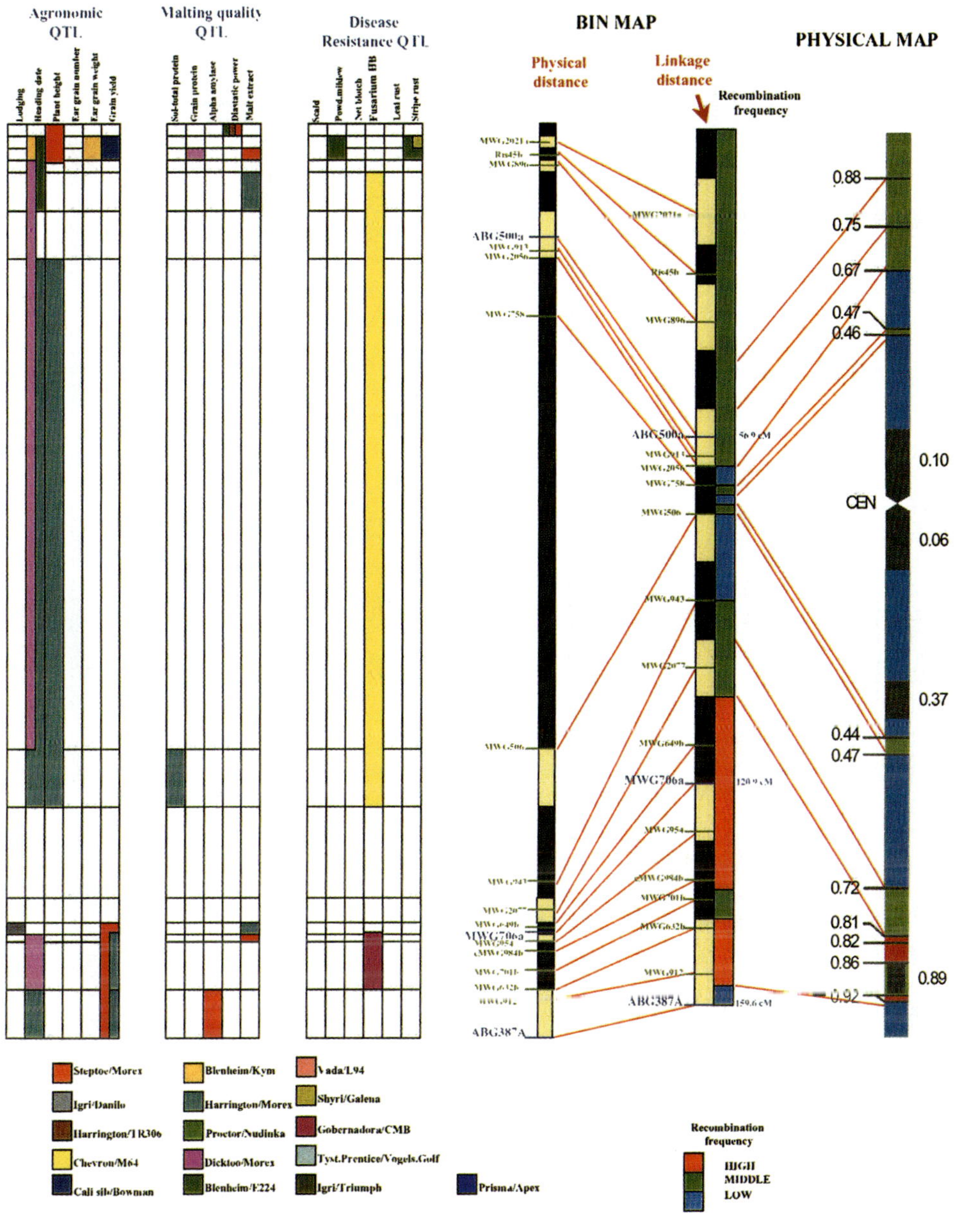

Chromosome 5H

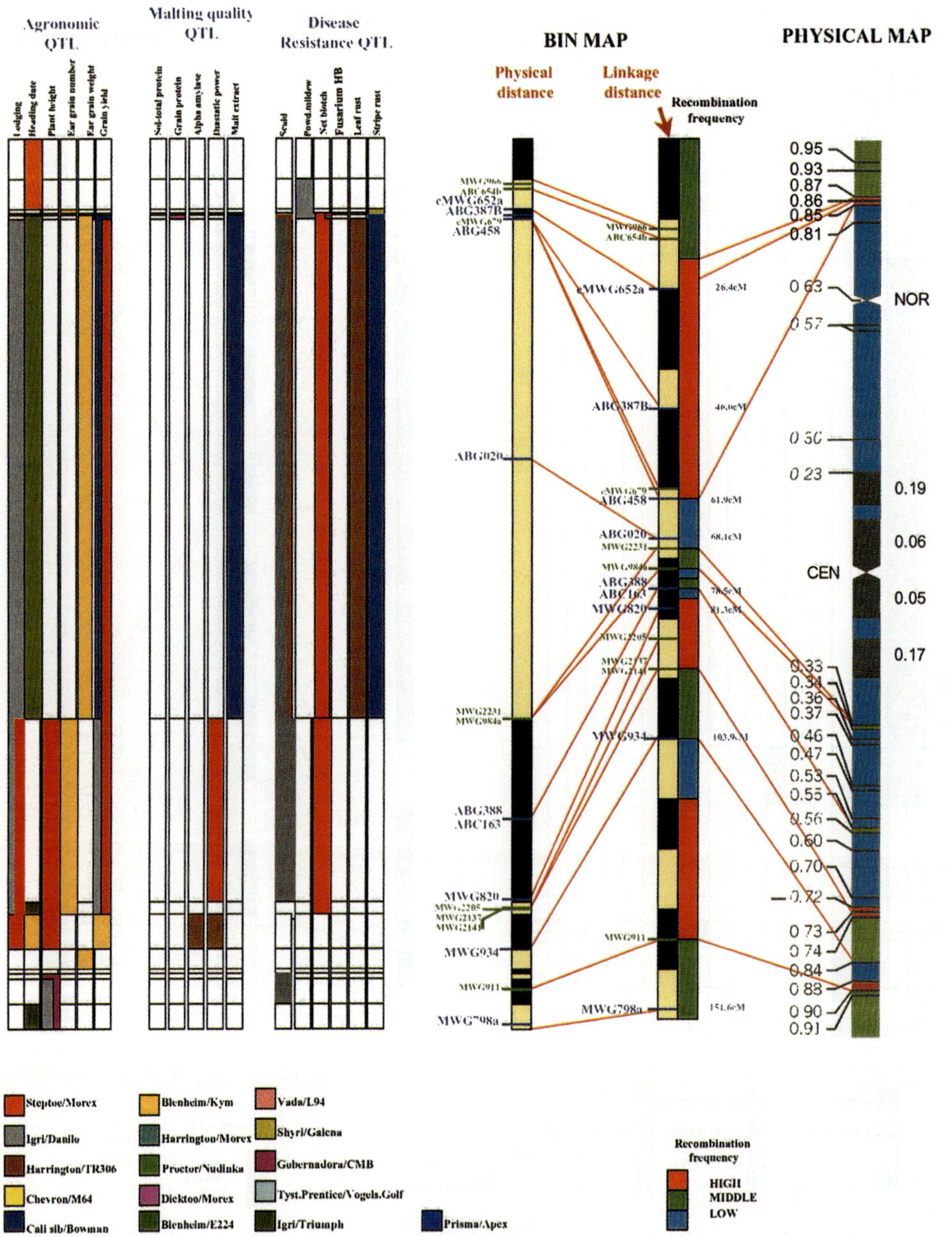

Chromosome 6H

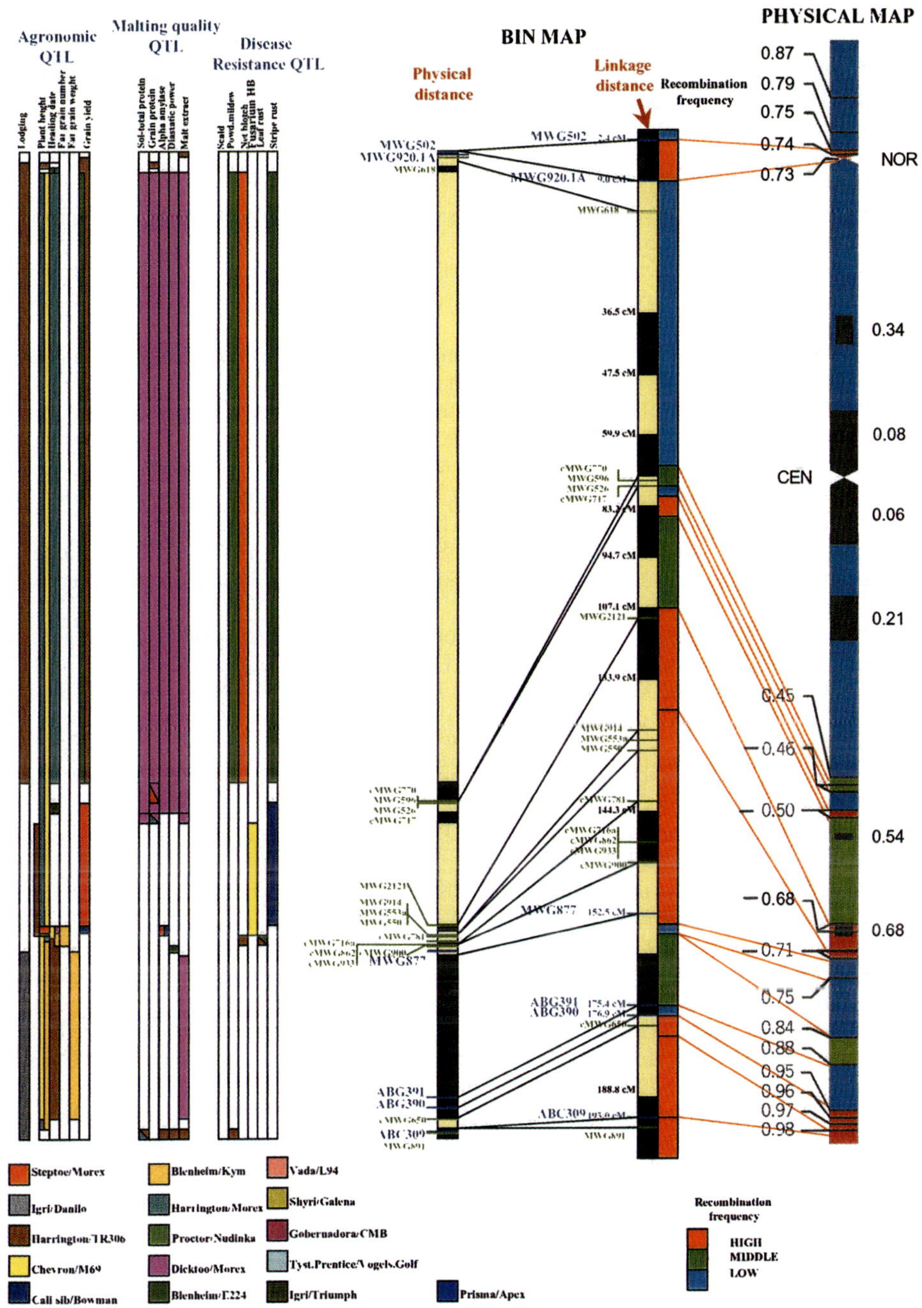

Chromosome 7H

References

Allard, R.W., 1988. Genetic changes associated with the evolution of adaptedness in cultivated plants and their wild progenitors. J. Hered. 79: 225-238.

Bamforth, C.W., and A.H.P. Barclay, 1993. Malting technology and the uses of malt. In: A.W. MacGregor and R.S. Bhatty (eds), Barley: Chemistry and Technology. Am. Assoc. Cereal Chemists, St. Paul, Minnesota, pp. 297-354.

Beldford, R.K., and R.H. Sedgley, 1991. Conclusions: ideotypes and physiology: tailoring plants for increased production. Field Crops Res. 26: 221-226.

Briggs, D.E., 1992. Barley germination: biochemical changes and hormonal control. In: P.R. Shewry (ed), Barley: Genetics, Biochemistry, Molecular Biology and Biotechnology. CAB International, UK, pp. 369-401.

Burger, W.C., and D.E. LaBerge, 1985. Malting and brewing quality. In: D.C. Rasmusson (ed), Barley, Am. Soc. Agron., Madison, Wisconsin, pp. 367-401.

Burton, R.A., X.Q. Zhang, M. Hrmova and G.B. Fincher, 1999. A single limit dextrinase gene is expressed both in the developing endosperm and in germinated grains of barley. Plant Physiol. 119: 859-871.

Cannell, M., A. Karp, P.G. Isaac and P. Shewry, 1992. Chromosomal assignment of genes in barley using telosomic wheat-barley addition lines. Genome 35: 17-23.

Castro, A.J., A. Corey, T. Filichkin, P. Hayes, J.S. Sandoval-Islas and H. Vivar, 2000. Stripe rust resistance QTL pyramids in barley. In: Barley Genetics VIII. Proc. 8th Int. Barley Genet. Symp. Adelaide, Australia, pp. 86-88.

Chen, F., D. Prehn, P.M. Hayes, D. Mulrooney, A. Corey, A. and H. Vivar, 1994. Mapping genes for resistance to barley stripe rust (*Puccinia striiformis* f. sp. *hordei*). Theor. Appl. Genet. 88: 215-219.

Debyser, W., W.J. Peumans, E.J.M. van Damme and J.A. Delcour, 1999. *Triticum aestivum* xylanase inhibitor (TAXI), a new class of enzyme inhibitor affecting breadmaking performance. J. Cereal Sci. 30: 39-43.

Dreccer, F., C. Grashoff and R. Rabbinge, 1997. Source-sink ratio in barley (*Hordeum vulgare* L.) during grain filling: effects on senescence and grain protein concentration. Field Crops Res. 49: 269-277.

Eglinton, J.K., P. Langridge and D.E. Evans, 1998. Thermostability variation in alleles of barley β-amylase. J. Cereal Sci. 28: 301-309.

Erkkila, M.J., R. Leah, H. Ahokas and V. Cameron-Mills, 1998. Allele-dependent barley grain β-amylase activity. Plant Physiol. 117: 679-685.

Evans, D.E., M.C. Sheehan and D.C. Stewart, 1999. The impact of malt derived proteins on beer foam quality. Part II: the influence of malt foam-positive proteins and non-starch polysaccharides on beer foam quality. J. Inst. Brew. 105: 171-177.

Fincher, G.B. and B.A. Stone, 1993. Physiology and biochemistry of germination in barley. In: A.W. MacGregor and R.S Bhatty (eds), Barley: Chemistry and Technology. Am. Assoc. Cereal Chemists, St. Paul, Minnesota, pp. 247-295.

Frary, A., T.C. Nesbitt, A. Frary, S. Grandillo, E. van der Knaap, B. Cong, J. Liu, J. Meller, R. Elber, K.B. Alpert, S.D. Tanksley, 2000. *fw2.2*: a quantitative trait locus key to the evolution of tomato fruit size. Science 289: 85-88.

Gallagher, J.N., P.J. Biscoe and R.K. Scott, 1975. Barley and its environment. V. Stability of grain weight. J. Appl. Ecol. 12: 319-336.

Giese, H., A.G. Holm-Jensen, H. Mathiassen, B. Kjær, S.K. Rasmussen, H. Bay and J. Jensen, 1994. Distribution of RAPD markers on a linkage map of barley. Hereditas 120: 267-273.

Hamblin, J., 1993. The ideotype concept: useful or outdated? In: D.R. Buxton, R. Shibles, A. Forsberg, B.L. Blad, K.H. Asay, G.M. Paulsen and R.F. Wilson (eds), International Crop Science I, CSSA, Madison, Wisconsin, pp. 589-597.

Harlan, J.R., 1995. The living fields: our agricultural heritage. Cambridge University Press, Cambridge, England and New York, NY, USA.

Harlan, H.V., 1957. One man's life with barley. Exposition Press, New York.

Hayes, P.M., A. Castro, A. Corey, L. Marquez-Cedillo, B. Jones, D. Mather, I. Matus, C. Rossi, and K. Sato, 2000. A summary of published barley QTL reports. http://www.css.orst.edu/barley/nabgmp/qtlsum.htm

Hayes, P.M., D. Prehn, H. Vivar, T. Blake, A. Comeau, I. Henry, M. Johnston, B. Jones and B. Steffenson, 1996. Multiple disease resistance loci and their relationship to agronomic and quality loci in a spring barley population. J. QTL. http://probe.nalusda.gov:8000/otherdocs/jqtl/index.html

Henson, C.A., and J.M. Stone, 1988. Variation in α-amylase and α-amylase inhibitor activities in barley malts. J. Cereal Sci. 8: 39-46.

Henson, C.A., and Z. Sun, 1995. Barley seed α-glucosidases: Their characteristics and roles in starch degradation. In: J.N. Saddler and M.H. Penner (eds), Enzymatic Degradation of Insoluble Carbohydrates. ACS Symposium, Series 618, pp. 51-58.

Hitchcock, A.S., 1971. Manual of the grasses of the United States, 2nd ed. (rev. by A. Chase). Dover Publications, New York.

Im, H., and C.A. Henson, 1995. Characterization of high pI alpha-glucosidase from germinated barley seeds: Substrate specificity, subsite affinities and active-site residues. Carbohydrate Research 277: 145-159.

Iyamabo, O.E., and P.M. Hayes, 1995. Effects of plot type on detection of quantitative trait locus effects in barley (*Hordeum vulgare* L.). Plant Breed. 114: 55-60.

Jefferies, S.P., A.R. Barr, A. Karakousis, J.M. Kretschmer, S. Manning, K.J. Chalmers, C. Nelson, A.K.M.R. Islam and P. Langridge, 1999. Mapping of chromosome regions conferring boron toxicity tolerance in barley (*Hordeum vulgare* L.). Theor. Appl. Genet. 98: 1293-1303.

Johnson, R., 1981. Durable resistance: definition of, genetic control, and attainment. Phytopathology 71: 567-568.

Jones, B.L., 1999. Malt endoproteinases and how they affect wort soluble protein levels. In: Proc. 9th Austral. Barley Techn. Symp., Pathway into the 21st Century, Melbourne, Australia. Barley Techn. Symp. Steering Committee, pp. 2.39.1-8.

Jones, B.L., and A.D. Budde, 1999. Endoproteinases and the hydrolysis of malt proteins during mashing. In: European Brewery Convention: Proc. 27th Congress, Cannes, France. European Brewery Convention, Zoeterwoude, Netherlands, pp. 611-618.

Jones, B.L., and L.A. Marinac, 2000a. Endogenous inhibitors of the endoproteinases and other enzymes of barley. In: T. Simoinen and M. Tenkanen (eds), Proc. 2nd Europ. Symp. Enzymes and Grain Processing, Helsinki, Finland, 1999. Techn. Res. Centre of Finland, Espoo, Finland, pp. 39-46.

Jones, B.L., and L.A. Marinac, 2000b. Purification and partial characterization of a second cysteine proteinase inhibitor from ungerminated barley (*Hordeum vulgare* L.). J. Agric. Food Chem. 48: 257-264.

Jui, P.Y., T.M. Choo, K.M. Ho, T. Konishi and R.A. Martin, 1997. Genetic analysis of a two-row × six-row cross of barley using doubled-haploid lines. Theor. Appl. Genet. 94: 549-556.

Kirby, E.J.M., and D.G. Faris, 1972. The effects of plants density on tiller growth and morphology in barley. J. Agric. Sci. 78: 281-288.

Kjær, B., and J. Jensen, 1996. Quantitative trait loci for grain yield and yield components in a cross between a six-rowed and a two-rowed barley. Euphytica 90: 39-48.

Kleinhofs, A., and A. Graner, 2000. The Barley BIN map (March 2000 posting). http://barleygenomics.wsu.edu/

Kleinhofs, A., A. Kilian, M.A. Saghai-Maroof, R.M. Biyashev, P. Hayes, F.Q. Chen, N. Lapitan, A. Fenwick, T.K. Blake and V. Kanazin, 1993. A molecular, isozyme and morphological map of the barley (*Hordeum vulgare*) genome. Theor. Appl. Genet. 86: 705-712.

Ko, H.L., and R.J. Henry, 1994. Identification of barley varieties using the polymerase chain reaction. J. Inst. Brew. 100: 405-407.

Ko, H.L., S. Weining and R.J. Henry, 1996. Application of primers derived from barley α-amylase genes to the identification of cereals by PCR. Plant Varieties and Seeds 9: 53-62.

Konishi, Y., A. Okamoto, J. Takahashi, M. Aitani and N. Nakatani, 1994. Effects of Bay m 1099, and α-glucosidase inhibitor, on starch metabolism in germinating wheat seeds. Biosci. Biotech. Biochem. 58: 135-139.

Künzel, G., L. Korzun and A. Meister, 2000. Cytologically integrated physical restriction fragment length polymorphism maps for the barley genome based on translocation breakpoints. Genetics 154: 397-412.

Langridge, P., A. Karakousis, J. Kretschmer, S. Manning, K. Chalmers, R. Boyd, C.D. Li, R. Islam, S. Logue, R. Lance and SARDI, 1996. RFLP and QTL analysis of barley mapping populations. http://greengenes.cit.cornell.edu/WaiteQTL/

Lauer, J.G., and S.R. Simmons, 1989. Canopy light and tiller mortality in spring barley. Crop Sci. 29: 420-424.

Li, C.D., X.Q. Zhang, P. Eckstein, B.G. Rossnagel and G.H. Scoles, 1999. A polymorphic microsatellite in the limit dextrinase gene of barley (*Hordeum vulgare* L.). Mol. Breed. 5: 569-577.

Loi, L., B. Ahluwalia and G.B. Fincher, 1988. Chromosomal location of genes encoding barley (1 → 3, 1 → 4)-β-glucan-4-glucanohydrolases. Plant Physiol. 87: 300-302.

Lusk, L.T., H. Goldstein and D. Ryder, 1995. Independent role of beer proteins, melanoidins and polysaccharides in foam formation. J. Am. Soc. Brew. Chem. 53: 93-103.

MacGregor, A.W., L.J. Macri, S.L. Bazin and G.W. Sadler, 1995. Limit dextrinase inhibitor in barley and malt and its possible role in malting and brewing. In: Proc. 25th Congr. Europ. Brewery Convention. Brussels, Belgium, pp. 185-192.

MacGregor, A.W., L.J. Macri, S.W. Schroeder and S.L. Bazin, 1994. Purification and characterisation of limit dextrinase inhibitors from barley. J. Cereal Sci. 20: 33-41.

MacLeod, L.C., R.C.M. Lance and A.H.D. Brown, 1991. Chromosomal mapping of the *Glb1* locus encoding (1 → 3, 1 → 4)-β-D-glucan 4-glucanohydrolase EI in barley. J. Cereal Sci. 13: 291-298.

Marquez-Cedillo, L.A., P.M. Hayes, B.L. Jones, A. Kleinhofs, W.G. Legge, B.G. Rossnagel, K. Sato, S.E. Ullrich, D.M. Wesenberg and the NABGMP, 2000. QTL analysis of malting quality in barley based on the doubled haploid progeny of two elite North American varieties representing different germplasm groups. Theor. Appl. Genet. 101: 173-184.

Marquez-Cedillo, L.A., P.M. Hayes, A. Kleinhofs, W.G. Legge, B.G. Rossnagel, K. Sato, S.E. Ullrich, D.M. Wesenberg and the NABGMP, 2001. QTL analysis of agronomic traits in barley based on the doubled haploid progeny of two elite North American varieties representing different germplasm groups. Theor. Appl. Genet. 103: 625-637.

Matus, I., A. Corey, T. Filichkina, P.M. Hayes, O. Riera-Lizarazu, K. Sato and M.I. Vales, 2000. In: Barley Genetics VIII. Proc. 8th Int. Barley Genet. Symp., Adelaide, Australia, pp. 86-88.

Mikkonen, A., I. Porali, M. Cercos and T.D. Ho, 1996. A major cysteine proteinase, EPB, in germinating barley seeds: structure of two intronless genes and regulation of expression. Plant Mol. Biol. 31: 239-254.

Munck, L., H. Mundy and P. Vaag, 1985. Characterization of enzyme inhibitors in barley and their tentative role in malting and brewing. Am. Soc. Brewing Chemists J. 43: 35-38.

Mundy, J., and J.C. Rogers, 1986. Selective expression of a probable amylase/protease inhibitor in barley aleurone cells: comparison to the barley amylase/subtilisin inhibitor. Planta 169: 51-63.

Oziel, A., P.M. Hayes, F.Q. Chen and B. Jones, 1996. Application of quantitative trait locus mapping to the development of winter habit malting barley. Plant Breed. 115: 43-51.

Parlevliet, J.E., 1983. Race-specific resistance and cultivar-specific virulence in the barley-leaf rust pathosystem and their consequences for the breeding of leaf rust resistant barley. Euphytica 32: 367-375.

Pennisi, E., 2000. *Arabidopsis* comes of age. Science 290: 32-35.

Powell, W., R.P. Ellis and W.T.B. Thomas, 1990. The effects of major genes on qualitatively varying characters in barley. III. The two row/six row locus (*V-v*). Heredity 65: 259-264.

Rasmusson, D.C., 1987. An evaluation of ideotype breeding. Crop Sci. 27: 1140-1146.

Rasmusson, D.C., 1991. A plant breeder's experience with ideotype breeding. Field Crops Res. 26: 191-200.

Rasmusson, D.C., and R.L. Phillips, 1997. Plant breeding progress and genetic diversity from *de novo* variation and elevated epistasis. Crop Sci. 37: 303-310.

Riggs, T.J., and E.J.M. Kirby, 1978. Developmental consequences of two-row and six-row ear type in spring barley. J. Agric. Sci. Camb. 91: 199-205.

Runeberg-Roos, P., K. Tormakangas and A. Ostman, 1991. Primary structure of a barley-grain aspartic proteinase. Eur. J. Biochem. 202: 1021-1027.

Shewry, P., 1993. Barley seed proteins. In: A.W. MacGregor and R.S. Bhatty (eds), Barley: Chemistry and Technology. Am. Assoc. Cereal Chemists, St. Paul, Minnesota, pp. 131-197.

Sissons, M.J., and A.W. MacGregor, 1994. Hydrolysis of barley starch granules by α-glucosidases from malt. J. Cereal Sci. 19: 161-169.

Sørensen, S.B., L.M. Bech, M. Muldbjerg, T. Beenfeldt and K. Breddam, 1993. Barley lipid transfer protein 1 is involved in beer foam formation. Master Brewers Assoc. Am. Tech. Quarterly 30: 136-145.

Sun, Z., and C.A. Henson, 1990. Degradation of native starch granules by barley α-glucosidases. Plant Physiol. 94: 320-327.

Sun, C., P. Sathish, S. Ahlandsberg and C. Jansson, 1999. Analyses of isoamylase gene activity in wild-type barley indicate its involvement in starch synthesis. Plant Mol. Biol. 40: 431-443.

Takahashi, R., J. Hayashi and I. Moriya, 1975. Basic studies on breeding barley by the use of two-rowed and six-rowed varietal crosses. In: Barley Genetics III. Proc. 3rd Int. Barley Genet. Symp., Thiemig, München, Germany, pp. 662-677.

Tanksley, S.D., and J.C. Nelson, 1996. Advanced backcross QTL analysis: a method for the simultaneous discovery and transfer of valuable QTLs from unadapted germplasm into elite breeding lines. Theor. Appl. Genet. 92:191-203.

Toojinda, T., E. Baird, L. Broers, X.M. Chen, P.M. Hayes, A. Kleinhofs, J. Korte, D. Kudrna, H. Leung, R.F. Line, W. Powell and H. Vivar, 2000. Mapping quantitative and qualitative disease resistance genes in a doubled haploid population of barley. Theor. Appl. Genet. 101: 580-589.

Vanderplank, J.E., 1978. Genetic and Molecular Basis of Plant Pathogenesis. Springer-Verlag, Berlin, Heidelberg, New York.

Venter, J.C. *et al.*, 2001. The sequence of the human genome. Science 291: 1304-1351.

Viëtor, R.J., A.G.J. Voragen and S.A.G.F. Angelino, 1993. Composition of non-starch polysaccharides in wort and spent grain from brewing trials with malt from a good malting quality barley and a feed barley. J. Inst. Brew. 99: 243-248.

Yamamoto, T., Y. Kuboki, S.Y. Lin, T. Sasaki and M. Yano, 1998. Fine mapping of quantitative trait loci *Hd*-1, *Hd*-2 and *Hd*-3, controlling heading date of rice, as single Mendelian factors. Theor. Appl. Genet. 97: 37-44.

Yano, M., Y. Harushima, Y. Nagamura, N. Kurata, Y. Minobe and T. Sasaki, 1997. Identification of quantitative trait loci controlling heading date in rice using a high-density linkage map. Theor. Appl. Genet. 95: 1025-1032.

Yano, M., Y. Katayose, M. Askikari, U. Yamanouchi, L. Monna, T. Fuse, T. Baba, K. Yamamoto and T. Sasaki, 2000. http://www.intl-pag.org/pag/8/abstracts/pag8819.html

Yasuda, S., 1964. Experimental studies on natural selection for time of heading and its inner factors in some barley hybrid populations. II. Genotypic constitutions of selection lines derived from the populations grown at different locations. Ber. Ohara Inst. Landw. Biol. Okayama Univ. 12: 197-215.

Yasuda, S., 1977. Effect of four different spring genes on earliness of heading, grain yield and some yield components in barley. In: Proc. 3rd Int. Cong. SABRAO 1: 30-32.

Yasuda, S., J. Hayashi and I. Moriya, 1992. Regional specialities in ecological agronomic characters of two-rowed barley cultivars from Japan and other countries. Bull. Res. Inst. Biores., Okayama Univ. 1: 61-74. (In Japanese with English summary).

Zhang, N., and B.L. Jones, 1995. Purification and partial characterization of a 31-kDa cysteine endopeptidase from germinated barley. J. Cereal Sci. 21: 145-153.

Zhu, H., L. Gilchrist, P. Hayes, A. Kleinhofs, D. Kudrna, Z. Liu, L. Prom, B. Steffenson, T. Toojinda and H. Vivar, 1999. Does function follow form? QTLs for *Fusarium* Head Blight (FHB) resistance are coincident with QTLs for inflorescence traits and plant height in a doubled haploid population of barley. Theor. Appl. Genet. 99: 1221-1232.

Section IV

Conservation and Future Utilization of Barley

Diversity in Barley *(Hordeum Vulgare)*
Roland von Bothmer et al (Editors).

Chapter 11

Detecting diversity – a new holistic, exploratory approach bridging phenotype and genotype

Lars Munck

The Royal Veterinary and Agricultural University, Rolighedsvej 30, DK-1958 Frederiksberg C, Denmark

Introduction

From the point of view of data acquisition, it is remarkable that nowadays it seems easier to tap data from the genotype by gene sequencing than from the phenotype using traditional biometric methods. The tacit question at hand in modern science is whether science itself can formulate and produce the necessary genetic variation. Is this a more viable and economic alternative to keeping genebanks for reasons of security to preserve the expected, yet undefined genetic variation, which may or may not be available? In this chapter supporting the genebank alternative, we will present experimental evidence of a new, interdisciplinarily inspired opportunity to explore biodiversity based on global spectroscopic and imaging data sets from plant phenotypes from which specific data could be extracted to match data from the genotype.

Plant breeding as a multidisciplinary technology

Today, the classical plant breeder seems to be one of the few surviving human cultivars who in his daily work uses the pragmatic age-old exploratory strategy of observation, drawing inductive conclusions *a posteriori* by enumeration of evidence through experience. This approach now seems obsolete in science. Hempel's view from 1968, which is still representative (Hempel, 1968) purports that induction is not feasible, because it would assume a brain of unrealistically gigantic capacity with regard to judging complex data. Instead, the strategy of science is now overwhelmingly based on deductive conclusions from *a priori* hypotheses after testing (Ziman, 1978).

Science and technology have been very successful in defining the underlying functional factors in nature at different levels of organisation such as atomic particles, atoms, molecules, genes and proteins, but are still at odds with how all these factors interact, for example, to create a whole plant. The simplified reductionistic paradigm of the physicists strictly applying a bivalent logic (either-or) has dominated and severely limited the biological sciences (Prigogine and Stengers, 1984; Atlan, 1986). We may now envisage a new degree of freedom in science in measuring and evaluating biological data from different levels of organisation based on new

---

Munck, L., 2003. Detecting diversity – a new holistic, exploratory approach bridging phenotype and genotype. In: R. von Bothmer, Th. van Hintum, H. Knüpffer and K. Sato (eds), Diversity in Barley (*Hordeum vulgare*), pp. 227-245. Elsevier Science B.V., Amsterdam, The Netherlands.

technology including the computer and an increasing understanding of how human cognition of such complex structures may work.

The ruling consensus in each scientific discipline was named "paradigm" by the physicist and historian Thomas Kuhn in his classic book "The Structure of Scientific Revolutions" published in 1962 and later discussed by him (Kuhn, 1977). Now and then consensus is threatened by scientific revolutions bringing new, fundamental evidence or new explanatory models either from researchers within a discipline or, more often, by short-circuiting the consequences of local paradigms between disciplines. The revolutionary concept is absorbed and modified in the form of a changed paradigm in a slow adaptation process which Kuhn calls "normal research".

Mendelian genetics caused such a scientific revolutionary paradigm shift in plant science at the beginning of the 20th century, with N.I. Vavilov as the front figure exploring biodiversity of cultivated plants in relation to the geographical regions of the world. He identified gene centres of diversity of wild forms of the cultivated plant species. This was also of fundamental importance in understanding the evolution of the pathogens involved (Vavilov, 1951). In the 1920s, germplasm collections were established by the U.S. Department of Agriculture in Beltsville, USA, and at the Vavilov Institute in St. Petersburg, Russia. Later, these were further expanded to form the present international network of plant genebanks of cultivated and wild species (Hawkes, 1981; Bothmer, 1992; Esquinas-Alcázar, 1993). In subsequent years, new, revolutionary themes were successively stated, altering and directing plant science and breeding. These include polyploidy breeding in the 1930s, mutation breeding in the 1930-50s, nutritional protein (lysine) breeding in the 1960s, the ecological conservationist approach in the 1960-70s, gene biotechnology in the 1970-90s, the "non-food" utilization concept of plant sources in the 1980s and information technology and bioinformatics in the 1990s. The latter was essentially supported by the new personal computer working on data from data banks provided by new gene sequencing techniques. In addition to genetic engineering, other new technologies, including new machinery, fertilisers, pesticides, herbicides and now even satellite-controlled management systems (Stafford, 1999), radically changed the image of agriculture in the industrialised countries from a "low tech" production to a "high tech" status where nature is thought to be controlled, manipulated and even designed by man.

It seems as if the genebanks of the world have until now primarily been used for "low tech", "normal research" in Kuhn's sense by plant breeders, plant pathologists, botanists and classical geneticists, e.g., to identify genes for disease resistance and salt tolerance, when needed. Inspired by the behaviour of the plant breeder as a model of selection, an improved complementary "high tech", exploratory research strategy for detecting biodiversity based on screening with spectroscopic sensors will be suggested, aiming at a dialogue with nature on the organisational level of the whole plant phenotype.

Screening methods for acquiring data banks on genotypic and phenotypic diversity

Data is not free in nature. It has to be revealed and measured often using elaborate, experimental, often destructive laboratory techniques. Such experimental procedures are slow, laborious and expensive, but provide valuable methodology for a basic understanding of the underlying functional factors at hand. In studying the complexity of nature outside the laboratory in order to detect plant biodiversity, special screening techniques have to be developed which are fast, reproducible, simple, cheap and preferably non-destructive.

In contrast to physics, modern instrumental techniques in biology, like DNA sequencing and

RFLP mapping, give rise to huge, complex, multifactorial data sets from individual objects that may be exploited by computers and multivariate data analysis (Laurie *et al.*, 1992; Baldi and Brunak, 1999). In analysing the phenotype, traditional, discrete, destructive and slow laboratory chemical analyses such as Kjeldahl protein, fat, ash and fibre analyses were developed by chemists in the 19th century, inspired by the paradigm of physics far from the level of biological organisation.

In the 1960s at the USDA laboratory in Beltsville, USA, Dr. Karl Norris (see Williams and Norris, 1987) had the remarkable idea of utilising the new computer facilities to evaluate thousands of wavelengths from a near infrared reflectance spectrometer (700-2500 nm) in milled wheat samples by calibrating it to the traditional slow analyses of protein and water. He used his own version of classical statistics in which he employed the first and second derivatives. In order to overcome the problem of dealing with such a large amount of data, he selected the 5-20 wavelengths that gave the highest correlation coefficients and utilised them in calibration models by multiple linear regression analysis (MLR). These data models could then be used directly to evaluate spectral information (x-data) from unknown samples by so-called "artificial intelligence" in "black box" models for protein and water (y-data) in wheat, the results being displayed on the computer within seconds. In cereals, non-destructive near infrared transmission (NIT) instruments measuring whole barley and wheat seeds are now state-of-the-art as "multimeters" in quite accurately determining water, protein and many other chemical components in agriculture and industry as well as in plant breeding in millions of analyses each year.

A new branch of multivariate analysis in chemistry called chemometrics (Martens and Næs, 1989; Massart *et al.*, 1997; Esbensen, 2000) is now integrated in the spectral agroindustrial screening methods as well as in a range of other chemical and industrial applications. Chemometric algorithms, in contrast to classical statistics, allow for data analysis of spectra with thousands of collinear wavelengths as a whole for classification or to be calibrated to the chemical reference methods for prediction. Examples of such algorithms are Principal Component Analysis (PCA, Wold *et al.*, 1987), Partial Least Squares Regression (PLSR, Geladi and Kowalski, 1986) and artificial neural nets (Massart *et al.*, 1997), the application of which will be discussed below. Parallel to the introduction of spectroscopic screening methods for analysis of agricultural grain commodities, new satellite systems including CCD camera sensors and position (GPS) techniques have created a new, rapidly growing field within agricultural engineering called "Precision Agriculture" (see Stafford, 1999). Spectroscopic near infrared and fluorescence "crop scanners" used in the field (Wiegand *et al.*, 1990; Bausch and Duke, 1996; Baret and Fourty, 1997) are now opening up new possibilities for the plant breeder to study biodiversity not only in seeds, but also on the developing canopy level.

In fact, both NIT spectrometers and crop scanners are already used by plant breeders in a normative way for a previously defined target, for instance, to measure protein in grains or the effect of nitrogen uptake on chlorophyll. It is seldom recognised, however, that the spectroscopic techniques produce "global" results within the limits of the performance of the instruments. That is, in principle they each grasp a holistic fingerprint of the state of the physics and chemistry of the sample where "protein" and "chlorophyll" are just two of the many physical and chemical signatures that are latent in the spectra. A large proportion of this potential information thus remains unexploited. The "revolutionary" scientific question at hand in Kuhn's (1977) sense is whether an explorative, inductive approach – "measuring first and validating afterwards" – could be rewarding in the land of undefined data. Is it possible to define what is "normal" and "deviating" barley by using these screening methods in a strategy of reversed engineering

(Pinker, 1997)? Is it at all feasible and "scientific" to apply non-invasive spectroscopic and imaging selection methods in, e.g., barley genebanks and mutation materials of unknown status, and thus select in a "global" way the genetically unique part of the variation of the chemical/physiological phenotype by such observation only?

Screening biodiversity of the barley endosperm proteome

In the 1960s and 70s a wide range of so-called "high-lysine barley genes" now known to regulate the endosperm proteome pattern were identified from natural and mutational gene sources by different research groups at Svalöf in Sweden and in Denmark, at Risø and Carlsberg (see review by Munck, 1992). They used the dye-binding screening method, which employed acilane-orange for estimating the total amount of basic amino acids (lysine, histidine and arginine). Let us take a new look at this unique collection of genes, which has now almost been forgotten, and exploit the spectroscopic and imaging screening methods and the data evaluation facilities of the late 1990s.

We measure the milled samples from the 128 barley lines with a NIR Systems 6500 VIS-NIR near infrared reflection spectrometer (Foss Electric A/S, Hillerød, Denmark) between the wavelengths 400 and 2500 nm, obtaining a maximum of 2,100 data points per spectrum. In our exploratory approach, we place our prior knowledge of the material "on the shelf", letting the spectroscopic data speak for themselves by a data selection process in the computer using the Unscrambler programme (Camo A/S Trondheim, Norway; Esbensen, 2000). With the help of the PCA and PLSR algorithms, we perform a data selection process of the spectroscopic material by data reduction into latent variables in several cycles. We validate the results as graphical patterns on the computer display using our *a priori* knowledge, by chemical reference analyses and by data experimentation where we divide the data material randomly into a calibration set and a test set.

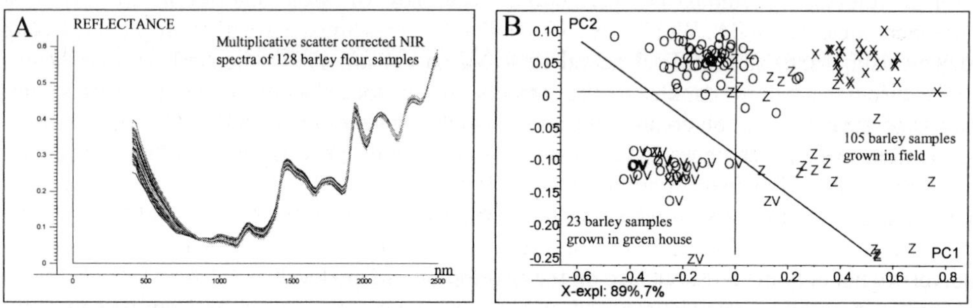

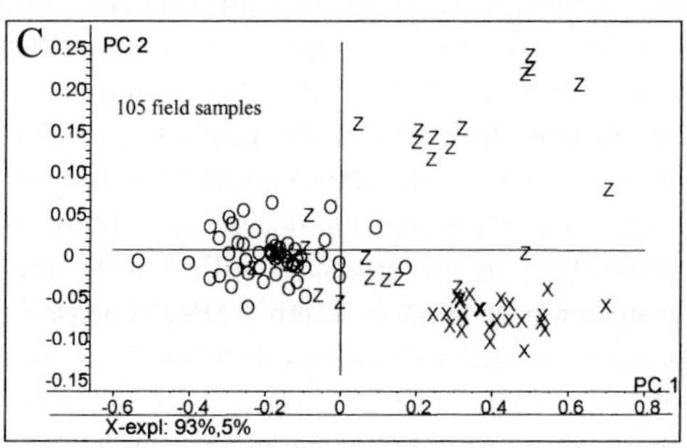

Figure 11.1. Barley seed material as measured by NIR (A) and evaluated by PCA (B and C). O = normal barley, X = high lysine *lys3a* gene, Z = other high lysine genes (see text), V = material grown in greenhouse.

The NIR spectra (from 400 to 2500 nm representing 2,100 wavelengths reduced to 1,050 data points for the 128 samples) are shown in Figure 11.1.A. The spectral measurements have a high degree of reproducibility. Therefore, even small differences in the spectral plot of samples, almost unrecognisable on visual inspection, may contain important reproducible information. In Figure 11.1.B, data from a whole spectrum is compressed to represent a point in a PCA score plot where *x*-axis Principal Component 1 (PC 1) represents 89% and *y*-axis PC 2 represents 7% of the variation. Spectra with similar patterns are placed near each other. Four main clusters are identified. Now we consult our identification list. Twenty-three samples in the left diagonal of the PCA plot were identified as having been grown in the greenhouse (marked V), while the other 105 were grown in the field. Most of the greenhouse samples were normal barley cultivars (marked O). Of the barleys grown in the field (right diagonal of the plot) in Figure 11.1.B, one group (above left) consists mainly of normal (marked O), one group (above right) is mainly crossings from Carlsberg with the extreme high-lysine gene *lys3a* (from Mutant Risø 1508) marked X. In addition, there is a smaller group with other lesser high-lysine genes marked Z below the two other groups as well as some X's intermediate to the O and X clusters.

In order to improve classification, we made a local PCA of the 105 spectra from the field lines (Figure 11.1.C). Within the three groups – normal (O), *lys3a* (X) and other mutants (Z) – there was a clear-cut classification between O and X. Some of the Z lines make their own cluster, but others are placed centrally in the plot, invading the O and X areas from left to right and from above to below, respectively. In evaluating the classification of the Z's again and consulting our field list we note that the Risø mutants 13, 16, 29, and 95 are clearly separated from the other classes, while the Carlsberg mutant in Nudinka No. 11 is located in the *lys3a* X area. Z samples in the outskirts of the classification of the normal barley area O are Svalöf gene selection *lys1* (Hiproly) and Risø mutant 7. There are numerous differentiation opportunities in this PCA investigation for plotting other principal component (PC) combinations in new diagrams in order to expose new aspects of classification.

The 128 barley lines represent widely differing protein contents and amino acid compositions due to genetics and environment. We expect that this will be reflected in NIR spectra by a varying content of nitrogen bonds such as R-NH2, AR-NH2, RNHR, CONHR and CONH2. In order to calibrate the NIR spectra to nitrogen data in a Partial Least Square Regression (PLSR) analysis, we decided to analyse the samples for nitrogen (N) and amide content (alkali volatile nitrogen), representing a general and a specific aspect of nitrogen bonds.

In the PLSR NIR analysis with protein ($N \times 6.25$) in Figure 11.2.A, the algorithm selects principal components in the form of mean spectra – *x* (loadings) representing different aspects for the prediction of protein (*y*). Validation is carried out by performing data experiments in the form of segmented cross-validation (Martens and Næs, 1989) and the regression coefficient *r* and error RMSEP are calculated. It is clear from Figure 11.2.A that two outliers from the prediction correlation can be identified. They are marked O and OV. The measured protein values were 7.5 and 8.3, while the predicted levels were 14.3 and 15.2. A new Kjeldahl analysis revealed an analytical error and the values were corrected to 15.1 and 15.7, which were near to the predicted results in Figure 11.2.A. The PLSR was, therefore, plotted again using the corrected figures (Figure 11.2.B). The correlation coefficient is improved and the error reduced. The possibility of critical outlier detection is fundamental in the establishment of a reliable multivariate analysis. One can see that greenhouse samples (V) are nicely included in the PLSR prediction of protein together with the field samples. The former have a high protein level.

A PLSR NIR correlation with the amide nitrogen (CONH2), also including small amounts of NH3, is presented in Figure 11.2.C showing a reasonable prediction. Clearly, the high-lysine

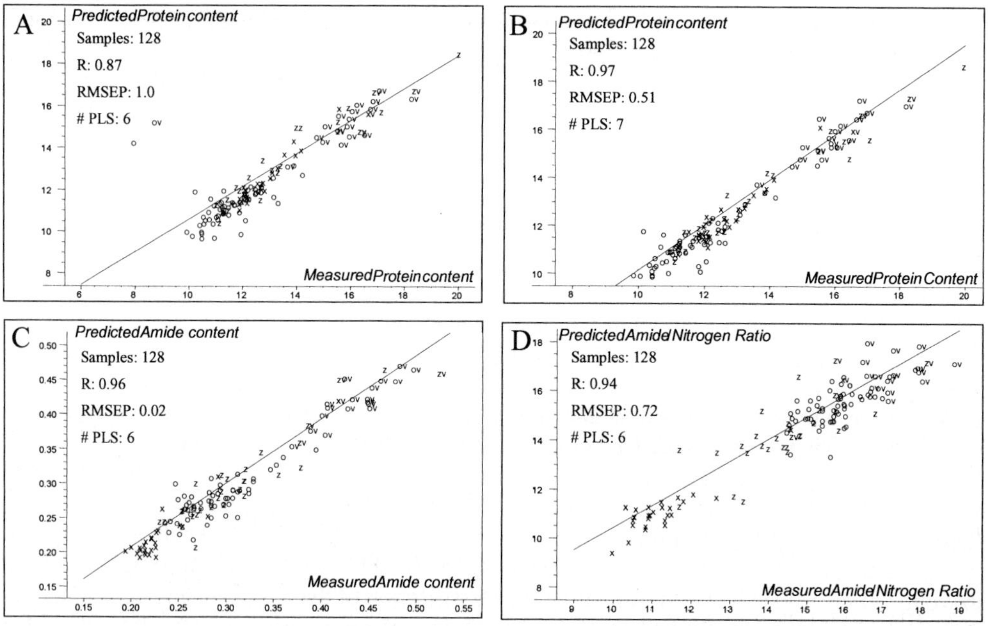

Figure 11.2. PLSR of the spectra in Figure 11.1 for prediction of protein (N×6.25) % d.m. (A and B), Amide N % d.m. (C) and Amide N to N ratio % d.m. (D).

lines and especially those with the *lys3a* gene (X) have low amide values compared to the normal lines (O). In order to investigate if a more clear-cut PLSR classification can be made in this material containing different high-lysine genes, a PLSR regression analysis was performed employing the amide nitrogen-to-nitrogen ratio as y and with NIR spectra as x. A similar result for classification of the barley lines was obtained in the PLSR plot in Figure 11.2.D as in the PCA analysis of the NIR spectra in Figure 1C. The *lys3a* lines (X) in Figure 11.2.D can be regarded as an extreme (here low in amide N/N ratio), with the other high-lysine lines (Z) as intermediate compared to the normal cultivars (O), which are high in this ratio.

It is clear that the amide content and especially the amide N/N ratio are reliable criteria for classifying the different classes of high-lysine and normal barley lines. Here we have shown that it is possible to obtain similar classifications with non-destructive near infrared reflection spectroscopy (NIR) which should reflect the propensity of different chemical groups (bonds), such as the amide residue. In trying to reveal how the NIR technique works, various methods can be applied to select wavelengths (x) that are especially highly correlated to y's like protein (N×6.25) and amide N. For this purpose, we carried out another experiment and extracted three submaterials: A. normal barley (n=72), B. *lys3a* barley (n=27) and C. normal + *lys3a* barley (n=99) and employed the (interval) i-PLS algorithm (Nørgaard *et al.*, 2000). We are thus testing wavelength regions in NIR spectra one by one for correlation with y. The algorithm then adapts to the region with highest correlations to the surroundings when trying to find the continuous area with the lowest error of prediction. It is noteworthy that errors in correlations with the *lys3a* submaterial are reduced by 50%, in spite of equal range in N and amide N.

We divided the spectra into 30 regions (1-30), each representing 35 data points and 70 nm. It

Table 11.1. Interval PLS identification of wavelength regions correlated to Nitrogen and Amide Nitrogen of the material in figures 11.1 to 11.4.

	Protein					Amide Nitrogen				
Material	Optimal interval (nm)	*r*	# PLS	RMSEP	Selected regions	Optimal interval (nm)	*r*	# PLS	RMSEP	Selected regions
Normal	1230-1302	0.98	4	0.45	13,20,19,26	1920-2004	0.97	4	0.016	23,19,26,20
lys3a	1434-1584	0.98	3	0.24	19,24,18,16	1934-2006	0.98	4	0.008	23,19,29,26
Normal + *lys3a*	1220-1285	0.97	4	0.47	13,20,26,19	1916-2008	0.98	4	0.014	23,19,20,16

was found (Table 11.1) that in all three submaterials with the NIR prediction of the specific form of nitrogen – amide, i-PLS selected very similar primary wavelength areas: 1920-2004 nm (A), 1934-2006 nm (B) and 1916-2008 nm (C). In all cases the number of principal components involved (3-4) was lower in the local than in the global calibration with the whole spectrum (6-7) in Figure 11.2.C, indicating a less complex model. In contrast, the wavelength area selected by i-PLS for best correlation to total nitrogen was highly variable between the high-lysine material *lys3a* (B) 1434-1584 nm on one hand and the normal (A) 1230-1302 nm and the combined material (C) 1220-1285 nm on the other. From Table 11.1 it is also evident that in the submaterials with different N composition PLSR relies on different orders of priority of the wavelength regions (here numbered 1-30) indicative of different forms of N in order to predict total N (protein). Of the four most important regions in the correlation with N, i-PLS has selected regions 24, 18 and 16, which are unique for the *lys3a* material, while regions 13 and 20 both have first priority in the two materials containing normal barley. The detection of amide N, which is more specific, is consequently simpler, in that all three materials rely on the same first and second priority wavelength area (regions 23 and 19). Region 19 is also involved in all three materials to predict total N.

The physical/chemical explanation and the underlying empirical basis to NIR spectroscopy are reasonably well established. Primary, clear-cut vibration bands characteristic of specific chemical bonds can be identified in the infrared region above 2500 nm. This information is repeated in the near infrared region as first, second, third and fourth as well as combination overtones with wavelengths decreasing from 2500 to 700 nm. The interpretation becomes more complex at higher overtones and decreasing wavelengths. Thus, in the region of 1916-2008 nm, which was selected by i-PLS for correlation to amide nitrogen, there are first and second overtones for amide bonds in protein such as carbonyl in amide at 1920 and 1950 nm and NH in amide at 1960, 1980 and 2000 nm. In finding priority areas for wavelength correlation for nitrogen, i-PLS selected additional primary information at lower wavelengths complementary to the amide information at higher wavelengths.

In a recent publication (Munck *et al.*, 2001) we have demonstrated that a small wavelength area between 2290 and 2360 nm provides unique spectroscopic signatures for the *lys3a* gene and the wildtype, while environmental effects are registered as offsets from the baseline. The alleles to the Risø 1508 mutant in the *lys3* locus Risø mutants 18 and 19 and Carlsberg mutant 1460 display similar spectral forms.

As we have demonstrated, covariate criteria and spectral signatures starting from simple ratios between chemical analyses at the micro level (e.g., between amide nitrogen and nitrogen) to global evaluation in a PCA plot representing correlations between thousands of spectral wavelengths at the macro level, are much more indicative of genetic variation and less dependent on environmental factors than univariate criteria like protein and starch content, as analysed with the classical statistical analysis of variance.

The upgraded plant breeder as a model for fundamental explorative strategy in science

In rational problem solving which requires the creation of a system of learning by exploration (Munck, 1972), e.g., from biodiversity, it is essential to systematically follow a cyclic sequence of events. One could call this the selection cycle (Figure 11.3; Munck, 1991, 1992, 1993), starting with a general survey (I) based on inventories including literature, interviews and observation in order to formulate a provisional primary selection hypothesis in the global area. Thus, a few scientists and plant breeders in Sweden and Denmark in the 1960-70s, inspired by the public discussion at that time, decided to breed for a better amino acid (lysine) balance in barley to improve its nutritional value. For this purpose, they created a screening method (II) with dye-binding, which was applied together with Kjeldahl analysis for selection (III) to the world barley collection and mutation materials. They found the *lys1*, *lys3a* and a range of other genes discussed above which were tested in yield trials (IV). However, the yields were low and seed quality poor; therefore, most breeders were discouraged from continuing or even starting. Scientists (see review by Nelson, 1979) observing that high-lysine genotypes generally had a strongly reduced level of storage proteins (prolamines) combined with reduced starch content

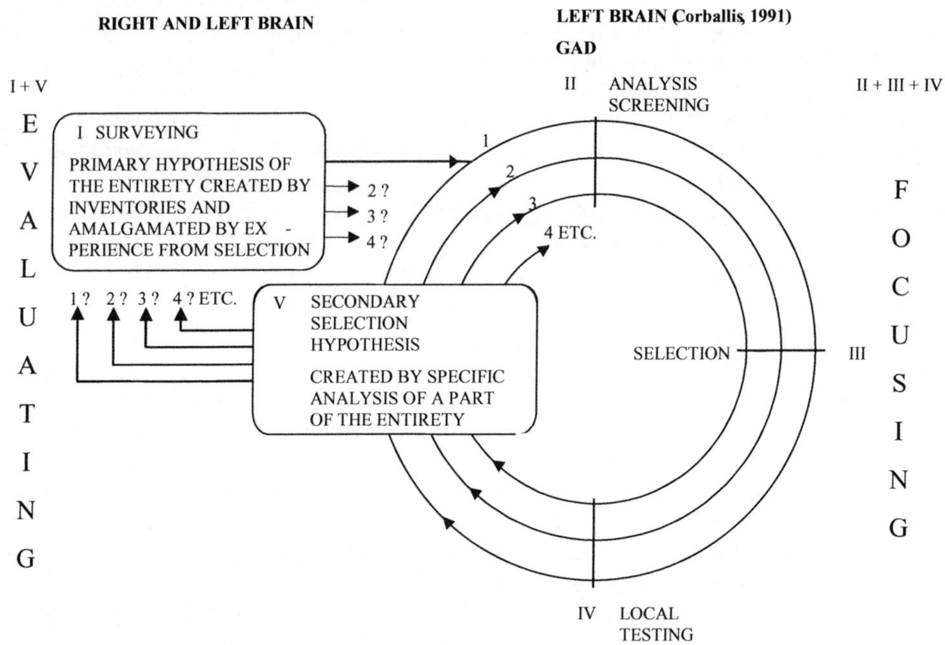

Figure 11.3. The selection cycle (Munck, 1991).

rationalised this drawback as a marker of an inevitably low yield. Mertz (1976) summarised the situation as "the high-lysine gene syndrome", implying an almost pathological situation.

Since then, two serious explorative attempts have been made to improve yield based on trust in the variability of nature to find genes which could correct the negative traits apparently pleiotropic to these genes. A secondary hypothesis (V) in the local area of the selection cycle was formulated, claiming that it was possible through cross-breeding to find gene backgrounds for correction of the yield depression and seed quality due to high-lysine genes. Such programmes were carried out in *lys3a* barley at Carlsberg (Munck, 1992) and in opaque-2 maize at CIMMYT (Vasal, 1999). In barley, at Carlsberg, traditional plant breeding methods based on extensive crossing in several selection cycles (Figure 11.4) were performed over a period of 15 years at a relatively low but steady input of one person per year. At the end of the 1980s, *lys3a* lines with equal or better yield than the control (a mixture of the four highest yielding commercial lines) were obtained in consecutive years (Munck, 1992). These lines had improved seed quality and starch content. Recently, in the 1999 harvest, the yield is about 10 percent lower than that of the controls due to lack of maintenance breeding since 1990. In opaque-2 maize breeding, high yielding competitive lines were obtained at the end of the 20th century. The soft opaque condition which provided less insect resistance has been changed to a hard endosperm type, still maintaining the improved amino acid composition trait (Vasal, 1999).

Thus, the flexibility of nature was confirmed, allowing corrections of the pleiotropic expression of a specific gene and still maintaining quality by empirically adopting a suitable gene background. However, now, years later, when consulting the primary hypothesis area in the selection cycle to find out whether the goals have been met, it seems that the protein issue of the 1960s has largely been forgotten. Protein and lysine additives are cheap today and there are no control methods for lysine used in the feed factories (Munck, 1992), even if such methods could easily be adapted to existing NIT spectrometers for analysing protein, which also work with the material presented here. However, there seems to be some interest in high-lysine maize in developing countries (Vasal, 1999).

In parallel, the idea of creating biodiversity by genetic manipulation working in the closed normative loop of the focused local secondary area of the selection cycle is attracting increased attention (Wettstein 1983; Shewry *et al.*, 1994). Thus, genes for new high-lysine proteins with 43% mol/mol lysine have been expressed in transgenic tobacco seeds (Keeler *et al.*, 1997), and in rapeseed the free content of lysine has been engineered to produce a 5-fold increase in total lysine content (Falco *et al.*, 1995). There is, however, still the question of how these results could be transferred to cereals and how much correction work has to be done by conventional plant breeding in order to obtain high yield with such genes which are not adapted to the cereal genome. Results from the breeding work in maize to include genetically engineered herbicide resistant genes in high yielding maize backgrounds demonstrate that such a project may be feasible, but that it will certainly take time and effort. In addition, one has to consider the public's increased scepticism towards transgenic plants, which should make inventories of genebanks by the new spectroscopic methods attractive. The spectroscopic screening methods should also work with transgenic breeding programmes for quality characteristics and substantially increase the efficiency of screening. It may be concluded that there was bias in the expectations and scope of the high-lysine programme in the 1960s, which contributed to the limited breakthrough of the research in practice, although there were important spin-off results.

The selection cycle thus embodies the necessity to secure a systematic dialogue between the primary global (society) area and the secondary local (laboratory) area, building up a chain of knowledge to optimise development with limited resources. The so-called "green revolution", which between 1972 and 1982 increased grain production in developing countries by 33% (Borlaug, 1981; Bosemark, 1993), is still one of the best, if not the best, example of the success of optimising this dialogue at agricultural, institutional, commercial and academic levels and where biodiversity has played an important role.

Corballis (1991), in addressing the neuro-psychology of the brain, has an explanation for the increasing dominance of normative research in present-day science. He considers it as a bias generated from the virtual reality of the brain, which we should take into account and compensate for. The vertebrate brain, which originally was symmetrical, has in humans evolved asymmetrically split functions in order to enlarge the specific computing capacity. The left brain embodies language by speech and sign (symbolic and propositional), reading, writing, recognition of man-made forms as well as rhythm and mathematics. The right brain records but is silent, and displays functions for pattern recognition of natural forms such as the texture of a plant canopy, spatial perception, non-symbolic mental imagery (which is analogue and picture-like) as well as a sense of intuition. Most human individuals follow this scheme, although there is a significant biological variation in humans with regard to the flexibility in the cooperation between the left and right brain. Our consciousness arises in a dialogue between these centres and with the whole brain. Corballis combines the skills of the left brain in the generalised concept of the **Generative Assembling Device (GAD)**. This is an analytical, forward-directed function, which makes decisions on exploiting our surroundings that allow a measure of purpose and planning. The influence of GAD may be seen in our theories of mind and matter where we can imagine real, potential and impossible objects. It helps us to generate a virtual reality. GAD is biologically programmed but to be developed it has to be taught. Its ruthless and restless construction activity and repetitivity means it opposes true creativity. Corballis claims that through the influence of GAD the natural environment is giving way to an artificial environment in the modern world. Nature is not as much harnessed as it is dominated. We can now identify GAD's place in the selection cycle (Figure 11.3) in the focused secondary local selection area in contrast to the primary global survey area in which there should be broader involvement of the right brain. The main issue now is whether it is possible to balance the normative focused bias of GAD localised in the left brain by supporting the analytical holistic skills of the silent right brain, by giving this region the ability to speak for itself using the tools of modern technology.

It is evident that man (and woman) as selector (Munck, 1991, 1993) in the selection cycle model is limited by physical resources, the abundance of (bio)diversity, the complexity of the selection criteria, the number of selection cycles (equal to experience), the efficiency of the testing process and communication with other selectors.

Here the ability of multivariate evaluation is crucial, as it involves the whole brain in a dialogue with nature. The PLSR algorithm is in itself a cyclic procedure (Massart *et al.*, 1997), which combines explorative x (spectra) and normative y (protein) elements. Holistic screening methods (e.g., spectroscopy) evaluated by the computer and exploratory software may thus be looked upon as an **Outer Exploratory Cognitive Device (OECD)** – an extension of the (plant breeder's) senses to trace (bio)diversity, a lifeline to the real world – with the task of keeping the bias of GAD under control when driving the selection cycle.

Our brain is a great achievement of nature with a capacity of one thousand billion neurons and one hundred thousand billion synapses, an enormous potential indicated by our limitless fantasy exhibited in GAD. However, we have not yet been able to fully utilise it in keeping in contact with the outer world (Corballis, 1991). The limits of the brain seem to reside on the input and output side and in the "software" which is necessary for communication and validation. Thus, the long-term memory of the human brain is able to store the meaning of 100,000 words passively and 10,000 actively (Corballis, 1991) while the short-term memory can only handle up to 20 letters per second (Nørretranders, 1991). When inspecting, for instance, a field plot of barley, we are receiving about hundreds of megabytes per second through our sight, which has the highest data load of all the senses. The brain is, therefore, constantly unconsciously involved in a substantial data reduction process, turning the overflow of data into information by using different forms of pattern recognition (Corballis, 1991; Pinker, 1997), which in our conscious generative mind – GAD – is turned into problem reduction with all the bias of too narrow a focus.

With the help of an OECD comprised of spectrographs and image analysers including data evaluation we might extend the vision of the scientist and breeder several thousand times on a wavelength basis by transferring several millions of bits per second to the memory of the computer. The crucial problem now is how the essence of data libraries with thousands of megabytes in the OECD can be explained to a brain with such a narrow bandwidth of communication and with a limited repertoire of recognizable linguistic symbols. The input from the OECD should be introduced in the first phase as a support to the observational holistic pattern recognition ability of the right brain, keeping prior knowledge "on the shelf". In the second phase, a validation dialogue is established with the generative executive function of the left brain through GAD. Obviously, the initial phase has to be done separately from GAD through the broadest channel – the sight – by data reduction into graphics by means of the computer, as in principal component analysis (PCA) demonstrated in our barley example above. The conclusions from the data then have to be translated into terms of scientific and common language. Known functional factors should thus be identified by name and new linguistic symbols created for the new ones by interpreting the latent principal components of the multivariate analysis. All applied mathematical operations should thus start and end with an exercise in language (Michod, 1999) to define and describe the probabilistic/deterministic character of the latent and manifest elements inherent in the multidimensional context, respecting the different levels of organisation and the context of the investigation (Atlan, 1986). In most cases at present, fragments of such a strategy are only tacitly included in the statistical analysis.

The mathematical languages of different scientific cultures

Before the computer, the limits of human cognition in handling complex multivariate problems discussed above had created a range of scientific cultures working at different levels of organisation of matter (Barrow, 1993). Consequently, each of those cultures are specialised with regard to mathematics and strategy (philosophy), most of them being based on problem reduction engaging GAD central in physics. The introduction of the computer with new software, enabling new ways of thinking, including data experimentation comparing measurements of different levels of organisation, is now changing this situation (Barwise and Etchemendy, 1998). However, this development is slow, because the computer is still most often programmed and used according to the inflexible way we had to think before we had the computer.

Here we can envisage at least three different cultures in applied mathematics within biology: (1) A probabilistic culture derived from game theory, insurance business and agricultural experimentation, which now prevails in genetics and bioinformatics (Baldi and Brunak, 1999) and is also a strong tendency in the social sciences (Gigerenzer and Murray, 1987). (2) A hard modelled culture based on fundamental determinate models (e.g., with differential equations) which is used in problem reduction in, for instance, agricultural engineering and precision agriculture (Stafford, 1999) and in environmental sciences together with the probabilistic strategy (Kingsland, 1995; Michod, 1999). (3) The third culture merging the two above includes multivariate analysis and chemometrics. It is basically a form of experimental mathematics in which the computer assists in focusing on covariate data structure, e.g., by using vector algebra displayed as a PCA on a graphic interface for exploration (Wold *et al.*, 1987) or as a PLS regression (Geladi and Kowalski, 1986; Martens and Næs, 1989) and Artificial Neural Net algorithms (Massart *et al.*, 1997), which is also used in bioinformatics (Baldi and Brunak, 1999).

The explorative strategy should be seen as an approach separate from these three cultures of applied mathematics. Hard and soft algorithms from all three cultures may be used with varying success in both an exploratory and a normative way, depending on the influence of an OECD or a GAD strategy. A few scientists associated with plant breeding (see review by Romagosa and Fox, 1993) have for a long time argued for a multivariate approach to model the covariate characteristics unique for each genotype, e.g., by PCA and the computer. The main problem in this approach has been in obtaining data of high quality, which is difficult when focusing on complex phenotypic environmentally dependent traits like genotypic stability of yield under favourable and stressed conditions. Earlier we indicated that an intermediate approach to data collection could be profitable in obtaining high quality data: the genetic information could be addressed at the phenotype level using a fingerprint of the physics and chemistry of the biological machinery via an OECD such as spectroscopy. Let us, therefore, explain in terms of language the tacit qualities inherent in the barley experiment described above.

Consider in a thought experiment two plots, each with 100 homozygotic plants of a normal barley 'Mandolin' and a derived *lys3a* segregant in F6 from a cross with the high-lysine mutant Risø 1508. The variation in earlier generations on the gamete and chromosome cross-over level is highly probabilistic, the evaluation of which involves algorithms like Chi Square. When analysing the seeds from 100 homozygotic plants from each plot for protein content which is highly environmentally influenced, we can see a typical normal distribution curve with a slightly different mean for each plot ideal for a variance analysis based on probability. However, when we analyse each plant for fibre, beta-glucan, and protein we find that even if each of these parameters are normally distributed within a plot, they are dependent upon each other. We can now identify two mini-patterns in a PCA analysis of the chemical parameters from the 2×100 plants with two main groups and some overlapping. In a classical multiple linear regression analysis (MLR) within each plot between protein, fibre and beta-glucan content of the different plants there are characteristically different slope coefficients for the two barley lines, apart from the fact that 'Mandolin' on the average is higher in beta-glucan and lower in fibre and protein compared to the *lys3a* segregant. After taking a fingerprint of the state of physics and chemistry of the seeds from the 2×100 plants using NIT spectroscopy, a PCA plot evaluating the information from whole spectra shows two distinct groups with no overlapping, signifying the genetic diversity at the chemical level. MLR is not applicable here in this large data set, unless a few variables are selected. Interestingly, when considering the issue of sampling, proper sampling is of paramount importance if the purpose is to obtain a representative number of plants reflecting the grain protein content of the seed yield from the plots; on the other hand, sampling is unimportant if the task is to identify using NIT or by multivariate chemistry if a plant is normal barley ('Mandolin') or a *lys3a* recombinant. Thus, in inspecting the same material, we have changed focus from a general, univariate and normal distributional view to a specific, multivariate and pattern recognition concept involving latent principal components. The mathematical tools (analysis of variance and principal component analysis), including the validation strategy, have changed correspondingly.

Modern statistics based on maximum likelihood cannot handle highly covariate data sets (like spectra) where the number of measurements (wavelengths) is high in comparison to the number of objects. Taking in many measurements in multivariate and multiway analysis often increases the specificity of the description of the object and facilitates classification of a normal population and its outliers (Bro, 1997, 1998, 1999; Munck *et al.*, 1998). This is the prerequisite for an explorative, inductive strategy with an OECD to be successful. Classical probabilistic statistics, e.g., analysis of variance, is based on normal distribution criteria and on an *a priori* convention of validation, which multivariate analysis lacks to date. Validation in PCA and PLSR is on an *a posteriori* process of data experiments based on internal relations in the data set between, for example, an arbitrarily chosen calibration set and test set to validate the model automatically chosen by the algorithm. There is no doubt that a skilled statistician and a chemometrician, both standing on their own platforms of tools, will be able to correct for much of the bias built into their respective methods. The risk rests with the users who might not have the insight for such flexibility and may turn to a fundamentalist approach. The PLS regression analysis in Figure 11.2, where protein is predicted employing whole NIR spectra, does contain tacit assumptions which could be regarded as a general hypothesis, e.g., the limitations in NIR spectroscopy, that Lambert Beer's law is valid, and that PLSR prescribes a linear model. In spite of these limitations, PLSR models are rather robust and are able to model non-multiplicative non-linearities by combining several combinations of principal components. In practice, all three mathematical cultures are rather liberal in trying out algorithms on data material using tacit assumptions, which are not always checked. The end result, if it makes sense, is more important than strictly upholding a formal basis founded on the original hypothesis.

PLSR implies a trust that the world is under indirect observation and that the data material can be reduced to latent principal components or functional factors, which may be identified by name (language). In introducing unknown samples measured by NIR, PLSR calculates the predicted protein value according to the model, implying by induction that the unknown sample follows the model. This may be checked by the operator in two ways. First, by the outlier control (Figure 11.2.B) where the multivariate NIR fingerprint is used to characterise whether the unknown sample belongs to the calibration, and secondly, by chemical analysis. Due to the multivariate, unique characteristics of an NIR spectrum cast in covariance, it is now possible in an OECD, including PLSR software and the computer, to obtain a **qualified inductive validation** (Munck *et al.*, 1998). This resembles the performance of Hempel's (1968) hypothesised brain for making trouble-free induction, discussed above, which was considered unrealistic at that time. The outlier control in which the computer automatically checks whether or not the characteristic multivariate fingerprint of a sample belongs to the model (the normal population) is the scientific basis for the revival of induction on a large scale now made possible by new software and the computer. This technology is now used worldwide to screen for quality in, for example, agricultural and pharmaceutical products, but it remains largely unused as an exploratory tool in other areas of science, such as in plant breeding and precision agriculture. It is now possible to acknowledge **the multivariate advantage** with regard to identifying biological objects that are more differentiated compared to those of quantum physics, and respect that these two levels of organisation must be represented and described mathematically in quite different manners.

In classical atomistic science, the establishment of a clear-cut cause-effect response based on bivalent logic is in focus. Such a strategy is necessary when lacking detailed data to define individual objects. However, this is not the case in biology. Respecting covariance at the organisational level of the phenotype means a new paradigm shift in a Kuhn sense (Kuhn, 1977) acknowledging a self-organising dynamic network (Atlan, 1986; Prigogine and Stengers, 1994) of morphological and biochemical components with primary, secondary, tertiary, etc., relationships leading to clusters of partial cause-effect responses which can now be explored by multivariate screening methods with the assistance of the computer. It is questionable whether it is possible or even desirable to develop a new polyvalent logic that is able to formalise and automate the selection patterns in, e.g., a PCA plot, when "global" data sets like spectroscopic fingerprints of the physics and chemistry of barley seeds are evaluated. This is now performed as a technology on an empirical, local, specific basis, in order to model covariance for establishing reliable cause–effect relationships between the different levels of organisation – the spectral phenotype, the chemical phenotype and the genotype – as demonstrated in this chapter. Barwise and Etchemendy (1998) have demonstrated a similar case based on visual interpretation, where data analysis is fundamental in revealing the everyday logic of dynamically explorative humans at work. Compared to interpretation by formal logic, such an investigation will offer a far richer scope and relevance in addressing the demands imposed by the level of organization and context of a complex biological data set.

It is now obvious that the hitherto neglected covariance between functional factors measured at different levels of organisation is essential in understanding the process of life and how to detect biodiversity.

Explaining and utilising the identified genetic diversity for new breeding goals in strengthening the production chains of barley

Recently, the focus on organisation in biotechnology is changing from gene sequencing to gene expression, as expressed in the concept of the proteome to describe the complement of proteins expressed by a cell. In a special section of the journal Nature (Dec. 16, 1999), this is described as "The post-genome revolution". We may thus conclude from the evidence collected in this chapter that this revolution will be facilitated by an OECD, implying new, non-destructive spectroscopic mass screening methods evaluated by multivariate analysis. This would imply a widening of the horizon of the plant breeder as well as the biotechnologist in detecting biodiversity on the proteome level.

But how can simple ratios and spectra reflect complex, genetically based changes in amino acid and protein composition? The answer lies in the covariate organisation of phenotypic life phenomena, as revealed in multivariate analysis of data from our barley example. As early as in 1930, Bishop in his nitrogen regulation principle revealed that the content of storage proteins (prolamines and glutelines), albumins and globulins could be predicted from the amount of protein (N×6.25) in barley. When the prolamines – low in lysine and essential amino acids but high in amides, glutamic acid and proline – increase as a percentage of total protein, the lysine content as a percentage of total protein declines (Postel, 1956). This phenomenon, which used to be considered almost as a natural law, is demonstrated by the regression line for normal barley in Figure 11.4.A. Eighteen barleys representative of the material presented earlier have been analysed here for lysine (mol %) and correlated to protein (N×6.25). However, as shown in the same figure, the introduction of the high-lysine genes falsifies this "law" by introducing several levels of outliers above the regression line for normal barley.

In the search for an explanation of how NIR spectroscopy and chemometrics may identify different high-lysine barley gene phenotypes, we can consult an NIR PLSR calibration with lysine percentage in Figure 11.4.B, demonstrating a correlation coefficient of $r = 0.96$. However, as shown in Figure 11.4.C, the correlation between the chemical analysis lysine mol % and the amide N to N ratio reveals an equally high correlation coefficient indicating that the NIR calibration for lysine is based on an indirect correlation with amide content. This observation is supported by the fact that the amide bond is a major source of information for spectral classification in this material, as indicated in the discussion of Table 11.1 and Figure 11.2.

It can be concluded that the amino acid pattern changes with increasing protein content within narrow limits in normal barley, although minor consistent variations on that theme can be detected among normal cultivars. The introduction of the high-lysine genes radically changes this pattern to other new consistent patterns. Thus, *lys1* (from Hiproly) and *lys3a* (from Risø 1508) show the same tendency of change in amino acid pattern relative to normal barley with regard to most amino acids – with increases, e.g., in lysine, aspargine, glycine and alanine and decreases in glutamic acid, proline and amide content (Munck, 1992; Munck *et al.*, 2001). The quantitative differences in relation to normal barley are, however, more pronounced with the *lys3a* genotype. There are, however, three amino acids, which behave qualitatively in different ways. In *lys1* isoleucine, leucine and phenylalanine are all increased relative to normal barley while they are decreased in *lys3a* lines.

The mechanism behind these small and large regularities of amino acid patterns due to single barley genes, most of which have been selected by the dye-binding method, can now be interpreted in detail as changes in protein patterns mediated by mutations in both regulatory and structural genes (see Mundy *et al.*, 1986; Leah *et al.*, 1991; Kreis and Shewry, 1992; Munck, 1992; and Blom Sørensen *et al.*, 1996). Thus, both *lys1* and *lys3a* are regulatory genes localised in chromosome 5H. They guide the synthesis of a range of proteins from structural genes in other chromosomes. As an example, the *lys3a* gene regulates a drastic reduction in both B and C hordeins low in lysine, resulting in a reduction of the protein bodies as seen in the electron microscope (Munck and Wettstein, 1976) and in a simultaneous increase in a range of different antimicrobial proteins such as chitinase, beta-glucanase, as well as subtilisin and protein synthesis inhibitors which are all high in lysine. In contrast, Risø mutant 56 is classified as a structural mutant. It has a deletion in the gene *Hor2* resulting in complete elimination of the B

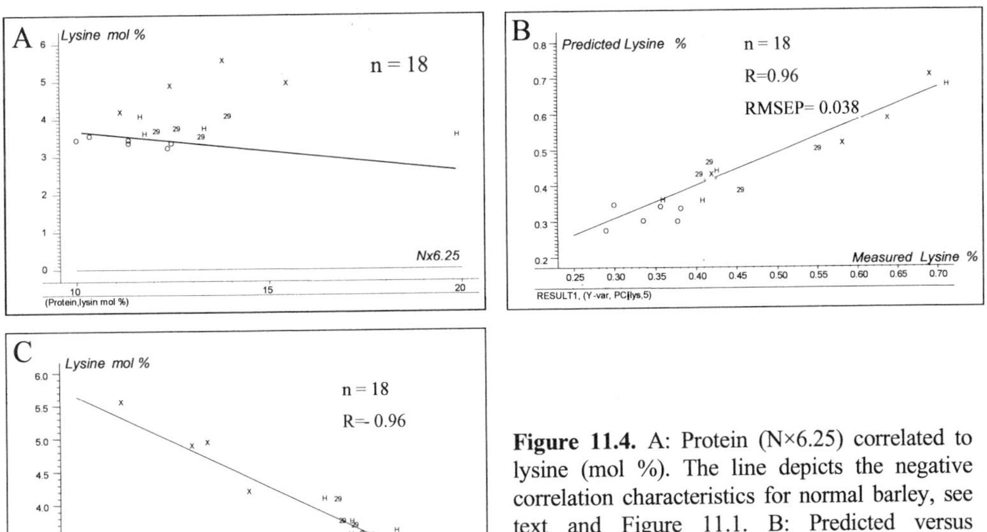

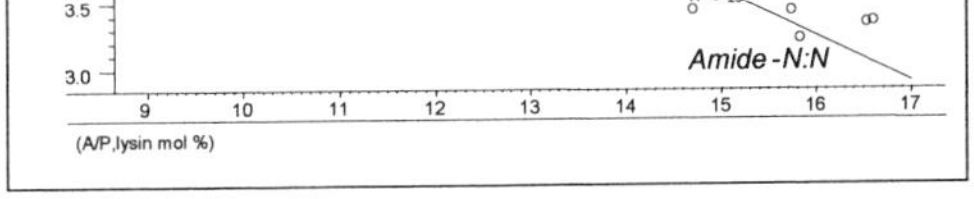

Figure 11.4. A: Protein (N×6.25) correlated to lysine (mol %). The line depicts the negative correlation characteristics for normal barley, see text and Figure 11.1. B: Predicted versus measured plot of a NIR calibration for lysine %. C: Ratio Amide N to N correlated to lysine (mol %).

hordeins.

Unfortunately, most of the high-lysine Mendelian genes selected at Risø and Carlsberg from mutation materials, using the dye-binding method still available, have not been studied as far as their molecular biology is concerned. No scientist to date seems to have motivated their GAD to utilise the unique endosperm model to study gene regulation in eucariotes, exploiting the fact that the endosperm tissue outside the germ line should be less sensitive to drastic mutations in regulating genes, which may not have survived in the germ line.

The barley endosperm could be looked upon as a cast form of proteome dynamics. When studying proteome development and regulation in, for instance, a canopy of barley plants, it might be more difficult to isolate the genetic from the environmental variation, also including direct effects due to changes in sunshine, climate and different pests. Looking at a 10 square meter barley plot using a fluorescence/near infrared spectrograph, data should basically contain a global fingerprint of the ongoing physics and chemistry which with modern chemometric methods could be calibrated to important morphological and chemical (proteome) characteristics of the plant (Munck *et al.*, 2001), assimilation efficiency, nutrient availability and infection of pests. A range of spectra collected during plant growth from the whole canopy and from different tissues could be calibrated to yield, as well as to scores from plant breeders and observations, and to physiological (proteome fingerprints) and genetic data. Spectrofluorometric landscapes displaying a high degree of specificity could be deconvoluted by "mathematical chromatography" into the excitation and emission spectra of the underlying unique fluorophores as interpretable loadings by the multiway algorithm PARAFAC (Bro, 1997, 1998, 1999; Munck *et al.*, 1998). In such a study in assembling an OECD, spectral libraries and supporting data could be built up in computers which would enable us, with the help of multivariate analysis, to explore and better understand gene regulation of plant growth processes in relation to yield and genotypic stability (Romagosa and Fox, 1993) and to optimise the use of biodiversity in plant breeding.

Conclusions and outlook

We have described a new, interdisciplinary technology of broad scientific relevance. The application of this technology in the detection of biodiversity in barley seeds is demonstrated in two steps, first by carrying out measurements using "global", highly reproducible spectroscopic screening analyses to create large covariate data sets, and then exploring the data with mathematical experiments using multivariate, chemometric algorithms in the computer.

The new approach is able to record biodiversity as a characteristic covariate, physical/chemical spectral fingerprint of the barley seed phenotype reflecting the genetic variation in chemical composition, including recording significant differences in proteome patterns and their consequences. Specific information on the biological organisational level (macro) from the "global" data set can be selected and modelled by calibration to the chemical organisational level (micro) for validation. The application of this technology on the canopy level to understand and control the realization of genetic diversity in plant growth and production would be a difficult but most rewarding challenge.

References

Atlan, H., 1986. A tort et a raison, intercritique de la science et du mythe. Éditions du Seuil, Paris.

Baldi, P., and S. Brunak, 1999. Bioinformatics. Massachusetts Institute of Technology, USA.

Baret, F., and T. Fourty, 1997. Radiometric estimates of nitrogen status of leaves and canopies. In: G. Lemaire (ed), Diagnosis of the Nitrogen Status in Crops. Springer, Berlin, pp. 201-227.

Barrow, J.D., 1993. PI in the Sky – Counting, Thinking and Being. Oxford Univ. Press.

Barwise, J., and J. Etchemendy, 1998. Computers, visualization, and the nature of reasoning. In: T.W. Bynum and J.H. Moor (eds), The Digital Phoenix: How Computers are Changing Philosophy. Blackwell Publishers, Oxford, pp. 93-116.

Bausch, W.C., and H.R. Duke, 1996. Remote sensing of plant nitrogen status in corn. Trans. Amer. Soc. Agric. Engineers 39: 1869-1875.

Blom Sørensen, M., M. Muller, J. Skerritt and D. Simpson, 1996. Hordein promoter methylation and transcriptional activity in wild-type and mutant barley endosperm. Mol. Gen. Genet. 250: 750-760.

Borlaug, N.E., 1981. Increasing and stabilising food production. In: K.J. Frey (ed), Plant Breeding II. Iowa State Univ. Press, Ames, pp. 467-492.

Bosemark, N.O., 1993. The need for a comprehensive plant breeding strategy. In: M.D. Hayward, N.O. Bosemark and I. Romagosa (eds), Plant Breeding – Principles and Prospect, Chapman and Hall, London, pp. 535-540.

Bothmer, R. von, 1992. The wild species of *Hordeum*: Relationships and potential use for improvement of cultivated barley. In: P.R. Shewry (ed), Barley Genetics, Biochemistry, Molecular Biology and Biotechnology. CAB International, Great Britain, pp. 3-18.

Bro, R., 1997. PARAFAC. Tutorial and applications. Chemom. Intell. Lab. Systems 38: 149-171.

Bro, R., 1998. Multi-way Analysis in the Food Industry – Models, algorithms and applications. Ph.D. thesis. Univ, Amsterdam, 228 p.

Bro, R., 1999. Exploratory study of sugar production using fluorescence spectroscopy and multi-way analysis. Chemom. Intell. Lab. Systems 46: 133-147.

Corballis, M.C., 1991. The Lopsided Ape – Evolution of the Generative Mind. Oxford University Press.

Esquinas-Alcázar, J.T., 1993. Plant genetic resources. In: M.D. Hayward, N.O. Bosemark and I. Romagosa (eds), Plant Breeding – Principles and Prospects. Chapman and Hall, London, pp. 33-55.

Esbensen, K.H., 2000. Multivariate Data Analysis in Practice (4th ed.). CAMO ASA, Oslo, Norway.

Falco, S.C., T. Guida, M. Locke, J. Mauvais, C. Sanders, R.T. Ward and P. Webber, 1995. Transgenic canola and soybean seeds with increased lysine. Biotechnology 13: 577-582.

Geladi, P., and B.R. Kowalski, 1986. Partial least square regression: A tutorial. Analytica Chimica Acta 185: 1-17.

Gigerenzer, G., and D.J. Murray, 1987. Cognition as Intuitive Statistics. Lawrence Erlbaum Associates, Hillsdale, N.Y., USA.

Hawkes, J.G., 1981. Germplasm collection, preservation and use. In: K.J. Frey (ed), Plant Breeding II. Iowa State Univ. Press, Ames, pp. 57-84.

Hempel, C., 1968. Philosophy of Natural Science. Prentice-Hall, Englewood Cliffs, N.J.

Keeler, S.J., C.L. Maloney, Y.P. Webber, C. Patterson, L.T. Hirata, S.C. Falco and J.A. Rice, 1997. Expression of *de novo* high-lysine alpha-helical coiled-coil proteins may significantly increase the accumulated levels of lysine in mature seeds of transgenic tobacco plants. Plant Mol. Biol. 34: 15-29.

Kingsland, S.E., 1995. Modelling Nature – Episodes in the History of Population Ecology. Univ. Chicago Press.

Kreis, M., and P.R. Shewry, 1992. The control of protein synthesis in developing barley seeds. In: P.R. Shewry (ed), Barley Genetics, Biochemistry, Molecular Biology and Biotechnology, CAB International, pp. 319-334.

Kuhn, T.S., 1977. The Essential Tension – Selected Studies in Scientific Tradition and Change. Univ. Chicago Press.

Laurie, D.A., J.W. Snape and M.D. Gale, 1992. DNA marker techniques for genetic analysis in barley. In: P.R. Shewry (ed), Barley Genetics, Biochemistry, Molecular Biology and Biotechnology. CAB International, pp. 115-132.

Leah, R., H. Tommerup, I. Svendsen and J. Mundy, 1991. Biochemical and molecular characterisation of three barley seed proteins with antifungal properties. J. Biol. Chem. 266: 1564-1573.

Martens, H., and T. Næs, 1989. Multivariate Calibration. John Wiley, Chichester.

Massart, D.L., B.G.M. Vandeginste, L.M.C. Buydens, S. DeJong, P.J. Lewi and J. Smeyers-Verbeke, 1997. Handbook of Chemometrics and Qualimetrics. Parts A and B. Elsevier, Amsterdam.

Mertz, E.T., 1976. Case histories of existing models. In: Genetic Improvement of Seed Proteins. Natl. Acad. Sci., Washington, D.C., pp. 57-70.

Michod, R.E., 1999. Darwinian Dynamics. Princeton Univ. Press, USA.

Munck, L., 1972. Improvement of nutritional value in cereals. Hereditas 72: 1-128.

Munck, L., 1991. Man as selector – a Darwinian boomerang striking through natural selection. In: Aa. Hansen (ed), Environmental Concerns. Elsevier Science Publishers, pp. 211-227.

Munck, L., 1992. The case of high-lysine barley breeding. In: P.R. Shewry (ed), Barley Genetics, Biochemistry, Molecular Biology and Biotechnology, CAB International, Great Britain, pp. 573-602.

Munck, L., 1993. On the utilization of renewable plant resources. In: N.D. Hayward, N.O. Bosemark and I. Romagosa (eds), Plant Breeding – Principles and Prospects. Chapman and Hall, London, pp. 500-522.

Munck, L., J. Pram Nielsen, B. Møller, S. Jacobsen, I. Søndergaard, S.B. Engelsen, L. Nørgaard and R. Bro, 2001. Exploring the phenotypic expression of a regulatory proteome-altering gene by spectroscopy and chemometrics. Anal. Chim. Acta 446: 171-186.

Munck, L., L. Nørgaard, S.B. Engelsen, R. Bro and C.A. Andersson, 1998. Chemometrics in food science – a demonstration of the feasibility of a highly exploratory, inductive evaluation strategy of fundamental scientific significance. Chemom. Intell. Lab. Systems 44: 31-60.

Munck, L., and D. von Wettstein, 1976. Effects of genes that change the amino acid composition of barley endosperm. In: Genetic Improvement of Seed Proteins. Natl. Acad. Sci., Washington, D.C., pp. 71-78.

Mundy, J., J. Hejgaard, A. Hansen, L. Hallgren, K.G. Jørgensen and L. Munck, 1986. Differential synthesis *in vitro* of barley aleurone and starchy endosperm proteins. Plant Physiol. 81: 630-636.

Nelson, O.E., 1979. Inheritance of amino acid contents in cereals. In: Seed Protein Improvement in Cereals and Grain Legumes. IAEA, Vienna, pp. 79-86.

Nørgaard, L., A. Saudland, J. Wagner, J.P. Nielsen, L. Munck and S.B. Engelsen, 2000. Interval partial least squares regression (iPLS): A comparative chemometric study with an example from near infrared spectroscopy. Appl. Spectroscopy 54: 413-419.

Nørretranders, T., 1991. Mærk Verden. Gyldendal, Copenhagen. (In Danish).

Pinker, S., 1997. How the Mind Works. Allen Lane, London.

Postel, W., 1956. Der Einfluß genetischer und ökologischer Faktoren auf den Eiweißhaushalt von Sommergersten unter besonderer Berücksichtigung der exogenen Aminosäuren. Züchter 26: 211-239.

Prigogine, I., and I. Stengers, 1984. Order out of Chaos – Man's New Dialogue with Nature. William Heineman Ltd., Great Britain.

Romagosa, I., and P.N. Fox, 1993. Genotype × environment interaction and adaptation. In: N.D. Hayward, N.O. Bosemark and I. Romagosa (eds), Plant Breeding – Principles and Prospects. Chapman and Hall, London, pp. 373-390.

Shewry, P.R., A.S. Tatham, N.G. Halford, J.H.A. Barker, U. Hannappel, P. Gallois, M. Thomas and M. Kreis, 1994. Opportunities for manipulating the seed protein composition in wheat and barley in order to improve quality. Transgenic Res. 3: 3-12.

Stafford, J.V. (ed), 1999. Precision Agriculture 99. Part I and II. Sheffield Academic Press.

Vasal, S.K., 1999. Quality Protein Maize Story in the Workshop "Improving Human Nutrition through Agriculture – The role of International Agricultural Research." Internat. Rice Res. Inst., Los Baños, Philippines, pp. 1-19.

Vavilov, N.I., 1951. The origin, variation, immunity and breeding of cultivated plants. Chronica Botanica 13: 1-366.

Wettstein, D. von, 1983. Genetic engineering in the adaptation of plants evolving human needs. Experientia 39: 687-804.

Wiegand, C.L., A.H. Gerbermann, K.P. Gallo, B.L. Blad and D. Dusek, 1990. Multisite analyses of spectral-biophysical data for corn. Remote Sens. Environ. 33: 1-16.

Williams, P.C., and K.H. Norris (eds), 1987. Near Infrared Technology in Agricultural and Food Industries. Amer. Assoc. Cereal Chemists, St. Paul, Minn., USA.

Wold, S., K. Esbensen and P. Geladi, 1987. Principal Component Analysis. Chemom. Intell. Lab. Systems 2: 37-52.

Ziman, J., 1978. Reliable Knowledge – An Exploration of the Grounds of Belief in Science. Cambridge Univ. Press.

Diversity in Barley *(Hordeum Vulgare)*
Roland von Bothmer et al (Editors).

Chapter 12

Diversity in *ex situ* genebank collections of barley

Theo van Hintum and Frank Menting

Centre for Genetic Resources The Netherlands (CGN), P.O. Box 16, NL-6700 AA Wageningen, The Netherlands

Introduction

The genetic resources of a crop serve as the raw material for both plant breeding and scientific research. Therefore, genetic resources have to be available for present users, and have to be conserved for future utilisation.

Following the early work of scientists such as Vavilov, Harlan and Frankel, conservation of plant genetic resources started to receive attention in the early sixties (Pistorius, 1997). Genetic erosion became an important issue and led to the 1967 FAO/IBP Technical Conference on Exploration, Utilisation and Conservation of Plant Genetic Resources, and the development of a global strategy (FAO, 1969). The international institutes funded by the Consultative Group on International Agricultural Research (CGIAR) were created, and started to collect systematically and conserve genetic diversity, notably landraces, of their various mandate crops. From the early seventies onwards, national, regional and institutional genebanks were established. According to the FAO Report on the State of the World's Plant Genetic Resources for Food and Agriculture (FAO, 1996), the number of genebanks has grown to more than 1,300 recorded collections storing and conserving more than six million accessions *ex situ* and more than half a million accessions held in field genebanks.

Barley is well represented in genebanks. According to the FAO Report on the State of the World's Plant Genetic Resources for Food and Agriculture (FAO, 1996), barley is the second largest crop. With 8% of all six million accessions worldwide, barley comes only after wheat (13%) and is followed by rice (7%), maize (5%), *Phaseolus* (4%), soybean and sorghum (both 3%) (see Table 12.1). FAO counts 485,000 barley accessions. In a more recent inventory by the authors, which is described below in some detail, the number of barley accessions was adjusted to 370,796. Such a huge number of accessions might create the impression that the matter, i.e., the conservation of the genetic resources of barley, has been properly taken care of. Peeters (1988) even considered the genebanks to be the new centres of diversity for barley. However, the number does not indicate whether these accessions represent the complete barley genepool or whether the material has been well preserved. Nor does it reveal if the accessions are accessible to the user, either with respect to their seeds, or general information about them.

Hintum, Th. van, and F. Menting, 2003. Diversity in *ex situ* genebank collections of barley. In: R. von Bothmer, Th. van Hintum, H. Knüpffer and K. Sato (eds), Diversity in Barley (*Hordeum vulgare*), pp. 247-257. Elsevier Science B.V., Amsterdam, The Netherlands.

Table 12.1. Numbers of accessions of 20 largest crops conserved *ex situ* in the world (source: FAO, 1996).

Crop	Total world accessions	Crop	Total world accessions
Wheat	784,500	Tomato	78,000
Barley	485,000 *	Chickpea	67,500
Rice	420,500	Cotton	49,000
Maize	277,000	Sweet potato	32,000
Phaseolus	268,500	Potato	31,000
Soybean	174,500	Faba bean	29,500
Sorghum	168,500	Cassava	28,000
Brassica	109,000	Rubber	27,500
Cowpea	85,500	Lentil	26,000
Groundnut	81,000	*Allium*	25,500

* this estimate is corrected in the current study to 370,796. It can be assumed that all numbers are overestimated.

Barley genetic diversity

When describing the state of barley genetic resources, the genetic diversity can be divided into the different genepools, and within the primary genepool in the wild and the cultivated barley. Even within the cultivated barley, a division needs to be made into cultivars, landraces and research material.

Genepools

The highest level at which the genetic resources can be described is the taxonomic level. If the focus is on cultivated barley, *Hordeum vulgare* ssp. *vulgare*, the first question should be about the relationship with other *Hordeum* taxa, i.e., the primary, secondary and tertiary genepool (Harlan and de Wet, 1971).

There is no crossing barrier between cultivated barley and wild barley, *H. vulgare* ssp. *spontaneum* (Asfaw and Bothmer, 1990). This makes wild barley part of the primary genepool. The secondary genepool of barley, material that can be crossed but with difficulties such as hybrid sterility, consists of only one species, *H. bulbosum*. This species has been used extensively in barley breeding and research for haploid production; hybrids of *H. vulgare* and *H. bulbosum* show chromosome elimination. But sometimes, apart from haploids, complete hybrids occur. The high meiotic pairing in these hybrids implies that recombination can occur (Kasha and Sadasivaiah, 1971; Bothmer *et al.*, 1987). All other wild *Hordeum* species can be considered part of the tertiary genepool. There are strict sterility barriers, and the genomes of these species are different from those of *H. vulgare* and *H. bulbosum.* Gene transfer is very difficult (see also Chapter 2).

Diversity in cultivated barley

The genetic diversity in cultivated germplasm is usually described in terms of breeding level, i.e., landraces versus cultivars versus research material, and/or in terms of geographical origin. All of these divisions are relevant to barley.

Some old European landraces are still available, but most of the landraces preserved in genebanks have been collected relatively recently, i.e., during the period 1950-1990, in traditional farming systems in Asia and North Africa. Many old cultivars, usually selections from landraces made in the first half of the 20th century, are still available. The modern cultivars, i.e., the commercial results of modern scientific plant breeding, are abundant in the genebanks. There is also a large amount of research material available in genebanks as a result of the role barley has played in scientific research. This category includes mutants, isogenic lines, addition lines, etc. (cf. Chapters 5 and 6).

Conservation methods

Seed storage

Like most genera in the Gramineae family, species in the genus *Hordeum* show orthodox seed storage behaviour, which means that the mature whole seeds not only survive considerable desiccation, but their longevity in air-dry storage is increased in a predictable way by reduction in seed storage moisture content and temperature. Data on seed storage behaviour are available for the species *bulbosum, jubatum, marinum, murinum, secalinum* and *vulgare* (both subspecies; Dickie *et al.*, 1990; Hong *et al.*, 1996).

Seeds of *H. vulgare* are not damaged by exposure to liquid nitrogen (Stanwood and Bass, 1981). If seeds are stored in the open in a temperate climate, the p50 value, i.e., the time in which half the seeds die, is about 7.2 years (Priestley, 1986). There is evidence that seeds tolerate desiccation to a moisture content of 1.8% (Hong *et al.*, 1996). Barley seeds in the 'Vienna sample' of 1877, which was sealed hermetically with a moisture content of 3.1%, showed a germination of 90% after 110 years of storage at ambient temperatures, which clearly illustrates the storability of barley (Steiner and Ruckenbauer, 1995).

The conditions for long-term storage in genebanks generally follow the recommendations of FAO/IPGRI (1994) for seeds of orthodox species: temperature between -20ºC and -15ºC and a moisture content of 3% to 7%. Under these conditions barley is expected to maintain high levels of viability over periods of 100 years and probably more.

Regeneration

Regeneration of cultivated barley is considered relatively easy. In general, regeneration is the most difficult operational aspect of collection management since the genetic integrity of the accessions needs to be maintained and therefore contamination, genetic drift and selection needs to be avoided. In the case of barley, the threat of pollen contamination is very low since it is a selfing crop, though this might not always be complete. Some cultivars can show, depending on the circumstances, a remarkably high rate of open flowering, i.e., tendency to cross-pollination (Hammer, 1975). The threat of seed contamination can be avoided using special sowing and threshing machines, or if they are not available, sowing and threshing manually. Furthermore, the post-harvest seed lots have to be handled carefully. These are standard procedures in genebanks. In addition, the danger of genetic drift can easily be avoided by using sufficiently large samples of the population. Intra-accession selection, however, is difficult to avoid in heterogeneous accessions, especially if they are regenerated outside their natural habitat.

Even for an 'easy' crop like barley, maintaining the genetic integrity of genebank accessions proves to be very difficult. Some genebanks, therefore, split heterogeneous accessions into morphologically distinct lines that are being maintained as separate accessions (Lehmann and Mansfeld, 1957). A recent study revealed that the effective population size in barley regenerations using an estimated 600 plants, was only 4.7 (Parzies *et al.*, 2000). Hintum and Visser (1995) showed that duplicate barley accessions had developed into quite different mixtures in different genebanks. Both studies looked at the results of procedures from the past, it is possible that they have improved since.

For the wild species there are more problems regarding regeneration. Sowing of, for example, ssp. *spontaneum* cannot be done mechanically, since the caryopses remain in the spikelets. There is the general problem with regenerating wild species of seed shattering; the rachis is brittle and as soon as the seeds become ripe, they fall to the ground, obviously making harvesting difficult. Another general problem is that of dormancy which can occur, causing problems with germination. Finally, not all *Hordeum* species are inbreeders, for example, *H. bulbosum* and *H. brevisubulatum* are almost obligate outbreeders. Sometimes it is not really known if a wild species is purely self-pollinating or partly or completely cross-pollinating. A wrongly assumed selfing nature will have considerable consequences for the genetic integrity, a wrongly assumed cross-pollinating nature has large effects on the costs of regeneration and therefore on the capacity of the programme. Bothmer *et al.* (1995) has provided relevant information on these points for many wild *Hordeum* species.

The category of research material has, in specific cases, its own regeneration problems. Material with male sterility will need maintainer lines; addition lines can lose the added chromosome; chlorophyll and other mutants can have problems in reaching the generative stage; lethal mutants can only be regenerated via heterozygotes, etc.

Collections

To describe the genetic diversity of barley in genebanks, an inventory of germplasm collections was made. Based on the information available in on-line accessible databases and from data sets kindly provided by curators of important germplasm collections, we could create an impression of the available genetic resources.

Data sources

The starting point of the survey was the IPGRI Directory of Germplasm Collections (IPGRI, 1999). The data in this directory were updated with data downloaded or received from institutes, universities and genebanks. In addition, a recent version of the European Barley Database was used (see Table 12.2).

Nomenclature

Since not all collections use the same nomenclature, the on-line accessible nomenclature of the Genetic Resources Information Network of the USDA (GRIN, 2000) was used as a source of synonymy to rename the species for the present study. If this list of synonyms was not sufficient, the monograph of Bothmer *et al.* (1995) was used. The most frequently occurring change was that from *Hordeum spontaneum* to *Hordeum vulgare* ssp. *spontaneum*, but the species in the tertiary genepool also frequently required the name to be changed.

Overview

The total number of barley accessions in the world is about 371,000. This number includes all taxa of *Hordeum*: *H. vulgare* ssp. *vulgare* (88%), *H. vulgare* ssp. *spontaneum* (10%), *H. bulbosum* (0.4%) and other wild species (1.7%). From all these accessions, 41% is maintained in Europe, 25% in North America and 12% in the Middle East. The remaining accessions are divided amongst Asia, South America, Africa and Australia with 9%, 7%, 4% and 2%, respectively.

Compared to *H. vulgare* ssp. *vulgare*, there are considerable differences in geographical distribution between the different groups of species. *H. vulgare* ssp. *spontaneum* can hardly be

Table 12.2. Sources of data used for the inventory and the duplication study.

Institute	Country, City	Acronym	Source type	Rank	% World total	% World *sponta-neum*	Reference
Plant Gene Resources of Canada	Canada, Saskatoon	PGRC	B	1	11.6	15.0	Diederichsen, 2000
USDA National Small Grain Collection	USA, Aberdeen	NSGC	A	2	7.3	4.5	PC-GRIN, 2000
International Centre for Agricultural Research in Dry Areas	Syria, Aleppo	ICARDA	B	3	6.8	5.5	Konopka, 2000
John Innes Centre, Norwich	UK, Norwich	JIC	A	4	6.4	44.2	JIC, 1999
Centro Nacional de Pesquisa de Recursos Genéticos e Biotec.	Brazil, Brasilia	CENARGEN	B	5	5.0	0.0	Oliveira, 2000
N.I. Vavilov Research Institute of Plant Industry	Russia, St. Petersburg	VIR	C	6	4.7	0.1	Kovalyova & Knüpffer, 2001
Institute of Crop Germplasm Resources	China, Beijing	CAAS	D	7	4.6	1.4	IPGRI, 1999
Institute of Plant Genetics and Crop Plant Research	Germany, Gatersleben	IPK	A	8	3.5	1.3	IPK, 1999
Biodiversity Conservation and Research Institute	Ethiopia, Addis Ababa	BCRI	D	9	3.4	0.0	IPGRI, 1999
Institute of Agroecology and Biotechnology	Ukraine, Kiev	IAB	D	10	2.2	0.0	IPGRI, 1999
Federal Centre for Breeding Research on Cultivated Plants	Germany, Braunschweig	BAZ	A	11	2.0	0.3	BAZ, 1999
Lieberman Germplasm Bank, Institute for Cereal Crops Improvement	Israel, Tel-Aviv	ICCI	B	12	1.9	20.1	Manisterski, 2000
Research Institute for Bioresources, Okayama University	Japan, Kurashiki	RIB	D	17	1.5	0.5	IPGRI, 1999
Centre for Genetic Resources, The Netherlands	Netherlands, Wageningen	CGN	E	23	0.9	0.1	CGN, 1999
Department of Crop Science, Swedish University of Agricultural Sciences	Sweden, Alnarp	SUAS	A	31	0.7	0.8	Bothmer, 2000
Institute for Agrobotany	Hungary, Tápiószele	ABI	C	32	0.7	0.0	Knüpffer, 1999
Institute of Evolution, University of Haifa	Israel, Haifa	Haifa	C	47	0.4	4.4	Knüpffer, 1999
Nordic Gene Bank	Sweden, Alnarp	NGB	A	50	0.3	0.0	NGB, 2000
Total					63.9	98.2	

Note: These data were used in addition to the IPGRI Directory of Germplasm Collections and the European Barley Database. The collections listed cover about 64% of the 370,796 global barley accessions and 98% of the world's 33,145 ssp. *spontaneum* accessions (in order of collection size). Data source: A= downloaded via Internet, B= received on request by Email, C= exported from the European Barley Database (November 1999), D= IPGRI Directory of Germplasm Collections (IPGRI, 1999), E= exported from database of authors.

found in collections in the continents that have the lowest number of barley accessions (Asia, Africa, South America and Australia). The Middle East, the centre of diversity for this subspecies, however, seems to focus on this group. It maintains 12% of all barley accessions but one-third of the *spontaneum* accessions. The majority of these accessions (6,652 acc.) are in the

Lieberman Germplasm Bank of the Tel-Aviv University in Israel, but ICARDA also has an important collection (1,816 accessions). The John Innes Centre in the United Kingdom holds nearly all the European *spontaneum* accessions (14,648 acc.). These accessions, which are derived from relatively few collecting sites, result from a joint UK-Israeli expedition but are not yet available for distribution. The North American *spontaneum* collections are in the possession of Plant Gene Resources of Canada, Saskatoon Research Centre in Canada (4,966 acc.) and the National Small Grain Collection USDA-ARS, United States (1,504 acc.).

H. bulbosum (1,370 acc.), the only representative of the secondary genepool, contributes only 0.4% to the global number of accessions of barley. Almost all of these accessions are maintained in Canada (786 acc.). The United States maintains 161, and Europe 411 accessions of this species.

The other wild species are very rare in genebanks, comprising only 1.7% of the world barley accessions. By far the most accessions of these species are maintained in Europe (48%) and North America (46%). The largest collections are those maintained at PGRC (41% of the world's accessions) and the Department of Plant Breeding Research of the Swedish University of Agricultural Sciences in Alnarp (39%). The best-represented species of the tertiary genepool are *H. murinum* (1,505), *H. marinum* (566), *H. jubatum* (337), *H. brevisubulatum* (335), *H. pubiflorum* (323), *H. chilense* (310) and *H. brachyantherum* (306). *H. guatemalense* and *H. erectifolium* are both listed once (in Alnarp). The total number of species from the tertiary genepool included in the inventory is 31.

From a considerable number of accessions (13%) the species name was not available for the current inventory. One can speculate that a relatively large proportion of these accessions are part of the secondary and tertiary genepool and not yet classified. However, it is also possible that some genebanks having only cultivated barley list barley or *Hordeum* as crop name, without specifying the, in their case, obvious species name, *vulgare*.

Duplication of germplasm

The huge number of 371,000 accessions suggests complete coverage of barley diversity. It has, however, already been indicated that the secondary and tertiary genepools are poorly represented. Nevertheless, despite the large number of accessions, the primary genepool also requires further investigation; under-representation of specific groups of material and excessive duplication within and between collections might change the picture.

To quantify the duplication within and between collections the data from a number of germplasm collections were analysed using the method proposed by van Hintum (2000), using random samples of size 100, resulting in errors of about 5% of the estimates for overlaps. Duplication was defined as being derived from the same original population. However, the decision concerning the number of duplicates was often arbitrary. For example, if a series of accessions within a collection were all derived from the same collecting site, each accession was considered to be duplicated once. Poorly documented accessions cannot be matched, and therefore had to be considered distinct. This means that a genebank with hardly any documentation appeared to be completely distinct.

In the analysis the three largest collections in the world were studied: the PGRC, the NSGC and the ICARDA collection, together accounting for 26% of the world's barley accessions (see Table 12.2). The analysis showed that the internal duplication was limited, 2%, 10% and 10%, respectively. However, the duplication between the collections was considerable. The overlap between the distinct accessions in PGRC and NSGC was estimated at over 19,000 (46% and

79%, respectively), between PGRC and ICARDA over 14,000 (34% and 63%), and between NSGC and ICARDA over 18,000 (76% and 81%). If all three collections were put together, an even more disturbing picture arose. The over 95,000 accessions maintained in these three collections appear to consist of only 51,000 distinct accessions (54%). The NSGC and ICARDA collections, which together maintain nearly 47,000 accessions, add only 9,000 distinct accessions to the PGRC collection (Figure 12.1). This extent of duplication could be the result of dedicated effort to mirror each other's collections for backup or availability reasons, however it sheds a new light on the huge numbers of barley accessions.

To get an impression of the role of smaller collections, a similar approach was followed for three European collections (IPK, BAZ and CGN), respectively 8th, 11th and 23rd in size, together maintaining 20,439 accessions, equivalent to 6% of the world's total. The internal duplication was rather high, 17%, 16% and 3%, respectively, mainly due to the accession management of the collections. For example, in the BAZ collection much duplication occurs in accessions like 'JIMMA V' (occurring 7 times), 'KAFFA II' (40 times), 'SHOA' (22 times) and 'WONDO III' (21 times). These are probably split-up samples because several infraspecific taxa occurred under the same accession identifier. In the analyses for internal duplication each of these accessions is assumed to be duplicated once.

The overlap with each other and the 'big ones', however, was relatively limited. The overlap with NSGC was 48%, 34%, and 40%, respectively. The mutual overlap was much less than that, despite their geographic proximity; when compared, the collections maintained about 17,500 distinct accessions, about 85% of the total number of accessions in these collections.

The general impression that arises from the analysis is that the large collections are

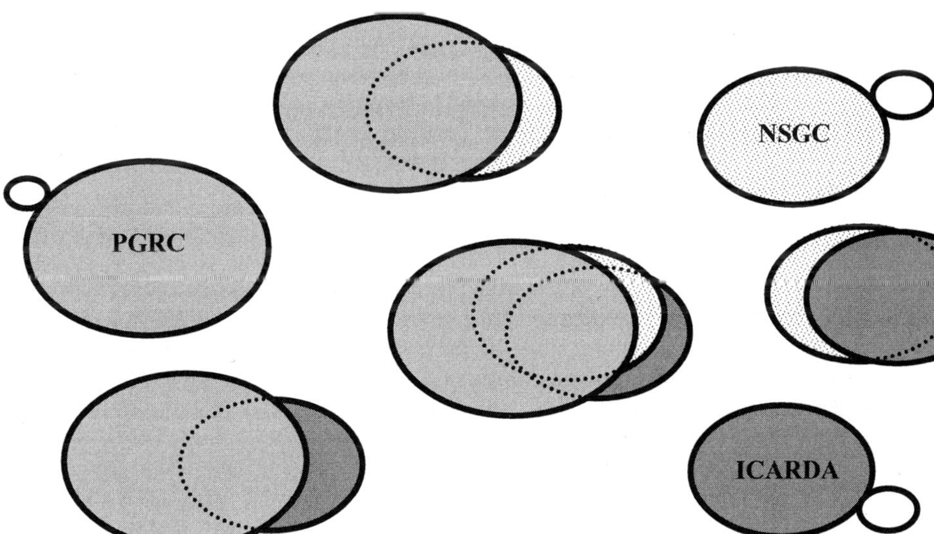

Figure 12.1. Visualisation of duplication within and between the three major barley collections. The area of the ellipses and their overlaps correspond to the number of accessions. The small white ellipses represent the internal redundancy. (PGRC: Plant Gene Resources of Canada, Saskatoon Research Centre, NSCG: USDA National Small Grain Collection, ICARDA: International Centre for Agricultural Research in Dry Areas).

duplicating each other, each adding only a small unique part. The smaller collections are partly duplicated by the large ones, but also contain considerable numbers of unique accessions.

Duplication of ssp. spontaneum

The situation for the smaller groups is quite different from the overall picture. The representation of the secondary and tertiary genepools is poor, even if there was no duplication.

The case of *spontaneum* was difficult to study since the concept of accession is used differently in different genebanks. In any case, the number of 32,977 accessions is a great overestimation.

The first cause of this overestimation is the number of accessions created from a single population. All 14,648 *spontaneum* accessions of JIC are derived from the initial multiplication of samples from only 213 sites (Ambrose, 1999). In addition, the accessions maintained at Haifa, ICCI, ICARDA and NSGC are probably samples from a few sites only. In 1997 Haifa maintained 1,300 genotypes from 33 Israeli populations, 500 genotypes from 20 Iranian populations, 450 genotypes from 20 Turkish populations, 700 accessions from 27 Jordanian populations and 550 genotypes for microgeographic studies (Maggioni *et al.*, 1999). In the inventory this collection is reduced to 1,442 accessions from 69 populations of Iran, Turkey and Israel based on the data in the European Barley Database (Knüpffer, 1999). The ICCI collection consists to a large extent of transect sampled lines from different natural populations (Maggioni *et al.*, 1999). Both in ICCI and Haifa the average number of accessions per site is 21, varying from 1 to 586 (accessions without listed collection site were excluded from the calculation).

The second cause of overestimation is the duplication of accessions via exchange between genebanks. The JIC reported to have sent their material to Svalöv for research purposes (this material was not included in the collection of the Swedish University of Agricultural Sciences) and partly to IPK, via a research station of BAZ in Aschersleben (Ambrose, 1999). Matching PI numbers showed that all NSGC *spontaneum* accessions are also maintained at PGRC and over three-quarters at ICARDA (based on a sample of 100 accessions). Furthermore, there is a strong overlap in collection sites of spp. *spontaneum* between Haifa, NSGC and ICARDA (the collection sites of PGRC were not available in the current study). Only the accessions in ICCI probably originate from another source.

Gaps in existing germplasm collections

The inventory of barley germplasm, as was used in this study, allows the analysis of redundancy in the form of duplication within and between collections. Finding the gaps in the collections is not possible on the basis of the inventory alone. In the cases of wild species or traditional crops, it is possible to compare the natural distribution area of the species or the cultivation area of the crop with the representation in the collections. This has not been done systematically for barley. For certain cultivars this is not possible, and good representation of this part of the primary genepool is a matter of subjective judgement. Despite these difficulties, identifying the gaps in the combined collections should receive more attention since these gaps represent the material potentially under threat of genetic erosion.

Access to the germplasm

The access to germplasm is determined by access to the information, and by the availability of the material.

Concerning the access to information, the last few years have shown a major improvement, as a result of the developments of information and communication technology. An increasing number of genebanks provide detailed information about the material in their collections on-line, accessible via the Internet. This is illustrated by the survey described above where much of the information was retrieved via the Internet. However, this is not yet complete, and some collections cannot be expected to become accessible on-line since they are considered an 'internal' resource.

The other aspect of access to collections is the availability of material. Genetic resources held in genebanks and public research institutes have traditionally always been freely accessible, unlike the material in collections held by private industries. However, this free access is no longer certain due to a number of developments.

First of all, recent years have shown a trend towards privatising breeding and research institutes. As a result, germplasm is considered an asset, and therefore free access is no longer obvious in all cases. Secondly, there is an ongoing discussion on the international level concerning access to plant genetic resources. The Convention on Biological Diversity (CBD) gave nations the jurisdiction over their biological diversity. However, the implications for crop genetic resources conservation programmes were not clear. Debates on this matter were held at different international platforms such as the Conference of Parties of the CBD (COP) and the FAO Commission on Genetic Resources for Food and Agriculture resulting in 2001 in a harmonised International Treaty on Plant Genetic Resources for Food and Agriculture allowing a relative free exchange of barley genetic resources, but only amongst signatories. Finally, the influence of Intellectual Property protection legislation (as formalised in the WTO TRIPS-agreement) on access to plant genetic resources is increasing. Again, the effects of this development are not yet completely clear.

Conclusions and outlook

Barley is relatively easy to conserve; the seeds store well, and regeneration is due to the selfing nature, easy. Nevertheless, care should be taken: the wild species need more attention, and even in the cultivated material studies have shown that the genetic integrity of barley genebank accessions is low. This can be due to incomplete selfing but also to genetic drift, selection or contamination.

The crop is well represented in genebanks; the number of accessions is very large. However:

- the secondary and tertiary genepools are hardly represented in *ex situ* collections,
- the ssp. *spontaneum* material, despite the relatively high number of accessions, can be shown to be derived from a limited number of populations,
- the cultivated material is very heavily duplicated.

Access to the information on material in genebanks is good and improving. The access to the material in the genebanks is good, but under threat due to international developments.

With the current genetic resources conserved in genebanks around the world and the increasing awareness of the value of these resources, there is no need for alarm. Only for the wild related *Hordeum* species, should care be taken that if habitat destruction threatens specific diversity, it needs to be sampled and conserved, since the likelihood that it is already represented in a germplasm collection is small.

Acknowledgements

The authors would like to thank Helmut Knüpffer for making the European Barley Database available, Axel Diederichsen and Eugene Timmermans (PGRC), José Caetano de Oliveira and Dijalma Barbosa da Silva (CENARGEN), Jan Valkoun and Jan Konopka (ICARDA), Fredrik Ottosson and Roland von Bothmer (SUAS) and Jacob Manisterski (ICCI) for sending their barley passport data. Furthermore we would like to thank Harold Bockelman (USDA-ARS), Lajos Horváth (ABI) and Mike Ambrose (JIC) for their remarks on, and help with the information about their barley collection and Loek van Soest and the editors of this book for their helpful comments on the manuscript.

References

Ambrose, M., 1999. Personal communication.

Asfaw, Z., and R. von Bothmer, 1990. Hybridization between landrace varieties of Ethiopian barley (*Hordeum vulgare* ssp. *vulgare*) and the progenitor of barley (*Hordeum vulgare* ssp. *spontaneum*). Hereditas 112: 57-64.

BAZ, 1999. Download from Federal Centre for Breeding Research on Cultivated Plants (BAZ), indigo3.dv.fal.de/bgrc/bgrc-g.html, on November 8, 1999.

Bothmer, R. von, 2000. Personal communication, data received on request via Email from Fredrik Ottosson and Roland von Bothmer on April 12, 2000.

Bothmer, R. von, N. Jacobsen, C. Baden, R.B. Jørgensen and I. Linde-Laursen, 1995. An ecogeographical study of the genus *Hordeum*. Systematic and Ecogeographical Studies on Crop Genepools 7. IPGRI, Rome.

CGN, 1999. Download from CGN's database GENIS, November 5, 1999.

Dickie, J.B., R.H. Ellis, H.L. Kraak, K. Ryder and P.B. Tompsett, 1990. Temperature and seed storage longevity. Ann. Bot. 65: 197-204.

Diederichsen, A., 2000. Personal communication, data received on request via Email from Axel Diederichsen and Eugene Timmermans (PGRC) on March 13, 2000.

FAO, 1969. Report of the Third Session of the FAO Panel of Experts of Plant Exploration and Introduction. 25-28 March 1969. Rome.

FAO, 1996. FAO State of the World's Plant Genetic Resources for Food and Agriculture. Rome.

FAO/IPGRI, 1994. Genebank Standards. Food and Agriculture Organization of the United Nation, Rome, International Plant Genetic Resources Institute, Rome. 13 p.

GRIN, 2000. Germplasm Resources Information Network taxonomy, on-line searchable Internet site: www.ars-grin.gov/npgs/tax/

Hammer, K., 1975. Die Variabilität einiger Komponenten der Allogamieneigung bei der Kulturgerste (*Hordeum vulgare* L. s.l.). Kulturpflanze 23: 167-180.

Harlan, J.R, and J.M.J. de Wet, 1971. Towards a rational classification of cultivated plants. Taxon 20: 509-517.

Hintum, Th.J.L. van, 2000. Duplication within and between germplasm collections. III. A quantitative model. Gen. Res. Crop Evol. 47: 507-513.

Hintum, Th.J.L. van, and D.L. Visser, 1995. Duplication within and between germplasm collections. II. Duplication in four European barley collections. Gen. Res. Crop Evol. 42: 135-145.

Hong, T.D., S. Linington and R.H. Ellis, 1996. Seed Storage Behaviour: a Compendium. Handbooks for Genebanks No. 4. International Plant Genetic Resources Institute, Rome.

IPGRI, 1999. IPGRI Directory of Germplasm Collections. www.cgiar.org/ipgri/doc/dbintro.htm (Database upload date: October 6. 1999, Report generation date: November 4. 1999).

IPK, 1999. Download from the Institute for Plant Genetics and Crop Plant Research (IPK), fox-serv.ipk-gatersleben.de/sbot.htm, on November 5, 1999.

JIC, 1999. Download from the John Innes Centre (JIC), www.jic.bbsrc.ac.uk/welcome.htm, on November 5, 1999.

Knüpffer, H., 1999. Data exported from the European Barley Database (EBDB), copy November 1999, compiled by H. Knüpffer.

Konopka, J., 2000. Personal communication, data received on request via Email from Jan Konopka (ICARDA) on February 14, 2000.

Kovalyova, O., and H. Knüpffer. 2001. Personal communication. Species composition of the VIR barley collection compiled February 2001.

Lehmann, C.O., and R. Mansfeld. 1957. Zur Technik der Sortimentserhaltung. Kulturpflanze 5: 108-138.

Maggioni, L., H. Knüpffer, R. von Bothmer, M. Ambrose, K. Hammer and E. Lipman (compilers), 1999. Report of a Working Group on Barley. Fifth Meeting, 10-12 July 1997, Alterode/Gatersleben, Germany. International Plant Genetic Resourses Institute.

Manisterski, J., 2000. Personal communication, data received on request via Email from Jacob Manisterski (ICCI) on April 20, 2000.

NGB, 2000. Download from the Nordic Gene Bank (NGB), 193.10.47.75/Databases/ on June 5, 2000.

Oliveira, J. de, 2000. Personal communication, data received on request via Email from José Caetano de Oliveira and Dijalma Barbosa da Silva (CENARGEN) on March 10, 2000.

Parzies, H.K., W. Spoor and R.A. Ennos, 2000. Genetic diversity of barley landrace accessions (*Hordeum vulgare* ssp. *vulgare*) conserved for different lengths of time in *ex situ* genebanks. Heredity 84: 476-486.

PC-GRIN, 2000. Data extracted from downloadable PC-GRIN (www.ars-grin.gov/npgs/pcgrin.html) on February 2, 2000.

Peeters, J.P., 1988. The emergence of new centres of diversity: evidence from barley. Theor. Appl. Genet. 76: 17-24.

Pistorius, R. 1997. Scientists, Plants and Politics – A History of the Plant Genetic Resources Movement. International Plant Genetic Resources Institute, Rome, Italy. 134 p.

Priestley, D.A. 1986. Seed Ageing. Cornell University Press.

Stanwood, P.C., and L.N. Bass, 1981. Seed germplasm preservation using liquid nitrogen. Seed Sci. Technol. 9: 423-437.

Steiner, A.M., and P. Ruckenbauer, 1995. Germination of 110-year-old cereal and weed seeds, the Vienna sample of 1877. Verification of effective ultra-dry storage at ambient temperature. Seed Sci. Res. 5: 195-199.

Diversity in Barley *(Hordeum Vulgare)*
Roland von Bothmer et al (Editors).

Chapter 13

Summarised diversity – the Barley Core Collection

Helmut Knüpffer[a] and Theo van Hintum[b]

[a]Institute of Plant Genetics and Crop Plant Research (IPK), D-06466 Gatersleben, Germany
[b]Plant Research International B.V., Centre for Genetic Resources, The Netherlands (CGN), PO Box 16, NL-6700 AA Wageningen, The Netherlands

Introduction

Genebank collections throughout the world comprise about 6.1 million accessions (FAO, 1996). The largest collection is of wheat, with barley occupying second place. The FAO estimated that about 485,000 barley accessions exist in *ex situ* germplasm collections such as genebanks, breeders' and research collections. Hintum and Menting (2000) corrected this to ca. 373,000 accessions.

To improve the accessibility of such large collections and to rationalise evaluation of plant genetic resources, the concept of core collections was developed (Frankel and Brown, 1984). This concept is now widely applied to genebank collections all over the world (Hamon *et al.*, 1995; Hodgkin *et al.*, 1995; Johnson and Hodgkin, 1999).

The International Barley Core Collection (BCC) has been developed since 1989 by an international consortium as a voluntary activity of the participating institutions (Hintum, 1993). It attempts to create a common set of barley genotypes for use mainly in research, allowing the compilation of a large set of data on the genetic diversity in barley.

This chapter describes the background of the BCC, its history, concepts, the network of collaborating institutions and individuals, and presents an account of the current composition, management and documentation aspects, and first results of its implementation and utilisation in research.

History

In 1989 the Barley Working Group of the present European Cooperative Programme for Crop Genetic Resources Networks (ECP/GR) recommended that an *ad hoc* working group should develop the concepts for setting up a European Barley Core Collection, BCC (IBPGR, 1989). Based on European genebank collections documented in the European Barley Database (Knüpffer, 1988), BCC accessions should be proposed ensuring a worldwide coverage of the barley genepool, and the possibility of extending the BCC at a later stage into an international collection was already envisaged (Anonymous, 1989). In, 1991, after three meetings and discussions with about 100 barley experts worldwide, the BCC task force presented this concept

---

Knüpffer, H., and Th. van Hintum, 2003. Summarised diversity – the Barley Core Collection. In: R. von Bothmer, Th. van Hintum, H. Knüpffer and K. Sato (eds), Diversity in Barley (*Hordeum vulgare*), pp. 259-267. Elsevier Science B.V., Amsterdam, The Netherlands.

at the Sixth International Barley Genetics Symposium in Helsingborg, Sweden (Hintum, 1992a). The Symposium recommended the involvement of non-European specialists and germplasm collections, in order to develop a truly International Barley Core Collection (Hintum, 1992b). An international BCC working group met in ICARDA, Syria, in 1992, and an international symposium on Core Collections was held in Brazil in the same year, where the BCC was presented, discussed and developed further (Knüpffer and Hintum, 1995). Since then, the International BCC Committee has organised business meetings and open workshops to monitor progress and coordinate activities, in connection with the International Barley Genetics Symposia in Saskatoon, Canada (1996) and Adelaide, Australia (2000).

Concepts

Definition

The BCC is a selected and limited set of accessions. It optimally represents the genetic diversity of cultivated barley and the wild species of *Hordeum*, covering the three genepools, and includes well-known genetic standards (Knüpffer and Hintum, 1995). This differs from the original definition given by Frankel and Brown (1984): “A core collection consists of a limited set of accessions derived from an existing germplasm collection, chosen to represent the genetic spectrum in the whole collection. The core should include as much as possible of its genetic diversity. The remaining accessions in the collection are called the reserve collection”. The BCC is a “synthetic core collection” (Brown, 1995), and it differs from the original definition in the following respects:

- BCC accessions are selected from all internationally available accessions, rather than from a single genebank collection.
- The BCC is a separate collection, rather than part of an existing genebank collection divided into a core and a “reserve collection”.

Objectives

The BCC is being developed in order to (1) increase the knowledge about the barley genepool; (2) increase the efficiency of evaluation and thus of utilisation of existing collections; (3) provide a manageable and representative, highly diverse selection of the available barley germplasm for use in research and plant breeding; (4) provide adequate standards, e.g., for studies of genetic diversity in barley.

Whenever a diverse set of barley accessions is needed for an investigation, the BCC or a subset of it can be used. Consequently, large numbers of diverse results of characterisation, evaluation and other research will be accumulated for a relatively small number of accessions. The BCC does not replace existing collections and does not make them superfluous. It is a key for better utilisation of the existing collections.

Structure and size

The barley genepool is hierarchically structured and can be described by a dendrogram (cf. Knüpffer and Hintum, 1995), each branch of which will be represented in the BCC by one or several accessions. The diversity of the barley genepool will be represented as completely as possible. The total number of BCC accessions should not exceed 2,000.

The genepool was initially divided into five main categories, namely, (1) cultivars (500 accessions), (2) landraces (800 accessions), (3) *Hordeum vulgare* ssp. *spontaneum* (150-200 accessions), (4) other *Hordeum* wild species (60-100 accessions, ca. two per species), and (5)

Table 13.1. Tentative sizes of regional subsets of the cultivar and landrace groups.

Subgroup	Cultivars	Landraces
West Asia, North Africa	15	300
South and East Asia	80	300
Ethiopia	5	100
Europe	200	80
North and South America	150	30
Australia, New Zealand, South Africa and other regions	35	0
Total	485	810

"genetic stocks" and reference material (max. 200 accessions). The further subdivision of the cultivated barleys and ssp. *spontaneum* follows ecogeographical criteria (Table 13.1), whereas the wild species of *Hordeum* are divided according to taxonomic and ecogeographical criteria.

Designated coordinators for subsets of the BCC, leading scientists in their field, selected candidate accessions from the world barley holdings, in cooperation with other experts for the particular region or taxonomic group.

Homogeneity of accessions

The BCC accessions are, as far as possible and when appropriate, homozygous and homogeneous lines derived from genebank accessions by techniques such as single seed descent or doubled haploids. Exceptions are two obligatory cross-pollinating wild species, namely, *H. bulbosum* and *H. brevisubulatum*. The homogeneity issue has been one of the most controversial issues within the BCC concept. The decisive advantages of homogeneity are the possibility of identical reproduction at different locations over a long period of time (stability of accessions), and the availability of identical material for various investigations (standards). A great disadvantage of homogeneity is, however, that a variable landrace is represented by only one of its lines although reference to the original landrace in the genebank collection is maintained. The access to the original genebank accession will, through the BCC documentation system, be available to anyone interested in studying the diversity within the accessions. In addition, the number of allele combinations is reduced, as compared to "normal", often heterogeneous genebank material.

Organisation and development

The BCC is developed in an international network, which consists of a coordinating committee and subset coordinators, responsible for the selection of BCC subsets, creation of BCC accessions via single seed descent, initial multiplication of BCC accessions and distribution to the 'active BCC centres' (Figure 13.1). These 'active BCC centres', major research institutions, are responsible for the distribution of BCC samples to *bona fide* users in their respective regions:

- the Institute of Plant Genetics and Crop Plant Research (IPK) in Gatersleben, Germany, for Europe,
- the International Centre for Agricultural Research in the Dry Areas (ICARDA) in Aleppo, Syria, for the WANA region,
- the Research Institute for Bioresources (RIB), Barley Germplasm Centre in Kurashiki, Japan, for South and East Asia,
- the USDA Small Grain Cereals Collection in Aberdeen, Idaho, U.S., for the Americas,
- the Australian Winter Cereals Collection (AWCC) in Tamworth, Australia, for Australia.

Furthermore, there is a base collection, with a back-up at two safety duplication sites. Users

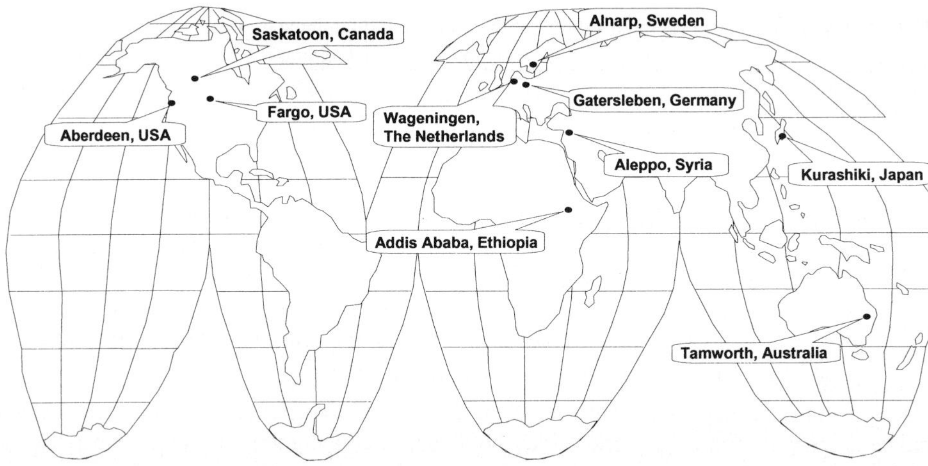

Figure 13.1. Network of cooperating institutions and their functions within the BCC network. USDA Small Grain Collection Aberdeen, Idaho, USA (subset Americas, active collection); International Centre for Agricultural Research in the Dry Areas, Aleppo, Syria (subset West Asia and North Africa, subset *H. vulgare* ssp. *spontaneum*, active collection, base collection); Swedish University of Agricultural Sciences, Alnarp, Sweden (chair, subset wild species); Nordic Genebank, Alnarp, Sweden, and North Dakota State University, Fargo, USA (both subset Genetic Stocks); Institute of Plant Genetics and Crop Plant Research, Gatersleben, Germany (subset Europe, active collection, documentation); Research Institute for Bioresources, Kurashiki, Japan (secretary, subset South and East Asia, active collection); Plant Gene Resources of Canada, Saskatoon, Canada (safety duplication); Australian Winter Cereals Collection, Tamworth, Australia (subset Oceania, active collection); Centre for Genetic Resources The Netherlands, Wageningen, The Netherlands (methodology, safety duplication). The Ethiopia part of the core collection is presently under development (Institute of Biodiversity Conservation and Research, Addis Ababa; subset Ethiopia).

are requested to submit a research proposal, or another indication of the intended use of the BCC entries, and a Material Transfer Agreement has to be signed.

The development of the different BCC subsets did not take place simultaneously. Some subsets have been made completely available, while others are still waiting to be created. In general, the development can be divided in four overlapping phases. In the first phase the concepts were developed and presented (1989-1992), in the second the BCC subsets were established (selection and initial multiplication of the accessions) (1992-1998). In the third phase, (parts of) the BCC have become operational, accessions are available for research and evaluation, multiplication, distribution and utilisation has been initiated (1996-). In the final and fourth phase, the BCC will be fully operational.

The selection and multiplication of BCC accessions by the designated subset coordinators is more or less complete. They are available for research and evaluation purposes from the respective subset coordinators. To a large extent, the missing parts will soon become available. The size of the BCC does not exceed 1,500 accessions at present. Given the maximum size of 2,000, this leaves room for future additions.

Subsets of the BCC

Landraces and cultivars from West Asia and North Africa (WANA)
The selection of 285 accessions of the WANA subset was initiated by J. Valkoun at ICARDA in 1993/1994. The accessions went through a cycle of single seed descent, after which they were multiplied. The material became available in 1996. Seeds of landraces from the WANA region were shipped to active centres for further distribution.

The material originates from 22 countries, with major contributions from Turkey (58 accessions) and Morocco (40 accessions). There is some geographical overlap between the WANA subset and the European subset, especially with respect to countries of the former Soviet Union, since the mandate region of ICARDA was extended to include the Caucasus and the Central Asian C.I.S. Republics. Some countries are still missing, and will be considered for inclusion in a later phase, especially Israel, Sudan, Tunisia, and Yemen.

Landraces and cultivars from South and East Asia
The Barley Germplasm Centre of the Okayama University in Kurashiki is responsible for selecting and managing East Asian accessions from Japan, Korea, China, Nepal, Bhutan and India (Sato and Takeda, 1997). The BCC material was initially selected by T. Konishi, to be continued later by K. Sato and K. Takeda. The Chinese material was selected in consultation with ICARDA. A selection of 300 landraces and 80 cultivars has been made from these seven countries, and material has been multiplied after a cycle of single seed descent. Material is available for distribution.

An overview of the basic characteristics of these accessions by country is given by Sato and Takeda (1997) and Liu *et al.* (1999). Covered, six-rowed and spring types are dominating. The East Asian BCC subset reflects the traditional barley types of the region, its composition differing from currently grown material in which the European malting type (covered, two-rowed spring barley) dominates. East Asian landraces have often been used directly for human food, or for food processing. Naked barleys predominate in Bhutan, the Tibetan plateau and parts of Nepal where barley is used as a source of food for human consumption. The complete East Asian subset was evaluated with respect to diversity, using 13 alleles at six isozyme loci (Liu *et al.*, 1999). Indian cultivars showed the highest diversity, followed by Korean and Chinese ones, while the landraces from Bhutan and Nepal had the lowest diversity. Cultivars showed higher diversity than landraces. All landraces clustered in one group, while cultivars from Japan, India and Korea formed independent groups. Hintum *et al.* (1995) tested different alternative sampling methods for selecting core entries in 96 cultivated barley accessions from China with reliable passport data, using isozyme systems. A stratification sampling based on collection site would give the best result (largest number of alleles preserved).

Landraces and cultivars from Ethiopia and Eritrea
For various reasons, an Ethiopian/Eritrean subset has not yet been completed. Attempts to involve the Ethiopian Biodiversity Institute (former Plant Genetic Resources Centre of Ethiopia) in a cooperative attempt to select the BCC candidate accessions from the Ethiopian germplasm collections failed, mainly due to the Ethiopian policy with regard to distribution of germplasm. Therefore, possible solutions are being sought, and recently a collaboration project between Ethiopia and Norway has started, aiming at selecting accessions for the international Barley Core Collection based on ecogeographical data and various markers.

Many studies on the diversity of Ethiopian barleys have been carried out, which could be the basis for selecting accessions for the BCC. For example, Demissie *et al.* (1998) assayed 43 Ethiopian barley landrace populations from a wide agro-geographical range using RFLPs with 31 probes distributed over all barley chromosomes and drew conclusions about optimal sampling strategies for selecting an Ethiopian core collection.

Landraces and cultivars from Europe

An initial set of 320 European barley landraces and cultivars was selected by G. Fischbeck and H. Knüpffer using the European Barley Database (Knüpffer, 1988), and requested from various European genebanks. After a cycle of single seed descent, the western European accessions were studied using RFLP analyses to adjust the selection (Hatz *et al.*, 1996; Hatz, 1997). From this study, it was concluded that more material from Spain and the Balkan area should be added, and some north-western material could be removed. The material was handed over to IPK for further multiplication and continued maintenance. Material from countries not sufficiently represented in the initial set were later selected from the IPK genebank on the basis of passport data, and passed the single-seed selection cycle at IPK. The set includes material from the Asian follower-states of the former Soviet Union, which are also represented in the WANA subset. It includes 298 accessions from 37 countries.

Liu *et al.* (2000) studied 79 accessions of the European subset with isoenzyme electrophoresis, using 26 alleles at ten loci. Most alleles known to occur in European barley are also observed in this European BCC subset, only five were absent. Nine of the 26 were rare (occurring in only two accessions), most of the rare ones were found in six-rowed winter barley. Greater diversity was found in six-rowed barley compared to two-rowed, winter was more diverse than spring barley.

Landraces and cultivars from the Americas

A first selection of American landraces and cultivars was made by H. Bockelman, in consultation with breeders and researchers from the U.S., Canada, Mexico and several South American countries. Criteria for the U.S. and Canadian barleys were the representation of various barley types grown in North and South America. Within these groups, the diversity was maximised by choosing accessions whose pedigrees differed as widely as possible. Most of the material was selected from the USDA Small Grain Collection (NSGC), but some, especially South American accessions, were selected and contributed to the BCC by foreign collaborators (H. Bockelman, pers. comm.). The material was multiplied after a cycle of single seed descent. The American subset is almost complete and has been distributed to the active centres; from here, it is available for research and evaluation purposes. Accessions from Argentina were added recently. Some material is of European origin and might be included in the European subset. The current set includes 155 accessions.

Liu *et al.* (2001) tested 151 accessions of the Americas subset for allozymic diversity, with the objective of providing valuable information for the further development of an optimal core collection in barley. A total of 25 alleles were found at the ten loci studied. Most significant differences in allelic frequency and genetic diversity values were found between spring (higher) and winter (lower) barley, the smallest difference was found between material from North and from South America.

Cultivars from Oceania and other parts of the world

A selection of ten Australian and one New Zealand cultivars has been made by M. Mackay. This subset includes material grown in southern Africa as, so far, predominantly Australian varieties were grown there, because of the similarity in natural conditions. Material was multiplied and handed over to the active BCC collections. Formerly this subset was referred to as "Other countries" or "rest of the world".

Hordeum vulgare *ssp.* spontaneum
A. Jaradat selected 152 entries of *H. vulgare* ssp. *spontaneum* from 16 locations in the early 1990s, and partly multiplied the material. Since the accessions were already single lines, single seed descent was not considered necessary. The material from Israel lacks passport data. The material has already been evaluated in plots and on a single line basis for a large number of descriptors. It is fully available. This original ssp. *spontaneum* subset was handed over to R. von Bothmer (Sweden) for continued maintenance, and multiplied. A set selected by J. Valkoun will be used to complement the existing set. At present, 70 accessions from 16 countries are available from this subset.

Wild Hordeum *species*
In the subset representing the wild species, excluding *H. vulgare* ssp. *spontaneum*, R. von Bothmer selected two entries from each species (when available). These samples were multiplied after a cycle of single seed descent. Due to problems in the multiplication of the cross-pollinating species *H. bulbosum* and *H. brevisubulatum*, material of these species is not yet available but the other 45 entries of 22 species are. Multiplication of wild species is difficult, and as a result, not all accessions have been reproduced to the extent requested. The wild *Hordeum* species subset was partly sent to the active centres. For *H. bulbosum* (which should be represented in the BCC by ca. 20 accessions), more well-documented candidate accessions are needed for the BCC.

Genetic stocks
The selection and preparation of the subset with genetic stock has experienced delay. U. Lundqvist and J. Franckowiak are involved in the selection of a basic list of candidate accessions, and arrangements for checking and regeneration have been made. In general, only material which is not too difficult to regenerate by genebank staff will be included in the BCC. It was decided not to include differential tester sets for resistance to various diseases.

Studies carried out on the BCC, use of the BCC, first results
The existing subsets of the BCC have been included in several research and/or evaluation programmes. The Barley Core Collection was the main object of evaluation for various biotic and abiotic stress factors in an EU-funded project "Evaluation and conservation of barley genetic resources to improve their accessibility to breeders in Europe" (Enneking *et al.*, 2002). Twenty-eight partners from EU member states and seven associated partners from non-EU countries participated in the project. Besides coordinated evaluation of the BCC using agreed protocols, the project included the multiplication of BCC accessions in order to provide enough seeds for these evaluations, characterisation of morphological characters, image documentation and the creation of herbarium specimens (spike samples). Access to the evaluation data from the project was restricted to the project partners for the first year, but these will later be made freely available through the BCC documentation system. The project included, for example, evaluation of leaf stripe (*Pyrenophora [Drechslera] graminea*) resistance in 150 BCC accessions from Europe, East Asia and the Americas. Twenty-nine accessions were found to be highly resistant.

Several research groups have used or are using the BCC for diversity studies employing isozyme and molecular markers. An EU project entitled "Rapid molecular screening of genetic diversity in cultivated and wild *Hordeum* spp." aimed at comparing different molecular characterisation methods, with respect to their applicability to barley. A project dealing with molecular analysis of the Gatersleben barley collection and parts of the BCC using both AFLP and microsatellite markers has started at IPK. The EU project "Analysis and exploitation of germplasm resources using transposable element molecular markers" uses the BCC to study novel molecular makers.

Conclusions and outlook

The BCC is the first large synthetic core collection, consisting of accessions selected from several genebank collections. Its success will, to a large extent, determine the future of this type of international cooperation in other crops. The BCC is an example of an international effort to make genetic resources management, research and utilisation more efficient.

The effort to create an international Barley Core Collection (BCC) is now reaching the final stages, even though some areas have still not been completely covered, such as *H. vulgare* ssp. *spontaneum*, landraces from Ethiopia and Eritrea, and genetic stocks. Furthermore, the documentation and web site still need to be organised.

'Bona fide' users can obtain seed material in small quantities on request from the distribution centres. Several research projects based on BCC accessions are in progress, but it is necessary to develop the BCC further, to increase awareness of the collection and to encourage further research on this unique, international collection.

Major questions still to be asked are: is the total diversity found in barley really reflected in the Barley Core Collection, should material be replaced, and should some material be added?

Acknowledgements

We thank our colleagues from the BCC coordinating committee for helpful information and comments. Mrs Gudrun Schütze (IPK) prepared the map.

References

Anonymous, 1989. Barley Core Collection (preliminary report). In: Report of a Working Group on Barley (Third Meeting). European Cooperative Programme for the Conservation and Exchange of Crop Genetic Resources. International Board for Plant Genetic Resources, Rome, pp. 59-61.

Brown, A.H.D., 1995. The core collection at the crossroads. In: T. Hodgkin, A.H.D. Brown, Th.J.L. van Hintum and E.A.V. Morales (eds), Core Collections of Plant Genetic Resources. Wiley & Sons, Chichester, pp. 3-19.

Demissie, A., Å. Bjørnstad and A. Kleinhofs, 1998. Restriction fragment length polymorphisms in landrace barleys from Ethiopia in relation to geographic, altitude, and agro-ecological factors. Crop Sci. 38: 237-243.

Enneking, D., E. Schliephake and H. Knüpffer, 2002. Documentation and evaluation of barley genetic resources in Europe. In: P. Hernández, M.T. Moreno, J.I. Cubero and A. Martín (eds), Triticeae IV. Proc. 4th Int. Triticeae Symp., Córdoba, Spain, pp. 183-186.

FAO, 1996. The state of the world's plant genetic resources for food and agriculture. Background documentation prepared for the International Technical Conference on Plant Genetic Resources, Leipzig, Germany, 17-23 June 1996. FAO, Rome, 510 p.

Frankel, O.H., and A.H.D. Brown, 1984. Plant genetic resources today: a critical appraisal. In: J.H.W. Holden and J.T. Williams (eds), Crop Genetic Resources: Conservation and Evaluation. Allen & Unwin, London, pp. 249-257.

Hamon, S., S. Dussert, M. Noirot, F. Anthony and T. Hodgkin, 1995. Core collections – complishments and challenges. Plant Breed. Abstr. 65: 1125-1133.

Hatz, B., 1997. Untersuchungen der genetischen Diversität innerhalb der Gattung *Hordeum* mit molekularen Markertechniken. Dissertation, München Tech. Univ.

Hatz, B.G., A. Jahoor and G. Fischbeck, 1996. RFLP-polymorphism among European accessions of barley core collection. In: Barley Genetics VII. Proc. 7th Int. Barley Genet. Symp, Saskatoon, Canada, pp. 176-178.

Hintum, Th.J.L. van, 1992a. Workshop summary: Barley Core Collection. In: Barley Genetics VI. Proc. 6th Int. Barley Genet. Symp., Helsingborg, Sweden, pp. 703-707.

Hintum, Th.J.L. van, 1992b. Organizational aspects of the Barley Core Collection. In: Barley Genetic Resources. International Crop Network Series No. 9. IBPGR, Rome, pp. 36-40.

Hintum, Th.J.L. van, 1993. The Barley Core Collection. In: E.A. Frison, M. Ambrose, F. Begemann and H. Knüpffer (eds), Report of a working group on barley (fourth meeting) held at the Institut für Pflanzengenetik und Kulturpflanzenforschung, Gatersleben, Germany, 10-12 May 1993. IBPGR, Rome, pp. 35-42.

Hintum, Th.J.L. van, R. von Bothmer and D.L. Visser, 1995. Sampling strategies for composing a core collection of cultivated barley (*Hordeum vulgare* s. lat.) collected in China. Hereditas 122: 7-15.

Hintum, Th.J.L. van, and F. Menting, 2000. Barley genetic resources conservation – now and forever. In: Barley Genetics VIII. Proc. 8th Int. Barley Genet. Symp., Adelaide, Australia, pp. 13-20.

Hodgkin, T., A.H.D. Brown, Th.J.L. van Hintum and E.A.V. Morales (eds), 1995. Core Collections of Plant Genetic Resources. Wiley & Sons, Chichester.

IBPGR, 1989. Report of a working group on barley (third meeting) held at the Zentralinstitut für Genetik und Kulturpflanzenforschung (ZIGuK), Gatersleben, Germany, 18-20 April 1989. International Board for Plant Genetic Resources, Rome, Italy, 61 p.

Johnson, R.C., and T. Hodgkin (eds), 1999. Core Collections for Today and Tomorrow. International Plant Genetic Resources Institute, Rome, Italy, 81 p.

Knüpffer, H., 1988. The European Barley Database of the ECP/GR: An introduction. Kulturpflanze 36: 135-162.

Knüpffer, H., and Th.J.L. van Hintum, 1995. The Barley Core Collection – an international effort. In: T. Hodgkin, A.H.D. Brown, Th.J.L. van Hintum and E.A.V. Morales (eds), Core Collections of Plant Genetic Resources, Wiley & Sons, Chichester, pp. 171-178.

Liu, F., R. von Bothmer and B. Salomon, 1999. Genetic diversity among East Asian accessions of the barley core collection as revealed by six isozyme loci. Theor. Appl. Genet. 98: 1226-1233.

Liu, F., R. von Bothmer and B. Salomon, 2000. Genetic diversity in European accessions of the Barley Core Collection as detected by isozyme electrophoresis. Genet. Res. Crop Evol. 47: 571-581.

Liu, F., G.L. Sun, B. Salomon and R. von Bothmer, 2001. Distribution of allozymic alleles and genetic diversity in the American Barley Core Collection. Theor. Appl. Genet. 102: 606-615.

Sato, K., and K. Takeda, 1997. Barley Core Collection in South and East Asia. In: Harmonizing Agricultural Productivity and Conservation of Biodiversity: Breeding and Ecology. Proc. 8th SABRAO General Congr. and Ann. Meet. Korean Breeding Soc., pp. 260-261.

Diversity in Barley *(Hordeum Vulgare)*
Roland von Bothmer et al (Editors).

Chapter 14

Barley diversity – an outlook

Kazuhiro Sato[a], Roland von Bothmer[b], Theo van Hintum[c] and Helmut Knüpffer[d]

[a]Barley Germplasm Center, Research Institute for Bioresources, Okayama University, Kurashiki, 710-0046, Japan
[b]Department of Crop Science, Swedish University of Agricultural Sciences, Box 44, SE-230 53 Alnarp, Sweden
[c]Plant Research International B.V., Centre for Genetic Resources, The Netherlands (CGN), PO Box 16, NL-6700 AA Wageningen, The Netherlands
[d]Institute of Plant Genetics and Crop Plant Research (IPK), D-06466 Gatersleben, Germany

Introduction

Different aspects of genetic diversity have been presented in this volume as reflected by various scientific disciplines. Several scientists have described the same thing from their own perspectives: barley diversity. Would it be possible to place these descriptions "on top of each other" and to obtain a clear-cut, unambiguous, general picture of the diversity in this crop and to establish what the future perspectives for research, breeding and conservation are?

Here an attempt will be made to develop a single model for the diversity in barley: how it emerged and how it acquired the current structure. A research agenda will be proposed aimed at filling the gaps in our current knowledge and resolving current discrepancies and controversies.

The final part of the chapter will examine the future role of barley as a model crop. Does the large genome size imply that barley will no longer play a major role as a model organism in these days of sequencing? Or are there, in fact, still important reasons for continuing to work with this beautiful crop?

Formation and structuring of genetic diversity

The immediate ancestor of cultivated barley, *Hordeum vulgare* ssp. *spontaneum*, is the result of successive steps of differentiation over millions of years of trial and error in the evolutionary process. This has resulted in a complex biological specialisation associated with a large genetic diversity.

The domestication process constituted a genetic bottleneck, which narrowed the diversity of the early cultivated forms even though introgression from ssp. *spontaneum* occurred. Early farmers proved to be skilful mutation breeders and selected agronomically important mutants, such as six-rowed, spring habit or hull-less types within the rather short period of one or a few

---

Sato, K., R. von Bothmer, Th. van Hintum and H. Knüpffer, 2003. Barley diversity, an outlook. In: R. von Bothmer, Th. van Hintum, H. Knüpffer and K. Sato (eds), Diversity in Barley (*Hordeum vulgare*), pp. 269-278. Elsevier Science B.V., Amsterdam, The Netherlands.

thousand years. Thus, domestication and more or less strict genetic isolation led to significant differences of the diversity spectra between the wild and the cultivated forms. In cultivated barley, the geographical distribution of diversity is significantly correlated with genes for adaptation to various areas and their ecological conditions, and to different uses, e.g., for food, feed, and malt. Four major sources can be postulated for the current genetic diversity in cultivated barley:

A) *Selection of mutations important for domestication* (Chapter 2), i.e., the transition from ssp. *spontaneum* to ssp. *vulgare*. The most important mutation was from a brittle to a non-brittle rachis, a character with an immense selective advantage for plants in cultivation. This mutation, along with a few others, probably only occurred a few times; and, due to the founder effect, in a relatively narrow genetic background.

B) *Spontaneous introgression from* ssp. *spontaneum to* ssp. *vulgare* (Chapters 2 and 3). This process might have played a considerable role in early cultivation since a mixture of wild, weedy and more or less domesticated forms often grew closely together. The early, non-brittle rachis types probably had higher outcrossing rates, similar to the higher allogamy of ssp. *spontaneum*, compared to modern barley types. There is a general evolutionary tendency for development from an allogamous breeding system in wild species towards autogamy in cultivated plants during the course of domestication.

C) *Spontaneous, beneficial or neutral mutations assembled in cultivated barley* (Chapters 3, 5-9, and 10). These mutations occurred at low frequencies, but over a long period (10,000 years). They contributed to increasing diversity, allowing for adaptation to new areas of cultivation.

D) *Induced mutation, recombination and introgression during the course of modern plant breeding* (Chapters 3 and 5). This process started relatively recently, about 150 years ago. The breeding efforts have been intense and the effects are specific, directed towards particular character combinations, such as yield and disease resistance, i.e., characters that are often quantitatively inherited.

Though basically similar, there is a distinct difference between mutations assembled in nature over millions of years of natural selection and the ones selected by man since the dawn of cultivation and breeding. In the former case the successful mutations created finely tuned mechanisms of prime importance for survival in nature such as seed shattering and seed dormancy. The mutations assembled and selected for during cultivation mainly caused disruption of gene functions (Lester and Daunay, 2002). These newly acquired traits are generally recessive to the wild types (Chapter 5). Based on the 'random' genetic diversity generated by mutation, recombination, and introgression, adaptation to a wide range of new environments was the result of natural and human selection (Chapter 4).

The current model of diversity

During the process of domestication, parts of the diversity of the wild progenitor were transferred into the cultivated form (Chapter 2). How much of the initial diversity of the wild progenitor has actually been lost during the domestication process, and how much has been retained has, however, not been satisfactorily elucidated. The answer to this question is particularly important since, in contrast to many other crops, both cultivated barley and its wild progenitor belong to the primary genepool, without any crossing barriers, making genetic exchange quite easy.

In earlier times, genetically variable landraces developed due to a combination of natural and human selection leading to a broad regional adaptation (Chapter 4). Modern material can usually be cultivated in a broader range of agricultural zones through accumulation of various genes for adaptation.

Since the beginning of the 20th century combination breeding has led to an increased variation in characteristics of specific human interest. The genetic diversity within a modern barley crop has generally decreased in comparison to older landraces as a result of the introduction of homogeneous and homozygous cultivars and the relatively narrow basis of plant breeding programmes (Chapters 3 and 7). Traits such as spring habit, malting quality and lodging resistance have been improved and thus diversified over the last century. Natural genetic variation in other characters might, however, have been reduced because of the selection for uniformity in agricultural production.

In many instances, the genetic basis of new cultivars has been narrowed due to the use of closely related parents in cross-breeding. In other cases, the genetic basis for cultivars has widened due to introgression of an increasingly wider range of genetic diversity from "exotic" sources and even wild material (ssp. *spontaneum*; Chapters 3 and 7).

Mutation is one of the key factors for creating diversity. Genetic diversity originates from the differentiation of DNA sequences caused mainly by amino acid changes in a gene, which may occur rather randomly in the genome. Mutants can be recognised when a phenotypic change from the wild type is visible. Changes of DNA sequences do not always result in phenotypic differences, but may lead to diversity at the molecular level only. Isozymes and random DNA markers are widely used to estimate genetic diversity since they are supposed to be neutral to selection. In addition, they give a good resolution for studies of allelic differences randomly distributed over the genome (Chapter 7).

For adaptation to new environments and for breeding purposes the mechanism of recombination is more important than mutation. Chromosomes constitute the framework for combination of genes (Chapter 6). They control recombination and contribute to allelic changes. They are not only carriers of genes but also principal coordinators of gene arrangement. The position of genes on a chromosome, viz., localisation in centromeric or in distal regions, affects the recombination frequency. After the discovery of synteny by comparing molecular maps of different species (Hammer and Schubert, 1994; Paterson *et al.*, 1995), it became obvious that the order of homoeologous genes is conserved among closely and even distantly related species, such as between barley and rice (Moore *et al.*, 1993; Moore, 1995).

A further important factor shaping diversity is natural selection. The plain selection pressure increases the fitness of specific phenotypes, such as those better adapted to colder environments. The increased fitness results in higher frequencies in the offspring, recombination of favourable types and phenotypes with an even greater ability to survive under a particular selection pressure. This mechanism is very common and occurs in response to abiotic stress (Chapter 9), and often in response to some biotic stresses. However, in the case of biotic stress (Chapter 8) another mechanism is also involved, i.e., co-evolution of host and pathogen. The plant population increases its fitness, as in natural selection, but the pathogen responds by adapting to the altered host. This causes a new selection pressure and thus a new adaptation by the host population.

Barley is a major temperate crop and adaptive traits are important for utilisation of diversity over large areas with varying environmental conditions. Since the genetic background for some agronomic traits, especially for yield, is not yet fully understood, newer approaches such as analysis of quantitative trait loci (QTLs) are systematically applied. This has led to a far better understanding of the diversity in agronomic and adaptive characters.

The genetic variability and the mechanisms of mutations are principally the same for genes related to quantitative inheritance as for characters that are under qualitative control. Both systems have been the target of natural and artificial selection (Chapter 10). Even in exotic material unexpected QTLs for cultivated traits can be found. Tanksley *et al.* (1996) and Tanksley and McCouch (1997) showed an excellent example in tomato where promising genes for yield not known in cultivated material were found in wild germplasm where they are, however, not expressed. These genes had simply been eliminated during the bottleneck of domestication.

There are many difficulties connected with the study of quantitative traits. Given the difficulty in quantifying quantitative traits, due to measuring errors and genotype × environment interactions, large sets of data are needed for analysis of quantitative traits. The study of quantitative traits is not possible without the help of powerful linkage detection software and these data handling systems are necessary for the estimation of genetic diversity in a wide range of germplasm. The example of Near Infrared spectroscopy and multivariate analysis (Chapter 11) describes the method of data mining, for the accommodation of a general paradigm shift from reductionistic experiments (question-experiment-answer) to holistic data gathering.

Collection and conservation of variation

The most important way of providing access to genetic diversity is through genebanks. These collections will never be complete, and the ‘gaps’ may comprise large sources of diversity. A monitoring system, which would study the threat of genetic erosion, such as that caused by habitat destruction, may contribute to reducing the risk of losing this diversity and improving the access to it (Chapter 12). The question whether the overall diversity in barley has decreased due to genetic erosion or increased due to mutation breeding and the use of a broader parental material (Chapter 3), has not yet been answered unambiguously. There are, however, more questions to be answered concerning the conservation of barley diversity, such as: what is or should be the role of *in situ* conservation of wild relatives, and what is the overlap between genebank collections. The conservation of material resulting from breeding activities, such as cultivars and breeders’ lines, is comparatively easy since it is usually homogeneous and homozygous and systematically organised. Landraces and wild forms are, however, much more heterogeneous and require other sampling strategies.

The Barley Core Collection (BCC; Chapter 13) is an efficient tool for monitoring the world barley diversity and could contribute to increased utilisation of barley germplasm. It has a high degree of homogeneity, represents an optimal size, and a fair representation of the diversity of the barley genepool. It will increase the overview of the diversity available and hence provide better access to the large amount of other material in the genebanks of the world. Large evaluations on a global scale may provide further useful information about the overall diversity present in the crop. A well-designed germplasm set, like the BCC, can be used as a model for developing data collection and processing strategies.

To increase the accessibility of diversity for breeding purposes, germplasm of cultivated barley should have a higher priority for conservation and utilisation than wild germplasm. Also within the heterogeneous group of “wild barley” embracing the primary, secondary and tertiary genepools, different priorities have to be applied. The progenitor of barley, *Hordeum vulgare* ssp. *spontaneum*, is more closely related to barley than other wild species and its genes are hence more easily accessible through normal recombination. There is a fair representation of ssp. *spontaneum* in the world genebanks (Chapter 12), but its documentation requires further elaboration.

A comparatively restricted amount of material of wild barley (apart from ssp. *spontaneum*) is currently available. Thus, more attention should be paid to the wild *Hordeum* species as far as collecting, conservation and research are concerned. As the knowledge of their habitat preferences and other edaphic conditions is limited, good sampling strategies need to be developed. This has to be done individually for species with different biological systems, such as perennials *vs.* annuals, or outcrossing *vs.* inbreeding forms (Bothmer and Seberg, 1995).

Evaluation of wild species is laborious and even the multiplication and maintenance require special skills and effort. For a targeted utilisation of this material in breeding, further evaluation is necessary, based on qualitative characters and including molecular analyses. Due to the difficulties caused by "wild" genes affecting the reproductive mechanisms, such as dormancy and seed physiological conditions, it is important to develop optimal long-term storage systems for seeds and DNA samples. This is also true when trying to maintain collections of mutations and other genetic stocks.

Analysis of variation

The earliest analyses of genetic variation in barley were based on the phenotype. Some traits, which both are agronomically important and good phenotypic markers, have been widely used as germplasm descriptors, such as row type, life form and major-gene disease resistance (e.g., Takahashi *et al.*, 1983; IPGRI, 1994). Since there is a restricted number of genes for phenotypical assessment of variation, different molecular and chemical marker systems have been developed. Obviously all methods have their advantages and shortcomings.

The geographical distribution of diversity should be further elucidated since knowledge about environmental conditions of a biotic and abiotic nature is inadequate. The diversity in agronomic traits should be evaluated under diverse growing conditions in major temperate areas in the world collection of barley. In order to achieve a systematic, worldwide evaluation system, the number of accessions should be reasonable to handle but also sufficient to cover major diversity groups. The Barley Core Collection is probably the most suitable approach for realising this (Chapter 13).

The results of diversity analysis should be well documented and easily available in searchable and processable computer databases. A careful description of variation may reduce the overlap between different analyses and support a systematic method of modelling diversity.

Revealing molecular polymorphism especially in coding regions is an area of high priority. Time and effort is needed to correlate the phenotype with the molecular polymorphisms. For this reason sampling of DNA from particular germplasms should be considered for both ssp. *vulgare* and wild *Hordeum* species. The following problems occur in this context:

1. The diversity is only confirmed by transformation of the gene into a contrasting genetic background;
2. Material necessary for transformation, i.e., genome libraries as well as transformation techniques, are not easily available in barley;
3. A gene-gene interaction is difficult to estimate.

There are difficulties with direct genetic comparisons between taxa. For this reason, backcrosses and substitution of ssp. *spontaneum* chromosomal segments into ssp. *vulgare* would be an ideal system for estimating gene functions. The disadvantage of this method is that comparatively few accessions can be handled due to the work involved in map and population development.

Induced mutations are excellent tools for visualising gene expression by a wide array of differentiated alleles (Chapter 5). However, these mutations are induced in a low number of genotypes and show a limited gene/cultivar interaction. Even though a considerable amount of research has been invested during the 20th century, induced mutations are not frequently utilised in breeding and, moreover, they are not even scientifically well known. A further problem is that agronomically important traits are controlled by several genes, and usually quantitatively inherited (Chapter 10). They are difficult to identify based on the selection of induced mutants.

Modelling of diversity

After the confirmation of genetic effects on a certain genetic background, one must check the effects on more generalised or targeted genetic backgrounds. This involves the following processes:

1. Estimation of the genetic basis for differences;
2. Confirmation of the genetic difference by studying the phenotype;
3. Verification of the genotype by a Mendelian genetic analysis or direct transformation of the gene into a target genotype;
4. Investigation of the environmental interaction of the genotype;
5. Studying of the gene-gene interaction.

Efficient data management is a key for access to the diversity in germplasm. The situation for germplasm documentation has changed considerably in recent years due to the introduction of computer databases, Internet communication and web browsing. Nowadays, most scientists have easy and cheap access to germplasm databases via the Internet. Holistic data-mining approaches will increase the efficiency of using these germplasm databases (Chapter 11). The combination of germplasm and DNA databases is gaining importance in the field of research into diversity in barley genes.

There are many and large sets of data for germplasm evaluation and for various genetic markers in the world. One of the major obstacles in this context is the lack of standardisation between the marker systems and the resulting data sets, preventing their integration into a global information system for general estimation of diversity (Chapter 7). The development of powerful interpretation methods, which can understand existing data sets, and translate them into comparable information for genetic variation, is a necessary prerequisite for achieving a holistic view of the genetic diversity.

Older information on different accessions can be of great value for estimating diversity. If all older information (both published and unpublished data) on germplasm were easily accessible and could be searched for electronically, it would add considerably to the understanding of diversity.

Application of barley diversity

Due to the economic importance of the crop and easy handling of seed samples, the number of barley accessions in the genebanks of the world is second only to wheat (Chapter 12). The basic analytical tools will soon be ready for large-scale diversity analysis of germplasm. These techniques may connect the phenotype, the source of variation, and DNA polymorphism via protein formation. Experimental material such as mutants and other genetic stocks will contribute to the identification of gene functions. A DNA bank or a DNA resource centre will possibly replace some of the present genebank activities. For example, it is difficult to multiply and maintain some of the subviable mutants but it is comparatively easy to keep samples of their DNA for future research purposes.

Genetic variation is the basis for breeding, but there is a certain conflict between modern agriculture based on the release of improved cultivars and the use of diversity. The introduction of modern cultivars promoted genetic erosion of the diversity present in former landraces.

Growing a few cultivars over large areas is advantageous for harvesting uniform agricultural products. However, the genetic uniformity may increase the vulnerability of agricultural production systems. The loss of diversity may increase the risk of serious damage due to environmental changes, such as severe diseases or extreme temperatures. The diversity available should thus be maximal in genebanks and in breeders' experimental trials but not necessarily on the farmers' fields.

The most effective strategy for the access to new resistance and tolerance sources is either to screen 'exotic' germplasm for 'non-host' resistance, or to screen material from geographical areas where the pathogen occurs naturally. In the case of abiotic stresses, the nature of the stress should be analysed, similar conditions or stress elements should be looked for in different geographical areas and material from these particular areas should be screened. If these prerequisites cannot be fulfilled, the alternative approach is to screen as much 'random' diversity as possible, to maximise the probability of identifying resistant or tolerant material. Marker-assisted selection (MAS) and advanced backcross techniques might accelerate the development allowing effective introgression of exotic genes.

The use of ssp. *spontaneum* in modern breeding programmes has been very limited, mainly because of the genetic load of wild genes resulting in a prolonged breeding cycle. Much effort has to be invested to exclude deleterious traits of wild forms such as dormancy or seed shattering, which are advantageous characters in nature. Exotic landrace material and the wild progenitor have better prospects for utilisation when contemporary breeding methods, such as marker-assisted breeding (MAS), become available.

General problems related to genetic diversity

Certain international political developments currently restrict access to genetic diversity (Chapter 12). International agreements such as the Convention on Biological Diversity (CBD) and the International Treaty for Plant Genetic Resources (at FAO) regulate the access to biological material. The national sovereignty tends to restrict the access to biodiversity and is a barrier to further collection and research on germplasm. The increasing private ownership of plant germplasm resources due to intellectual property rights make the situation even more complicated. Improved access to germplasm in existing *ex situ* collections and due to increased collecting efficiency will improve both plant breeding and research.

Barley as a model crop for genetic research

The domestication of barley and wheat occurred simultaneously in the same area. Migration of the two crops and their subsequent adaptation to new areas followed similar routes. Apart from the polyploid system in wheat, several prerequisites for breeding strategies are thus rather similar in the two crops, such as, for example, diversity for stress tolerances. Compared to wheat, barley has a great advantage due to its diploid level of ploidy. It is easy to produce mutants and to carry out genetic analysis. The large chromosomal syntenies make barley an ideal model crop for the whole tribe, Triticeae.

The great interest in barley has promoted the research and screening of mutants, with, at present, more than 10,000 documented mutants (Chapter 5). Molecular genetics using mutants produces a significant impact also on research into *Arabidopsis* (Kalantidis *et al.*, 2000) but there are more well characterised mutants in barley than in *Arabidopsis*. In this sense, crops are more suitable for mutation research than other species. The results obtained in one crop species can also prove informative for other species, making breeding strategies based on the genome analysis more efficient. If the species are closely related, the chances of sharing the results become more realistic. For example, agronomically important QTLs in barley can be identified in other crop species even if the locus is not identified.

Barley has been widely used for cytogenetic research as it is a diploid organism with large chromosomes. However, the larger genome size makes molecular studies more problematic. There are some genome sequencing research programmes in model species such as rice or *Arabidopsis*, but because of the larger genome size, there is no ongoing project for genome sequencing in barley. One approach would be to use cDNA analysis or to perform a sequencing analysis for gene-rich regions. Genome sequencing programmes are not able to include many genotypes due to high costs and time. Studies of a single genotype are sufficient for gene identification and for estimation of the genomic constitution. After these initial sequencing programmes, studies of polymorphism or diversity within species will be the next target.

Conclusions and outlook

As visualised in previous chapters of this book, the picture of diversity in barley is highly complex. Is it at all feasible to draw any general conclusions with the regard to diversity, that is – is the entire elephant visible or not? The full answer must be "*no*". There are still many empty spaces and too complex a pattern to allow easy visualisation. However, with the current state of knowledge, it is possible to show some major, clearly visible tendencies of development and distribution of diversity and it is also possible to show where the gaps are and indicate ways of tackling the problems.

General conclusions on diversity

The process of domestication was a genetic bottleneck, leading to loss of diversity. The first cultivated forms were genetically depauperated in comparison to the wild progenitor. *H. vulgare* ssp. *spontaneum* shows a larger diversity in supposedly neutral markers.

A large amount of marker data has been generated, but individual data sets include only a restricted number of accessions and are usually based on a unique marker set. Hence, it is not yet possible to pool data into a common database in order to further complete the picture of the diversity structure in the genepool.

Already in the earliest cultivated forms of barley, the mechanisms of mutation and recombination were actively starting to create "new" diversity, which was continuously subject to natural as well as artificial selection. These processes have led to an immense diversification in morphological, agronomical, and other adaptive traits as well as in important quality characters for different uses. Thus, the entire gene pool of cultivated barley in the world has a far larger diversity in these traits than its wild progenitor (cf. Lester, 1989; Lester and Daunay, 2003). In addition, for these sets of characters it is difficult to obtain a good estimate of the diversity, since continuous breeding efforts have caused the formation of regional and temporal diversity patterns, reflecting several factors, such as cropping system (winter *vs.* spring types), end use (feed, food or malt) and the strategy of individual breeders to rely on distinct progenitors.

In general, one can observe a tendency of decreasing diversity over the last century from the earlier landraces to modern cultivars. However, due to use of exotic germplasm, in some breeding material the diversity has again increased over the last decades.

Modern plant breeding in barley was originally based on selections from earlier landraces. During the last century of combination breeding, it is mainly elite material or new cultivars that have been used as crossing partners. The basis for diversity in the breeding material is thus rather limited and the use of exotic germplasm can be much improved.

Barley as a model crop

As convincingly shown in the earlier chapters, barley offers several advantages for studies of diversity and as a major model organism in biological research. This is due to a number of factors:

1. The diploid ploidy level of the crop;
2. The inbreeding habit and the diploid system, which make inheritance studies easy to perform;
3. Large chromosomal synteny in the Triticeae and in the entire grass family. The results obtained in one species can be applied to other species;
4. The wild progenitor belongs to the primary genepool of the crop and crosses are hence easy to perform. The large number of closely and distantly related species in the Triticeae to other grasses makes barley a central organism for studies of evolution and relationships. The entire tribe is also a gigantic genepool for crop improvement (wheat, barley, rye and forage grasses);
5. A large number of well-documented mutations and other genetic stocks are available;
6. Most genotypes are homozygous, which makes it easy to repeat experiments;
7. The uniqueness of barley is largely due to its long history as a crop, and accentuated by the multitude of uses and the development of many different, often quantitatively inherited quality characters;
8. A large number of well-documented accessions are available in germplasm collections. The abundant genetic resources will certainly support the continuous and systematic research of barley diversity and eventually further genetic analyses. Seed samples are cheap to handle and easy to keep for a long period.

References

Bothmer, R. von, and O. Seberg, 1995. Strategies for the collecting of wild species. In: L. Guarino, V. Ramanatha Rao and R. Reid (eds), Collecting Plant Genetic Diversity. CAB International, UK, pp. 93-111.

Hammer, K., and I. Schubert, 1994. Are Vavilov's law of homologous series and synteny related? Genet. Res. Crop Evol. 41: 123-124.

IPGRI, 1994. Descriptors for barley (*Hordeum vulgare* L.). International Plant Genetic Resources Institute, Rome, 46 p.

Kalantidis, K., L.G. Briarty and A.A. Wilson, 2000. *Arabidopsis* mutant characterization; microscopy, mapping, and gene expression analysis. In: Z.A. Wilson (ed), *Arabidopsis*: a practical approach. Oxford Univ. Press, pp. 77-104.

Lester, R.N., 1989. Evolution under domestication involving disturbance of genetic balance. Euphytica 44: 125-132.

Lester, R.N. and M.-C. Daunay, 2003. Diversity of African vegetable *Solanum* species and its implications for a better understanding of plant domestication. In: H. Knüpffer and J. Ochsmann (eds), Rudolf Mansfeld and Plant Genetic Resources. ZADI, Bonn, Germany, pp. 136-151.

Moore, G., 1995. Cereal genome evolution: pastoral pursuits with 'Lego' genomes. Curr. Opinion Genet. Develop. 5: 717-724.

Moore, G., M.D. Gale, N. Kurata and R.B. Flavell, 1993. Molecular analysis of small grain cereal genomes: current status and prospects. Biotechnol. 11: 584-589.

Paterson, A.H., Y.R. Lin, Z. Li, K.F. Schertz, J.F. Doebley, S.R.M. Pinson, S.C. Liu, J.W. Stansel and J.E. Irvine, 1995. Convergent domestication of cereal crops by independent mutations at corresponding genetic loci. Science 26: 1714-1718.

Takahashi, R., S. Yasuda, J. Hayashi, T. Fukuyama, I. Moriya and T. Konishi, 1983. Catalogue of Barley Germplasm Preserved in Okayama University. 217 p.

Tanksley, S.D. and S.R. McCouch, 1997. Seed banks and molecular maps: unlocking genetic potential from the wild. Science 277: 1063-1066.

Tanksley, S.D., S. Grandillo, T.M. Fulton, D. Zamir, Y. Eshed, V. Petiard, J. Lopez and T. Beck-Bunn, 1996. Advanced backcross QTL analysis in a cross between elite processing line of tomato and its wild relative *L. pimpinellifolium*. Theor. Appl. Genet. 92: 213-224.

Keyword list

The list contains selected keywords for each chapter. The first page of the corresponding chapter in which the keyword appears is indicated.